Climate Change and Environmental Sustainability-Volume 5

Climate Change and Environmental Sustainability-Volume 5

Editors

Bao-Jie He
Ayyoob Sharifi
Chi Feng
Jun Yang

MDPI • Basel • Beijing • Wuhan • Barcelona • Belgrade • Manchester • Tokyo • Cluj • Tianjin

Editors
Bao-Jie He
Chongqing University
China

Ayyoob Sharifi
Hiroshima University
Japan

Chi Feng
Chongqing University
China

Jun Yang
Northeastern University
China

Editorial Office
MDPI
St. Alban-Anlage 66
4052 Basel, Switzerland

This is a reprint of articles from the Topics published online in the open access journal *Atmosphere* (ISSN 2073-4433), *Buildings* (ISSN 2075-5309), *Land* (ISSN 2073-445X), *Remote Sensing* (ISSN 2072-4292), *Atmosphere* (ISSN 2073-4433) (available at: https://www.mdpi.com/topics/Climate_Environmental).

For citation purposes, cite each article independently as indicated on the article page online and as indicated below:

LastName, A.A.; LastName, B.B.; LastName, C.C. Article Title. *Journal Name* **Year**, *Volume Number*, Page Range.

ISBN 978-3-0365-3795-5 (Hbk)
ISBN 978-3-0365-3796-2 (PDF)
161 k words.

Contents

About the Editors

Bao-Jie He is a Research Professor of Urban Climate and Built Environment at the School of Architecture and Urban Planning, Chongqing University, China. Prior to joining Chongqing University, Bao-Jie He was a PhD researcher at the Faculty of Built Environment, University of New South Wales, Australia. Bao-Jie is working on the Cool Cities and Communities and Net Zero Carbon Built Environment project. Bao-Jie has strong academic capability and has published around 80 peer-reviewed papers in high-ranking journals and given oral presentations at reputable conferences. Bao-Jie acts as the Topic Editor-in-Chief, Leading Guest Editor, Associate Editor, Editorial Board Member, Conference Chair, Sessional Chair, and Scientific Committee of a variety of international journals and conferences. Dr. He received the Green Talents Award (Germany) in 2021 and National Scholarship for Outstanding Self-Funded Foreign Students (China) in 2019. Dr. He was ranked as one of the 100,000 global scientists (both single-year and career top 2%) by the Mendeley, 2021.

Ayyoob Sharifi works at the Graduate School of Humanities and Social Sciences, Hiroshima University. He also has a cross-appointment at the Graduate School of Advanced Science and Engineering. Ayyoob's research is mainly at the interface of urbanism and climate change mitigation and adaptation. He actively contributes to global climate change research programs such as the Future Earth and is currently serving as a lead author for the Sixth Assessment Report (AR6) of the Intergovernmental Panel on Climate Change (IPCC). Before joining Hiroshima University, he was the Executive Director of the Global Carbon Project (GCP)—a Future Earth core project—leading the urban flagship activity of the project, which is focused on conducting cutting-edge research to support climate change mitigation and adaptation in cities.

Chi Feng received his joint PhD training in South China University of Technology (China) and KU Leuven (Belgium). He is now a research professor in the School of Architecture and Urban Planning, Chongqing University (China) and is leading a research group of more than 10 members. His research topics cover coupled heat and moisture transfer in porous building materials, as well as the hygrothermal performance of building envelopes and built environments. He has led eight international and national research projects, including the China–Europe round robin campaign on material property determination (nine countries participated), the National Natural Science Foundation of China, and the National Key R&D Program of China. He has published more than 60 peer-reviewed journal/conference papers at home and abroad. He is drafting two Chinese standards and participating in another nine international/national ones.

Jun Yang is working at the Urban Climate and Human Settlements Lab, Northeastern University (Shenyang China). His research expertise involves urban climate zones, urban ecology, urban human settlements, and sustainability. As PI or Co-PI, he has been involved in 50 research projects, receiving a total of 15 million RMB from EGOV.CN (e.g., NFC, MOST, and MOE) since 2002. He has authored and co-authored more than 160 papers and book chapters and published more than 50 English papers and more than 110 Chinese papers in academic journals. He is now the Associate Editor of the *Social Sciences* section of the *International Journal of Environmental Science and Technology*, sits on the Editorial Board of *PLoS ONE*, *PLOS Climate*, and *Frontiers in Built Environment*, is a Lead Guest Editor of Complexity, and a Guest Editor of *IEEE Journal of Selected Topics in Applied Earth Observations and Remote Sensing*.

Preface to "Climate Change and Environmental Sustainability-Volume 5"

The Earth's climate is changing; the global average temperature is estimated to already be about 1.1 °C above pre-industrial levels. Indeed, we are now living in conditions of a climate emergency. Climate change leads to many adverse events, such as extreme heat, flooding, bushfire, drought, and many other associated economic and social consequences. Further warming is projected to occur in the coming decades, and climate-induced impacts may exceed the capacity of society to cope and adaptive in a 1.5 °C or 2 °C world. Therefore, urgent actions should be taken to address climate change and avoid irreversible environmental damages.

Climate change is interrelated with many other challenges such as urbanisation, population increase and economic growth. For instance, cities are now the main settlements of human being and are major sources of greenhouse gas emissions that are key contributors to climate change. Moreover, rapid and unregulated urbanisation in some contexts further causes urban problems such as environmental pollution, traffic congestion, urban flooding and heat island intensification. In the absence of well-designed measures, increasing urbanisation trends in the next two–three decades are likely to further aggravate such problems. Overall, climate change and many other challenges have deteriorated the sustainable development of the world.

The framework also calls for the support and engagement of all societal stakeholders. To support the achievement and implementation of the framework, this book focuses on climate change and environmental sustainability by covering four key aspects, including climate change mitigation and adaptation, sustainable urban–rural planning and design, decarbonisation of the built environment in addition to climate-related governance and challenges. Climate change mitigation and adaptation covers topics of greenhouse gas emissions and measurement, climate-related disasters and reduction, risk and vulnerability assessment and visualisation, impacts of climate change on health and well-being, ecosystem services and carbon sequestration, sustainable transport and climate change mitigation and adaptation, sustainable building and construction, industry decarbonisation and economic growth, renewable and clean energy potential and implementation in addition to environmental, economic and social benefits of climate change mitigation.

Sustainable urban–rural planning and design deals with questions of climate change and regional economic development, territorial spatial planning and carbon neutrality, urban overheating mitigation and adaptation, water-sensitive urban design, smart development for urban habitats, sustainable land use and planning, low-carbon cities and communities, wind-sensitive urban planning and design, nature-based solutions, urban morphology and environmental performance in addition to innovative technologies, models, methods and tools for spatial planning. Decarbonisation of the built environment addresses issues of climate-related impacts on the built environment, the health and well-being of occupants, demands on energy, materials and water, assessment methods, systems and tools, sustainable energy, materials and water systems, energy-efficient design technologies and appliances, smart technology and sustainable operation, the uptake and integration of clean energy, innovative materials for carbon reduction and environmental regulation, building demolition and material recycling and reusing in addition to sustainable building retrofitting and assessment. Climate-related governance and challenges concerns problems of targets, pathways and roadmaps towards carbon neutrality, pathways for climate resilience and future sustainability, challenges, opportunities and solutions for climate resilience, the development and challenges climate change governance coalitions (networks), co-benefits and synergies between adaptation and mitigation measures, conflicts and trade-offs between adaptation and mitigation measures, mapping, accounting and trading carbon emissions, governance models, policies, regulations and programs, financing urban climate change mitigation, education, policy and advocacy of climate change mitigation and adaptation in addition to the impacts and lessons of COVID-19 and similar crises.

Overall, this book aims to introduce innovative systems, ideas, pathways, solutions, strategies, technologies, pilot cases and exemplars that are relevant to measuring and assessing the impact of climate change, mitigation and adaptation strategies and techniques in addition to public participation and governance. The outcomes of this book are expected to support decision makers and stakeholders to address climate change and promote environmental sustainability. Lastly, this book aims to provide support for the implementation of the United Nations Sustainable Development Goals and carbon neutrality in efforts aimed at achieving a more resilient, liveable and sustainable future.

This volume of Climate Change and Environmental Sustainability covers topics on greenhouse gas emissions, climatic impacts, climate models and prediction, and analytical methods. Issues related to two major greenhouse gas emissions, namely of carbon dioxide and methane, particularly in wetlands and agriculture sector, and radiative energy flux variations along with cloudiness are explored in this volume. Further, climate change impacts such as rainfall, heavy lake-effect snowfall, extreme temperature, impacts on grassland phenology, impacts on wind and wave energy, and heat island effects are explored. A major focus of this volume is on climate models that are of significance to projection and to visualise future climate pathways and possible impacts and vulnerabilities. Such models are widely used by scientists and for the generation of mitigation and adaptation scenarios. However, dealing with uncertainties has always been a critical issue in climate modelling. Therefore, methods are explored for improving climate projection accuracy through addressing the stochastic properties of the distributions of climate variables, addressing variational problems with unknown weights, and improving grid resolution in climatic models. Results reported in this book are conducive to a better understanding of global warming mechanisms, climate-induced impacts, and forecasting models. We expect the book to benefit decision makers, practitioners, and researchers in different fields and contribute to climate change adaptation and mitigation.

Prof. Bao-Jie He acknowledges Project NO. 2021CDJQY-004, supported by the Fundamental Research Funds for the Central Universities. We appreciate the assistance from Mr. Lifeng Xiong, Mr. Wei Wang, Ms. Xueke Chen, and Ms. Anxian Chen at the School of Architecture and Urban Planning, Chongqing University, China.

Bao-Jie He, Ayyoob Sharifi, Chi Feng, and Jun Yang
Editors

MDPI

Article

Greenhouse Gas Emissions from Agriculture in EU Countries—State and Perspectives

Paulina Mielcarek-Bocheńska [1,*] and Wojciech Rzeźnik [2]

[1] Institute of Technology and Life Sciences, 05-090 Raszyn-Falenty, Poland
[2] Institute of Environmental Engineering and Building Installations, Poznan University of Technology, 60-965 Poznań, Poland; wojciech.rzeznik@put.poznan.pl
* Correspondence: p.mielcarek@itp.edu.pl

Abstract: Agriculture is one of the main sources of greenhouse gas (GHG) emissions and has great potential for mitigating climate change. The aim of this study is to analyze the amount, dynamics of changes, and structure of GHG emissions from agriculture in the EU in the years 2005–2018. The research based on data about GHG collected by the European Environment Agency. The structure of GHG emissions in 2018 in the EU is as follows: enteric fermentation (45%), agricultural soils (37.8%), manure management (14.7%), liming (1.4%), urea application (1%), and field burning of agricultural residues (0.1%). Comparing 2018 with the base year, 2005, emissions from the agricultural sector decreased by about 2%, which is less than the assumed 10% reduction of GHG emissions in the non-emissions trading system (non-ETS) sector. The ambitious goals set by the EU for 2030 assume a 30% reduction in the non-ETS sector. This will require a significant reduction in GHG emissions from agriculture. Based on the analysis of the GHG emission structure and available reduction techniques, it was calculated that in this period, it should be possible to reduce emissions from agriculture by about 15%.

Keywords: greenhouse gases; agriculture; climate change; mitigation

Citation: Mielcarek-Bocheńska, P.; Rzeźnik, W. Greenhouse Gas Emissions from Agriculture in EU Countries—State and Perspectives. *Atmosphere* **2021**, *12*, 1396. https://doi.org/10.3390/atmos12111396

Academic Editors: Baojie He, Ayyoob Sharifi, Chi Feng and Jun Yang

Received: 30 September 2021
Accepted: 24 October 2021
Published: 25 October 2021

1. Introduction

The phenomenon of the natural greenhouse effect is positive for living conditions on the Earth. Thanks to this, the temperature of Earth surface is increased by 20–34 °C. Without the greenhouse effect, the average temperature of the Earth would be around −19 °C [1]. Gases that absorb radiation in the range emitted by the Earth's surface cause the greenhouse effect and are called greenhouse gases (GHGs). The main GHGs are water vapor (H_2O), carbon dioxide (CO_2), methane (CH_4), nitrous oxide (N_2O), ozone (O_3) [2]. Measurements carried out in recent decades have shown that the level of radiation escaping into space is getting smaller; so, the heat is accumulated on the Earth, and energy balance is disturbed. Therefore, it is observed the intensification of the greenhouse effect (global warming), which is caused by the growing concentration of greenhouse gases in the atmosphere [3–5].

The main anthropogenic GHG, carbon dioxide, is responsible for about 81% of global GHG emissions in the European Union (EU) (according to data for 2018); the next are methane and nitrous oxide, accounting for 10% and 6%, respectively [6]. Relatively high shares of methane and nitrous oxide in GHG emission, despite their low concentration in the atmosphere, are connected to global warming potential, which compares the ability of 1 kg of each gas to capture heat over a 100-year perspective. Methane has 21–36 times greater potential than CO_2, and nitrous oxide 265–310 times greater than CO_2 [7,8]. Another important parameter is the remaining time of gases in the atmosphere. Carbon dioxide does not break down easily and remains in the atmosphere for several centuries; nitrous oxide remains for about 121 years; and for methane, it is about 12 years [8,9].

The methane emission from agriculture is about 54% of total emissions of this gas in the EU, and for N_2O, it is nearly 79% [6]. Therefore, despite the relatively small share of the agricultural sector in the EU's global GHG emissions of around 10%, agriculture has great potential and is an important link in the strategy of reducing greenhouse gas emissions and mitigating climate change.

The intensification of the greenhouse effect and climate changes, for which the consequence is a necessity of GHG reduction, are currently some of the key issues in the EU's environmental policy. One of the goals of the EU 2020 climate and energy package is to reduce greenhouse gas emissions by 20% compared with 1990 levels [10]. In sectors not covered by the emissions trading system (non-ETS sectors), reduction targets have been set individually for each member state [11]. On the other hand, according to the EU 2030 climate and energy framework, greenhouse gas emissions are to be reduced by at least 40% compared with those in 1990. In sectors covered by the emissions trading system, emission reductions of 43% compared with those in 2005 are assumed, and in non-ETS sectors, by 30%. The land use, land-use change, and forestry sector (LULUCF) were included in the UE 2030 climate and energy framework [12]. Moreover, at the EU meeting in December 2020, the target of GHG reduction was increased to at least 55% by 2030.

Considering the ambitious GHG reduction targets set by the EU, mainly those for 2030, their implementation requires comprehensive knowledge about the amount of emissions and their structure from each non-ETS area. Additionally, knowledge about the status of implementation of the GHG reduction levels in each EU country may be helpful and constitute a significant contribution to the planning of mitigation strategies. It will make it easier to take the necessary initiatives: defining the measures, changing legal regulations, etc. To realize such a high reduction level of GHG emissions, it will be necessary to act in each of the emission areas, including agriculture.

The aim of the study is to analyze the amount, dynamics of changes, and the structure of GHG emissions from agriculture in the EU in the years 2005–2018.

2. Materials and Methods

The study analyses the period from 2005 to 2018. The research is based on data about GHG collected by the European Environment Agency [13]. These annual data are reported by each EU state, which is obligatory under the United Nations Framework Convention on Climate Change (UNFCCC). These emissions are reported according to classification and in the Common Reporting Format (CRF) for five main categories: 1. Energy; 2. Industrial processes and product use; 3. Agriculture; 4. Land use, land use change and forestry (LULUCF); and 5. Waste. They are calculated according to the methodology published by the Intergovernmental Panel on Climate Change [14]. These guidelines allow for the estimation of emissions at various levels of detail, depending on the availability of national methods as well as emission parameters and indicators.

The analyses based on GHG emission data, expressed in CO_2 equivalent, without the LULUCF sector, and emissions from international aviation and international maritime transport. The research covers all EU countries, including the United Kingdom (EU-28).

GHG EU's emissions were reviewed in total from agriculture and individually for each member state. The dynamics of emission changes was calculated, assuming that the base year 1990 in the case of total GHG emissions and the base year 2005 in the case of GHG emissions from agriculture.

The structure of GHG emissions from the agriculture sector was studied for EU and individual member states. These are the included structures: enteric fermentation, manure management, agricultural soils, field burning of agricultural residues, liming, and urea application. For each agricultural source, the dynamics of changes in GHG emissions was calculated, assuming the base year of 2005. The trends in GHG emissions from individual sources in agriculture in the period 2005–2018 were also analyzed.

3. Results and Discussion

3.1. GHG Emissions in the EU

In the analyzed period (1990–2018), GHG emissions, compared with the base year 1990, showed a downward trend. However, between 1990 and 2004, GHG emissions in the EU fluctuated. There were both periodic increases and decreases in emissions. Since 2005, GHG emissions have been characterized by a constant downward trend. The largest decreases (over 10%) were observed over the last ten years (Figure 1). In 2018, the total GHG emission without the LULUCF sector was 4225.97 Tg CO_2 eq. (CO_2—81.4%, CH_4—10.3%, N_2O—5.6%) and was lower by 25.2% compared with that in 1990. With the LULUCF sector, the level of reduction was 26.8%. Considering the dynamics of GHG emissions in the analyzed period, it can be projected that the 20% GHG emission reduction will be achieved.

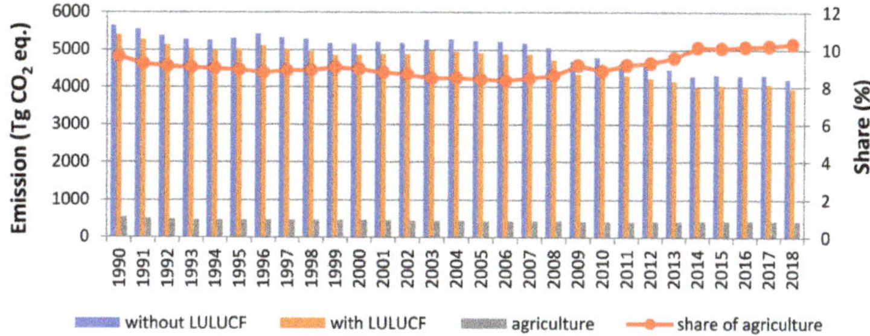

Figure 1. Total greenhouse gas (GHG) emissions and agricultural GHG emissions in the EU for 1990–2018.

The values of GHG emissions in the EU countries in 2018 are presented in Figure 2.

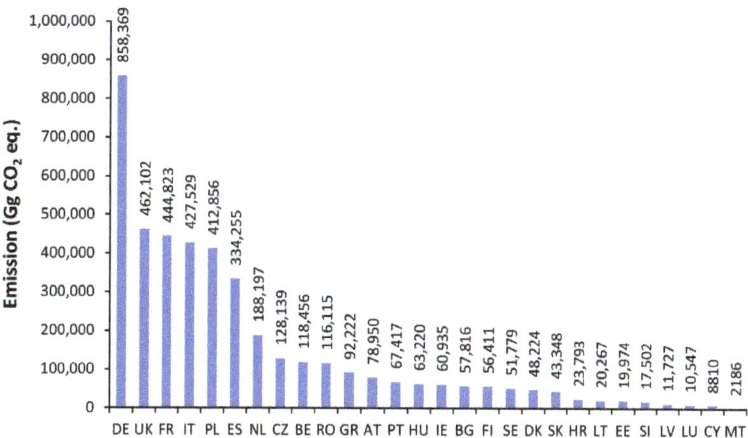

Figure 2. GHG emissions in the EU countries in 2018. Explanation: DE—Germany, UK—United Kingdom, FR—France, IT—Italy, PL—Poland, ES—Spain, NL—The Netherlands, CZ—Czech Republic, BE—Belgium, RO—Romania, GR—Greece, AT—Austria, PT—Portugal, HU—Hungary, IE—Ireland, BG—Bulgaria, FI—Finland, SE—Sweden, DK—Denmark, SK—Slovakia, HR—Croatia, LT—Lithuania, EE—Estonia, SI—Slovenia, LV—Latvia, LU—Luxembourg, CY—Cyprus, MT—Malta.

From the analysis of the changes of GHG emissions in the EU countries, both decreases and increases were observed. The greatest reductions in GHG emissions, compared with those in 1990, were recorded in countries with a relatively small share in the total GHG emissions in the EU. These were Lithuania (−57.8%), Latvia (−55.5%), Romania (−53.2%), and Estonia (−50.4%). Probably, this is related to the economic changes taking place in these countries. In the countries with the highest share of GHG emissions in the EU, the reductions were, respectively, as follows: Germany (−31%), United Kingdom (−41.8%), France (−18.9%), Italy (−17.2%), and Poland (−13.1%) (Figure 3). The GHG emissions were higher in Cyprus (+55.0%), Spain (+15.5%), Portugal (+15.0%), Ireland (+9.9%), and Austria (+0.6%). Apart from Spain, countries in this group have a small share in the total EU's GHG emissions, which practically does not affect the total emission.

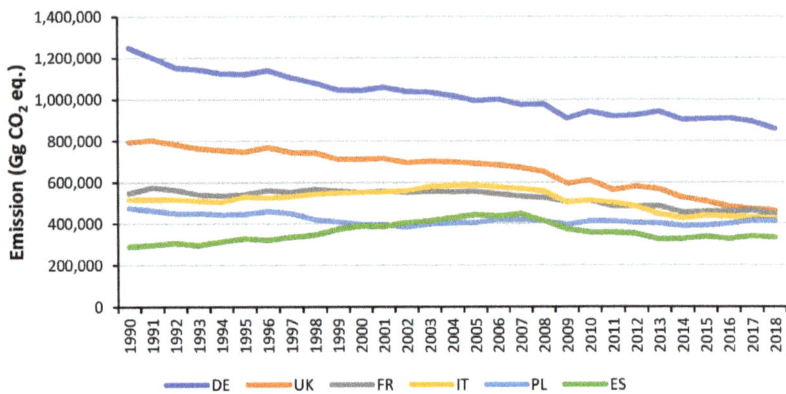

Figure 3. Total GHG emission in selected countries in 1990–2018.

3.2. *Agricultural GHG Emission in the EU*

One of the sources of GHG emissions is agriculture. Its average share in total GHG emissions (without LULUCF) in the EU in 2005–2018 was 9.3%. The share of this sector in total GHG emissions in the EU has increased from 8.4% in 2005 to 10.3% in 2018 (Figure 1). Changes in GHG emissions from agriculture are not in line with the trend in total GHG emissions in the EU. The agricultural GHG emissions showed a downward trend in the years 2005–2012, whereas since 2013, GHG emissions increased slightly from 427.6 Tg CO_2 eq. up to 440.8 Tg CO_2 eq. in 2017. Compared with the previous year, only in 2018, it was recorded a slight decrease in emissions by 1.3%. The analysis of the dynamics of emission changes in relation to the base year 2005 generally showed a decrease in GHG emissions in 2006–2018. In 2018, GHG emissions from agriculture in the EU amounted to 435.3 Tg CO_2 eq. (CO_2–2.6%, CH_4–53.7%, N_2O–43.7%) and were lower by 1.2% compared with those in 2005.

Almost 60% of the GHG emissions from agriculture in the EU come from France, Germany, the United Kingdom, Spain, and Poland. Meanwhile, the countries with the lowest agricultural GHG emissions—Luxembourg, Cyprus, and Malta—were responsible for 0.3% of total EU GHG emissions. The share of agriculture in total GHG emissions differed in each country. It was the largest in Ireland (32.7%), Denmark (22.9%), Latvia (22.3%), and Lithuania (21.1%), and the lowest in Malta (3%), Cyprus (5.7%), Slovakia (6.3%), Luxembourg (6.5%), and the Czech Republic (6.7%) (Figure 4).

Based on the analysis of changes in GHG emissions from agriculture in the years 2005–2018, it was noticed that there was a generally downward trend in France. The situation was opposite in Romania and Spain. Meanwhile, in Ireland and the Netherlands, emissions decreased until about 2011 and then emissions increased. In other countries, the emission was characterized by high variability without a clear upward or downward slope (Figure 5).

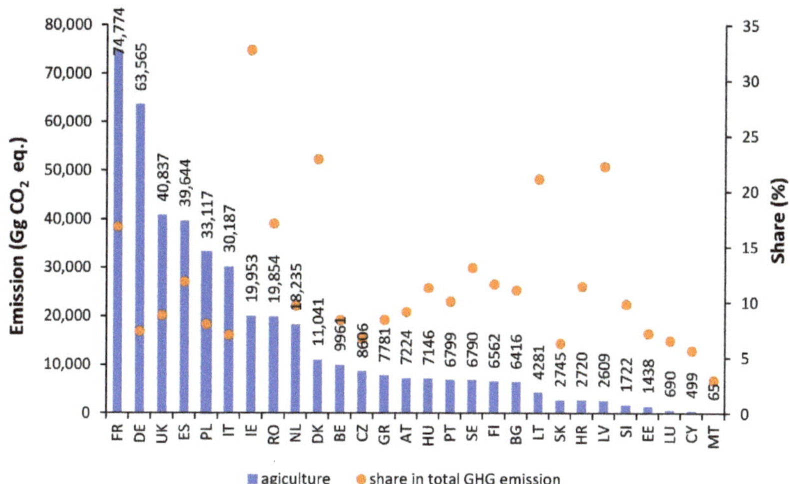

Figure 4. GHG emissions from agriculture and its share of national total GHG emissions in 2018.

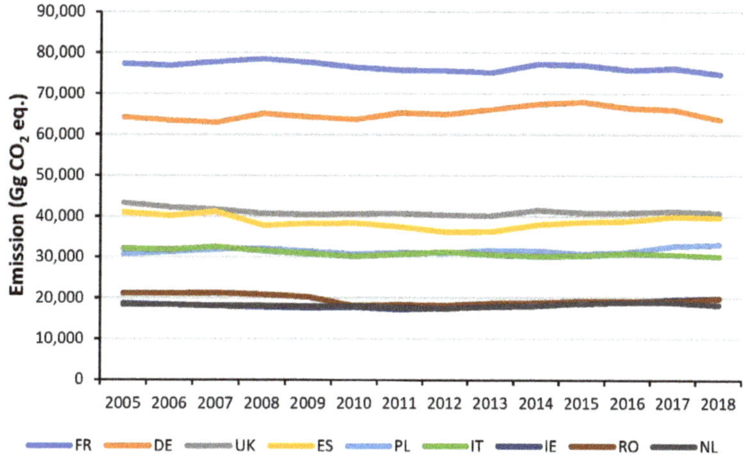

Figure 5. GHG emissions from agriculture in selected countries in 2005–2018.

3.3. *State of Reduction of Agricultural GHG Emissions in the EU*

The assumed reduction targets of GHG emissions from agriculture in the EU are related to 2005 (base year). The achievements of these targets in the EU countries are presented in Table 1.

Table 1. Agricultural greenhouse gases (GHG) emission changes in EU countries in 2005–2018.

Country	Agricultural GHG Emission in 2018 (Gg CO_2 eq.)	Changes in Agricultural GHG Emissions for 2005–2018 (%)	GHG Emission Limits in 2020 Compared with 2005 GHG Levels (%)	Reduction of GHG Emissions in 2030 Compared with 2005 GHG Levels (%)
FR	74,774.04	−3.3	−14	−37
DE	63,564.89	−1.0	−14	−38
UK	40,837.00	−5.6	−16	−37
ES	39,643.76	−3.2	−10	−26
PL	33,117.07	7.9	14	−7
IT	30,186.58	−5.8	−13	−33
IE	19,953.07	6.5	−20	−30
RO	19,854.03	−6.1	19	−2
NL	18,234.55	−0.6	−10	−36
DK	11,041.26	−2.0	−20	−39
BE	9960.88	−3.8	−15	−35
CZ	8606.50	5.1	9	−14
GR	7781.50	−13.1	−4	−16
AT	7224.35	3.3	−16	−36
HU	7145.64	16.5	10	−7
PT	6798.76	1.6	1	−17
SE	6790.17	−3.6	−17	−40
FI	6562.49	0.3	−16	−39
BG	6415.69	24.1	20	0
LT	4280.66	3.3	15	−9
SK	2745.29	4.5	13	−12
HR	2720.30	−17.5	-	−7
LV	2609.40	12.3	17	−6
SI	1721.71	−0.6	4	−15
EE	1437.79	20.4	11	−13
LU	690.44	9.8	−20	−40
CY	499.40	−6.3	−5	−24
MT	65.46	−13.7	5	−19

Explanation: red color—EU countries that have not yet reached the 2020 GHG limit in 2018; green color—EU countries that have reached the 2020 GHG limit in 2018. Source: [12,14].

According to decision 406/2009/EC of the European Parliament and of the Council of 23 April 2009 on the effort of Member States to reduce their greenhouse gas emissions to meet the Community's greenhouse gas emission reduction commitments up to 2020, Member States have set GHG emission limits for 2020 compared with 2005 in non-ETS sectors, which include agriculture. Depending on the size of the country, the structure of agriculture, and the assumed changes in this sector, some of them are to reduce emissions, and some may increase them within the assumed limit. Comparing the assumed GHG limits for 2020 with the GHG emissions for 2018, 11 of the EU Member States meet the GHG emission limits (green color in Table 1). These are mainly countries with a small share in the EU GHG emissions, accessed to EU after 2004. The countries, which have not yet reached the GHG limits for 2020 (red color in Table 1) are mainly old EU Member States, and among them, the countries with the highest share in GHG emissions from agriculture in the EU.

Comparing the agricultural GHG emissions in the EU in 2005 and 2018, there was a 2% reduction in emissions during this time. This is lower than the limit for the non-ETS sector, which is 10% (base year 2005). It should be noted that the assumed 2020 GHG emission limits apply to all non-ETS sectors together, not only agriculture. On the other hand, the GHG reduction limit for 2030 is much higher and, for non-ETS sector, assumed 30%. It seems that to achieve this goal, it may be necessary to reach a significant reduction in each area of the non-ETS sector.

Studying the GHG limits established for 2030, it may be concluded that three countries should not have problems with reaching these limits: Malta, Greece, and Croatia. Other countries should take decisive steps to implement measures to reduce GHG emissions (Table 1).

3.4. GHG Emissions from Main Agricultural Sources

According to the IPCC methodology [14], the main sources of GHG emissions from agriculture are enteric fermentation (CH_4), manure management (CH_4, N_2O), agricultural soils (N_2O), field burning of agricultural residues (CH_4, N_2O), liming (CO_2), and urea application (CO_2). The structure of GHG emissions in 2018 in the EU is shown in Figure 6. The largest sources were enteric fermentation (45%) and agricultural soils (37.8%), the emissions of which are mainly related to the use of natural and mineral fertilizers. Compared with the analysis conducted by Syp [15] for GHG emissions in the EU in 2014 in the agricultural sector, the shares of the two main emission sources were 45% agricultural soils and 38% enteric fermentation.

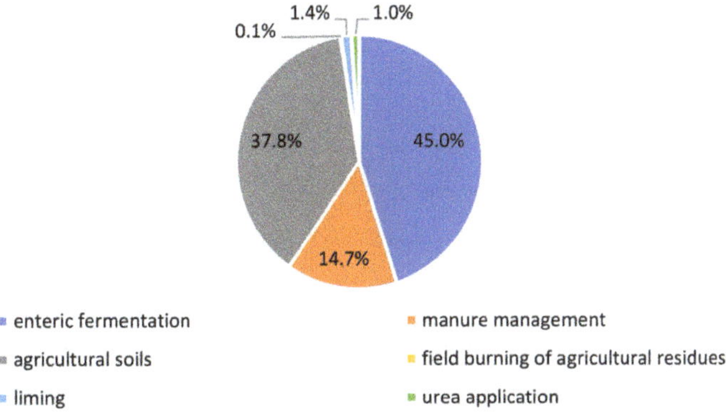

Figure 6. GHG emissions from agricultural sources in the EU in 2018.

Based on an analysis of GHG emissions from agricultural sources for 2005–2018 in the EU, a downward trend was recorded in emissions from enteric fermentation, manure management, field burning of agricultural residues, and liming. Only the emissions from urea application increased. For agricultural soils, emissions decreased until 2009, after which an upward trend was observed (Figure 7).

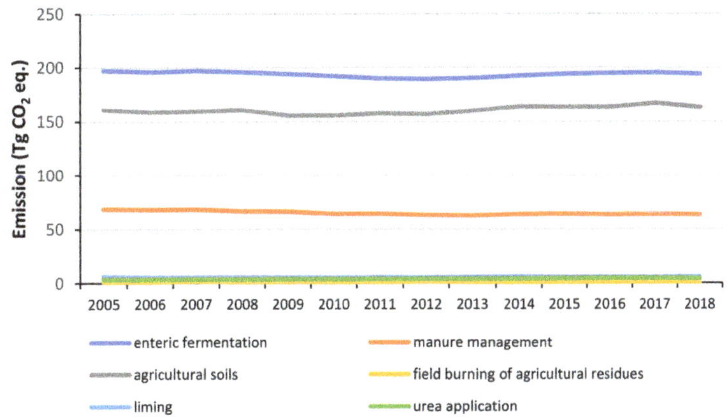

Figure 7. GHG emissions from agricultural sources in the EU in 2005–2018.

3.4.1. Enteric Fermentation

In 2018, GHG emissions from enteric fermentation in the EU amounted to 194 Tg CO_2 eq. (CH_4–100%). Analyzing the share of GHG emissions from enteric fermentation in 2018 in EU countries, it was noticed that over 40% of EU GHG emissions from enteric fermentation came from three countries: France, Germany, and the United Kingdom with shares of 17.7%, 12.9%, and 10.9%, respectively. Luxembourg, Cyprus, and Malta had the lowest emissions from enteric fermentation (Figure 8). The GHG emissions from enteric fermentation are related primarily to the amount of livestock production, resulting from the size, geographic location, and policies of the country. This has a significant impact on the share of enteric fermentation in the total GHG emissions from agriculture. In eight countries, enteric fermentation was over 50% of GHG emissions from agriculture: Luxembourg (58.4%), Ireland (57.9%), Austria (57.0%), Romania (54.6%), Slovenia (53.9%), Cyprus (52.4%), the United Kingdom (51.8%), and Portugal (51.4%). Only in two countries were the emissions from this source below 30%: Bulgaria (23.2%) and Hungary (28.7%).

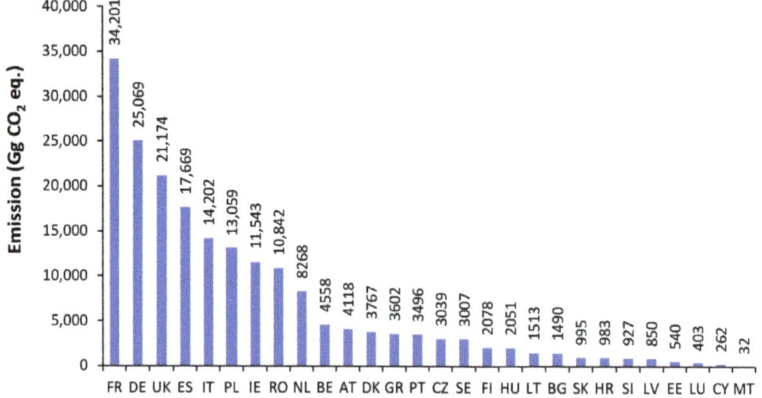

Figure 8. GHG emissions from enteric fermentation in EU countries in 2018.

The analysis of changes in GHG emissions from enteric fermentation in the period 2005–2018 showed a downward trend in the countries that are the largest sources (France, Germany, the United Kingdom). An upward trend was observed in Poland and Italy. In Ireland, emissions decreased until 2011 and then emissions increased.

The dynamics of changes in GHG emissions from enteric fermentation in relation to the 2005 base year showed emission reductions in France (−3.46%), Germany (−3.32%), the United Kingdom (−6.49%), Spain (−6.26%), and Romania (−8.71%). Increases in emissions were recorded in Poland (+10.87%), Ireland (+6.46%), and Italy (+3.60%) (Figure 9). This may be due to an increase in the cattle population.

In 2018, GHG emissions from enteric fermentation in the EU were lower by 1.75% compared with those in 2005. In this area, the majority of GHG emissions are related to dairy and beef cattle farming. It has a relatively high reduction potential for GHG emissions by 2030. The ongoing genetic and breeding work improve efficiencies such as feed conversion-to-milk yield ratio, which significantly reduce the amount of GHG emissions by 5–15% [16–18]. Further, modifications to nutrition by reducing fibre levels can lower emissions by 5–10% [18–20]. Additionally, the increase in the share of pasture feed in nutrition and the reduction in TMR consumption may result in reductions in GHG emissions [21,22]. In summary, the abovementioned actions may lead to a reduction in GHG emissions of up to 10%.

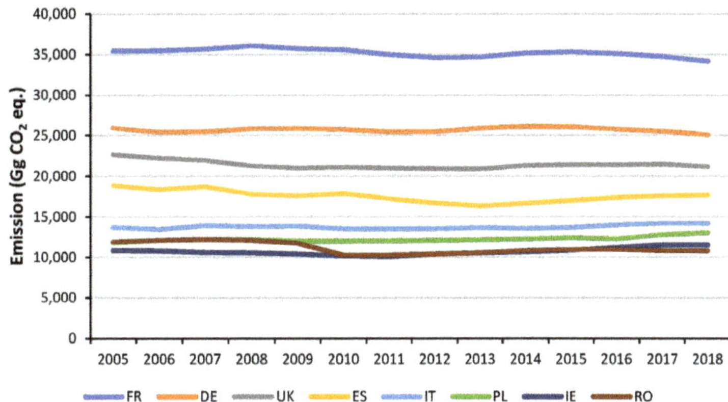

Figure 9. GHG emissions from enteric fermentation in selected EU countries in 2005–2018.

3.4.2. Manure Management

In 2018, GHG emissions from manure management in the EU amounted to 63 Tg CO_2 eq. (CH_4–65%, N_2O–35%). The biggest sources of GHG emissions from manure management were Germany (14.8%), Spain (13.8%), the United Kingdom (11.1%), and France (10%); Cyprus, Luxembourg, and Malta had the lowest emissions (Figure 10).

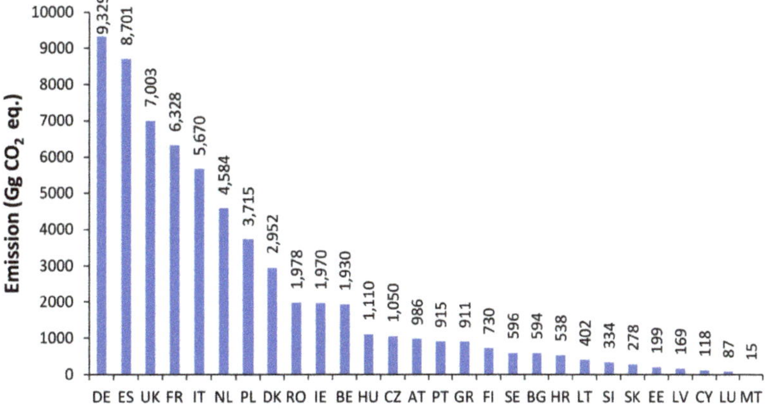

Figure 10. GHG emissions from manure management in EU countries in 2018.

Only in five countries, emissions from manure management in GHG emissions from agriculture in these countries was over 20%: Denmark (26.7%), the Netherlands (25.1%), Cyprus (23.7%), Malta (22.7%), and Spain (21.9%); meanwhile, in six countries, the share of emission from this source was below 10%: Ireland (9.9%), Lithuania (9.4%), Bulgaria (9.3%), Sweden (8.8%), France (8.5%), and Latvia (6.5%).

The analysis of GHG emissions from manure management in 2005–2018 showed that in Germany and Denmark, there was a clear downward trend in this period. In the Netherlands there was a clear upward trend. Emissions remained constant in the United Kingdom and France (Figure 11).

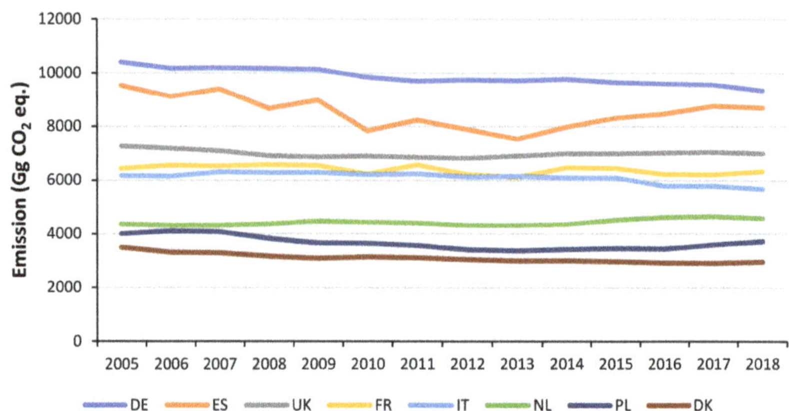

Figure 11. GHG emissions from manure management in selected EU countries in 2005–2018.

The dynamics of changes in GHG emissions from manure management in relation to the 2005 base year showed significant reductions in emissions in Denmark (−15.3%), Germany (−10.3%), Spain (−8.7%), Italy (−8.2%), and Poland (−7.2%). In the Netherlands, there were increases in emissions (+5.1%).

In 2018, GHG emissions from manure management in the EU were lower by 7.94% compared with those in 2005. Ongoing intensification of livestock production leads to increasing volumes of manure to be managed and may increase GHG emissions from the manure management area. At the same time, it changes the level of specialization and mechanization of European livestock production, which may reduce emissions. In larger farms, implementation of GHG emission reduction techniques is cheaper per animal, but complicated. One method of reducing GHG emissions is by improving or changing the housing system. It may result, depending on the type of animal, up to a 30% reduction of GHG emissions [23–26]. Another effective method is covering manure or slurry storage and closing the slurry channel, producing a 10% GHG reduction [27,28]. Total GHG emission reduction potential is quite low at about 10%. Changing the housing system is quite an effective way to reduce GHG emissions and is relatively cheap but only in new buildings. In existing livestock buildings, it often requires a change in construction, which is expansive. On the other hand, covering manure and slurry storages is not so expensive and may be forced by legal acts.

3.4.3. Agricultural Soils

In 2018, GHG emissions from agricultural soils in the EU amounted to 163 Tg CO_2 eq. (N_2O–100%). GHG emissions from agricultural soils in 2018 were the highest in France, Germany, and Poland whose shares were 19.7%, 15.1%, and 9.4%, respectively. Significant sources of emissions were also Spain (7.6%) and the United Kingdom (7.0%). The lowest emissions were again noted in Luxembourg, Cyprus, and Malta (Figure 12).

In six countries, the share of emissions from agricultural soils in the GHG emissions from agriculture in these countries was over 50%: Bulgaria (64.9%), Latvia (59.3%), Lithuania (54.7%), Finland (53.9%), Hungary (52.5%), and Slovakia (50.7%). In the next six countries, this share was over 40%: the Czech Republic (49.1%), Estonia (47.3%), Poland (46.4%), Sweden (45.0%), France (42.9%), and Croatia (41.3%); in nine countries, the share of GHG emissions from agricultural soils was below 30%: Ireland (29.5%), the Netherlands (29.3%), Malta (28.9%), the United Kingdom (27.9%), Austria (27.7%), Italy (27.6%), Luxembourg (27.3%), Slovenia (25.5%), and Cyprus (23.9%).

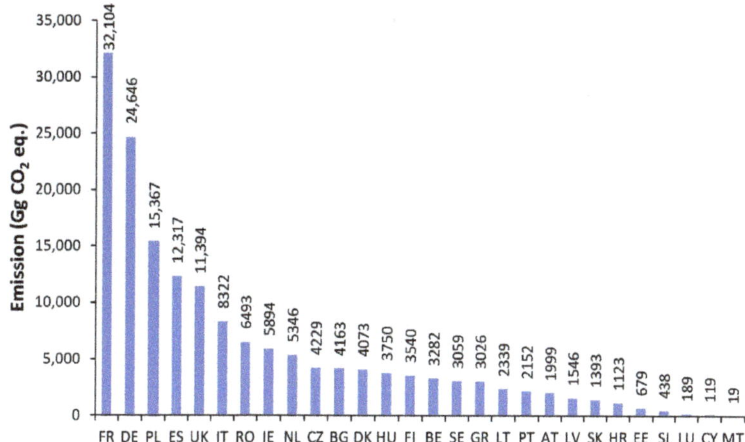

Figure 12. GHG emissions from agricultural soils in EU countries in 2018.

Changes in GHG emissions from agricultural soils in 2005–2018 were similar in the United Kingdom and Ireland; emissions remained constant. A downward trend was observed in Italy and France while the upward trend was in Spain, Poland, and Romania (Figure 13).

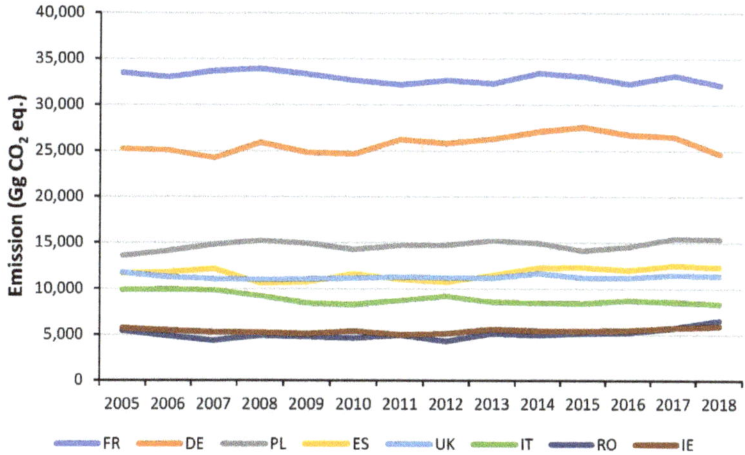

Figure 13. GHG emissions from agricultural soils in selected EU countries in 2005–2018.

The analysis of the dynamics of changes in GHG emissions from agricultural soils in relation to the 2005 base year showed a clear increase in emissions in Poland (+13.1%) and Romania (+20%). These increases may result from the stabilization of the agricultural market and the decline in the fallow land. Lower increases in emissions were observed in Spain (+5.7%) and Ireland (+3.8%). Decreases occurred in Italy (−15.6%), France (−4.1%), the United Kingdom (−3.1%), and Germany (−2.2%).

In 2018, GHG emissions from agricultural soils in the EU were higher by 1.36% compared with those in 2005. The natural and synthetic fertilization is the main source of GHG emissions in the area of agricultural soils. The best way to reduce these emissions is to optimize the process of fertilization, understood as a precise selection of the fertilizer dose. The dose selection is closely related to the method of fertilization. To minimize the fertilizer dose and GHG emissions, the main direction is the direct land application of fertilizers.

These solutions primarily reduce the time of contact between fertilizers and air by their covering by soil shortly after application in the field or their application directly into the soil. Using this method, it was observed that the GHG emissions reduced by 20% [29–31].

3.4.4. Field Burning of Agricultural Residues

In 2018, GHG emissions from field burning of agricultural residues in the EU amounted to 0.6 Tg CO_2 eq. (CH_4–73%, N_2O–27%). In 2018, GHG emissions from field burning of agricultural residues were noted in only thirteen countries; it was the largest in Romania and accounted for nearly 57% of EU emissions in this source, followed by France (9.5%) and Portugal (8.2%). Austria (0.1%), Hungary (0.06%), and Cyprus (0.04%) had the smallest share in GHG emissions from field burning of agricultural residues in the EU (Figure 14).

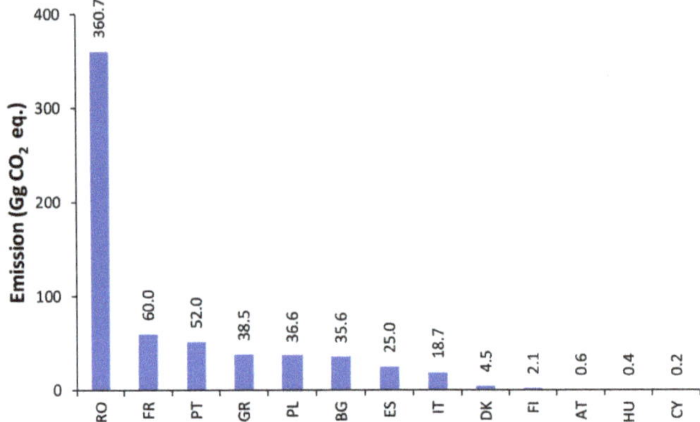

Figure 14. GHG emissions from field burning of agricultural residues in EU countries in 2018.

The highest share of emissions from field burning of agricultural residues in the total emissions from agriculture was in Romania and was 1.8%. The next three countries were Portugal (0.76%), Bulgaria (0.55%), and Greece (0.49%). In other countries, this share was marginal.

According to the analysis of GHG emissions from field burning of agricultural residues in 2005–2018, it was noticed that in Poland and Bulgaria, emissions slowly increased; a downward trend was observed in Romania, Greece, and Spain. In other countries, emissions remained relatively constant (Figure 15).

The analysis of the dynamics of changes in GHG emissions from field burning of agricultural residues in relation to the base year 2005 showed significant increases in emissions in Poland (+29.1%) and Bulgaria (+75.1%). It may be due to increase in the share of crops, which generate the field burning residues. On the other hand, significant decreases in emissions from this source were noted in Romania (−49.9%), Spain (−39.8%), and Greece (−23.1%).

In 2018, GHG emissions from field burning of agricultural residues in the EU were lower by 36.56% compared with those in 2005. Field burning of agricultural residues is fast and economical, but it is highly unsustainable, as it produces large amounts of air pollutants. The best way to reduce GHG emissions from this area is through the legal prohibition of such practices and alternative agricultural residues management. It may be converted to formed fuel (brick, pellets) or combusted directly for energy [32,33]. Another method to reduce emissions from this area is to compost the residues for fertilization and incorporation into the soil [34,35]. Projected reduction level of GHG emissions in this area may be as much as 70%.

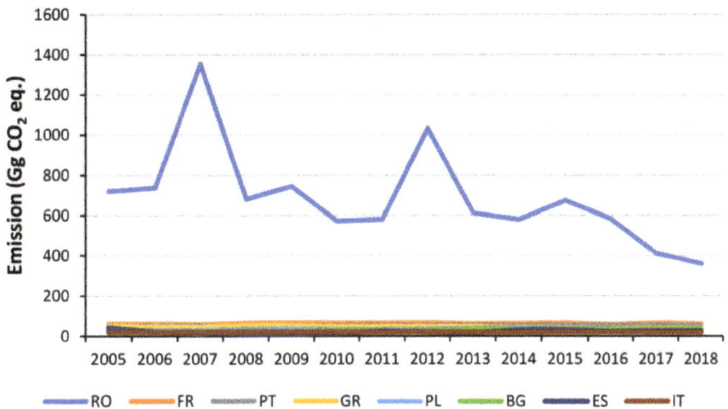

Figure 15. GHG emissions from field burning of agricultural residues in selected EU countries in 2005–2018.

3.4.5. Liming

In 2018, GHG emissions from liming in the EU amounted to 6 Tg CO_2 eq. (CO_2–100%). GHG emissions from liming in 2018 were inventoried in 24 countries. The largest source was Germany, responsible for almost 36% of emissions in the EU. The next were the United Kingdom and France, whose shares were 15.4% and 12.2%, respectively. Slovakia, Slovenia, Lithuania, Luxembourg, Croatia, Hungary, and Portugal had a small share in GHG emissions from liming in the EU (Figure 16).

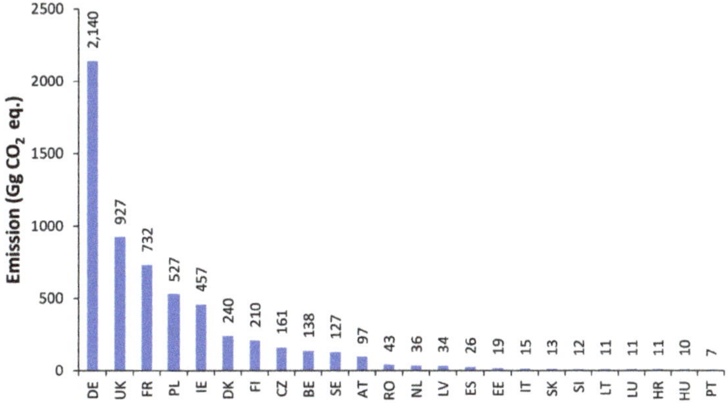

Figure 16. GHG emissions from liming in EU countries in 2018.

The largest share of emissions from liming in the total emissions from agriculture was in Germany and Finland, which amounted to 3.4% and 3.2%, respectively. In Denmark, Ireland, and the United Kingdom, the share was slightly above 2%. In contrast, a marginal share of 0.1% was noted in Portugal, Spain, Hungary, and Italy.

The analysis of GHG emissions from liming in 2005–2018 showed a clear upward trend in Denmark. On the other hand, a clear downward trend was observed in the United Kingdom. In other countries, the emissions were characterized by high variability without a clear upward or downward trend (Figure 17).

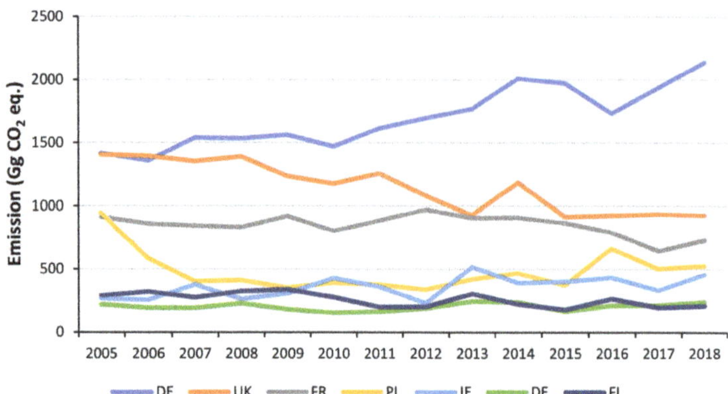

Figure 17. GHG emissions from liming in selected EU countries in 2005–2018.

The dynamics of changes in GHG emissions from liming in relation to the base year 2005 showed reductions for France (−20%), Finland (−27.5%), the United Kingdom (−34.1%), and Poland (−44.2%). Germany and Ireland noted an increase in emissions from liming (+51%) and (+71.5%), respectively.

In 2018, GHG emissions from liming in the EU were lower by 3.35% compared with those in 2005. The reduction of GHG emissions from liming is a complex process. The change in soil pH during liming contributes to the reduction of N_2O emissions from acid soils. However, this reduction effect is counterbalanced by CO_2 emissions during the chemical dissolution of calcium carbonate. Actually, research is carried out on the chemical and biological processes taking place during liming and the release of CO_2, N_2O, and CH_4 from the soil. There are a small number of studies on the effect of liming on GHG emissions due to changes in biological processes in soil, limiting the possibility of including these processes in the GHG emission modelling process [36–38]. The level of GHG emission reduction in the next 10 years can be assumed at the level of 0–5%.

3.4.6. Urea Application

In 2018, GHG emission from urea application in the EU amounted to 4 Tg CO_2 eq. (CO_2–100%). GHG emissions from urea application in 2018 were observed in 25 countries. France, with a share of over 30% of EU emissions, was the largest source. Subsequently, there were Germany, Spain, and Poland, whose shares amounted to 13.4%, 11.1%, and 9.7%, respectively. Sweden, Finland, Denmark, Cyprus, and Estonia had a small share in GHG emissions from urea application in the EU (Figure 18).

The largest shares of urea application in the total agricultural emissions were in Croatia and Slovakia, which was 2.4% in both countries. In six countries, this share was above 1%, they are France (1.72%), Hungary (1.58%), the Czech Republic (1.46%), Italy (1.34%), Poland (1.25%), and Spain (1.19%). On the other hand, a marginal share (0.04% and less) was recorded in Cyprus, Denmark, Estonia, Finland, and Sweden.

Based on the GHG emissions from urea application in the period 2005–2018, a clear upward trend in emissions was observed in France. An upward trend was also noted in the United Kingdom and Spain. In other countries, emissions were characterized by high variability without a clear upward or downward trend (Figure 19).

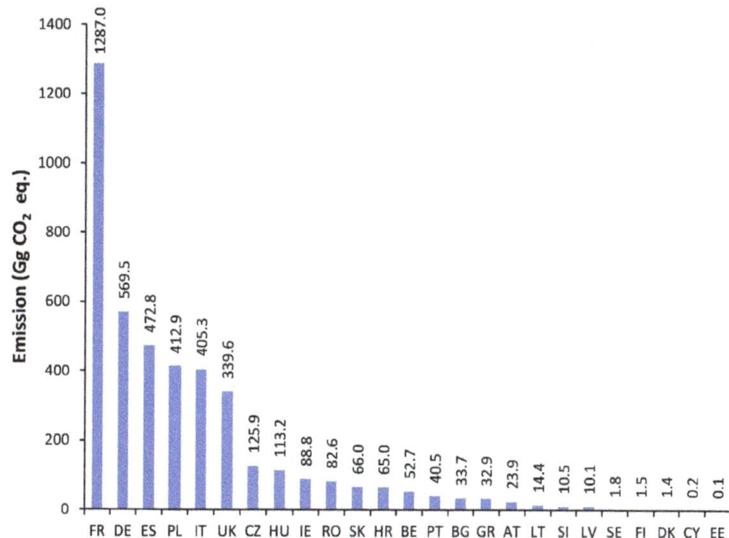

Figure 18. GHG emissions from urea application in EU countries in 2018.

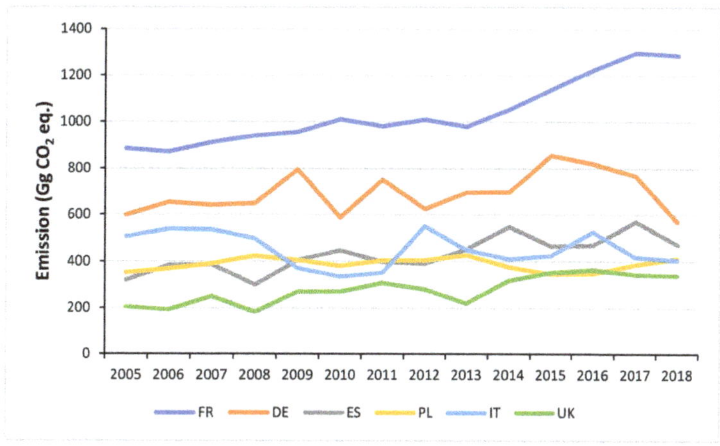

Figure 19. GHG emissions from urea application in selected EU countries in 2005–2018.

The analysis of the dynamics of changes in GHG emissions from urea application compared with the base year 2005 showed the largest emission increases in France (+45.3%), Spain (+48.3%), and the United Kingdom (+66.7%). Only in Italy, the GHG emissions significantly decreased (−20.1%).

In 2018, GHG emissions from urea application in the EU were higher by 25.72% compared with those in 2005. It is possible to significantly reduce GHG emissions in this area. One way is to replace urea with other ammonium nitrate fertilizers. For a long time, the potential of urease and nitrification inhibitors in reducing greenhouse gas emissions because of the massive use of agricultural fertilizers has also been recognized. Another method is by polymeric coating urea granules. The total GHG reduction potential of these methods ranges from 60 to 90% [39–42].

4. Conclusions

Climate change is a global problem, therefore, only the efforts of many countries—especially the largest ones—can bring measurable benefits in the form of stabilization and then reduction of anthropogenic GHG emissions into the atmosphere. For this reason, measures taken by individual EU countries should be coordinated, because then one can expect significantly beneficial effects of the policy, as a result of revealing synergistic effects. The proposed measures and instruments to reduce GHG emissions and to mitigate climate change result from the level of development of EU countries and their economic situation.

Comparing 2018 with the base year 2005, emissions from the agricultural sector decreased by about 2%, which is less than the assumed 10% reduction of GHG emissions in non-ETS sector. The ambitious goals set by the EU for 2030 assume a 30% reduction in the non-ETS sector. This will require a significant reduction of GHG emissions from agriculture. Based on the analysis of the GHG emission structure and available reduction techniques, it was calculated that in this period, it should be possible to reduce emissions from agriculture by about 15%. The concentration and intensification of agriculture in the EU, considered as a threat to the environment, may contribute to the reduction of GHG emissions. This is related to the areas where emissions may be significantly reduced, in particular, enteric fermentation, manure management, and agricultural soils, which together account for about 98% of GHG emissions. Reduction of emissions from these areas is associated with the rebuilding of existing buildings or the construction of new facilities, the purchase of equipment for the precise preparation of feed and the application of fertilizers, and the employment of qualified personnel. In large farms, the implementation of such techniques generates lower investment and operating unit costs. In smaller farms, the application of reduction techniques is simply unprofitable financially.

Reducing GHG emissions requires the involvement of significant human resources, changes in legal regulations, financial outlays, as well as organizational and technical changes. However, the level of reduction is difficult to predict because the changes in the livestock population and the crop structure for the next 10 years, which have a direct impact on GHG emission levels, are difficult to determine.

Author Contributions: Conceptualization, P.M.-B. and W.R.; methodology, P.M.-B. and W.R.; formal analysis, P.M.-B. and W.R.; investigation, P.M.-B. and W.R.; resources, P.M.-B. and W.R.; writing—original draft preparation, P.M.-B. and W.R.; writing—review and editing, P.M.-B. and W.R.; visualization, P.M.-B. and W.R.; supervision, P.M.-B. and W.R. All authors have read and agreed to the published version of the manuscript.

Funding: This research received no external funding.

Institutional Review Board Statement: Not applicable.

Informed Consent Statement: Not applicable.

Data Availability Statement: The data on greenhouse gas emissions are available online at https://www.eea.europa.eu/data-and-maps/data/data-viewers/greenhouse-gases-viewer, accessed on 15 February 2021.

Conflicts of Interest: The authors declare no conflict of interest.

References

1. Kweku, D.; Bismark, O.; Maxwell, A.; Desmond, K.; Danso, K.; Oti-Mensah, E.; Quachie, A.; Adormaa, B. Greenhouse Effect: Greenhouse Gases and Their Impact on Global Warming. *J. Sci. Res. Rep.* **2018**, *17*, 1–9. [CrossRef]
2. Cofała, J.; Amann, M.; Asman, W.; Bertok, I.; Heyes, C.; Höglund-Isaksson, L.; Klimont, Z.; Schöpp, W.; Wagner, F. Integrated Assessment of Air Pollution and Greenhouse Gases Mitigation in Europe. *Arch. Environ. Prot.* **2010**, *36*, 29–39.
3. Le Treut, H.; Somerville, R.; Cubasch, U.; Ding, Y.; Mauritzen, C.; Musset, A.; Peterson, T.; Prather, M. Historical Overview of Climate Change. In *Climate Change 2007: The Physical Science Basis. Contribution of Working Group I to the Fourth Assessment Report of the Intergovernmental Panel on Climate Change*; Solomon, S., Qin, D., Manning, M., Chen, Z., Marquis, M., Averyt, K.B., Tignor, M., Miller, H.L., Eds.; Cambridge University Press: Cambridge, UK, 2007.
4. Valipour, M.; Bateni, S.M.; Jun, C. Global Surface Temperature: A New Insight. *Climate* **2021**, *9*, 81. [CrossRef]

5. Nijsse, F.J.; Cox, P.M.; Huntingford, C.; Williamson, M.S. Decadal Global Temperature Variability Increases Strongly with Climate sensitivity. *Nat. Clim. Chang.* **2019**, *9*, 598–601. [CrossRef]
6. Eurostat. The Source Data for GHG Emissions. 2020. Available online: http://ec.europa.eu/eurostat/data/database (accessed on 12 December 2020).
7. Schimel, D.; Alves, D.; Enting, I.; Heimann, M. Chapter 2: Radiative Forcing of Climate Change. In *Climate Change 1995: The Science of Climate Change. Contribution of Working Group I to the Second Assessment Report of the Intergovernmental Panel on Climate Change*; Houghton, J.T., Meira Filho, L.G., Callander, B.A., Harris, N., Kattenberg, A., Maskell, K., Eds.; Press Syndicate of the University of Cambridge: Cambridge, UK, 1996.
8. Myhre, G.; Shindell, D.; Bréon, F.-M.; Collins, W.; Fuglestvedt, J.; Huang, J.; Koch, D.; Lamarque, J.-F.; Lee, D.; Mendoza, B.; et al. Anthropogenic and Natural Radiative Forcing. In *Climate Change 2013: The Physical Science Basis. Contribution of Working Group I to the Fifth Assessment Report of the Intergovernmental Panel on Climate Change*; Stocker, T.F., Qin, D., Plattner, G.-K., Tignor, M., Allen, S.K., Boschung, J., Nauels, A., Xia, Y., Bex, V., Midgley, P.M., Eds.; Cambridge University Press: Cambridge, UK, 2013.
9. Prather, M.J.; Holmes, C.D.; Hsu, J. Reactive greenhouse gas scenarios: Systematic exploration of uncertainties and the role of atmospheric chemistry. *Geophys. Res. Lett.* **2012**, *39*, L09803. [CrossRef]
10. COM. Communication from the Commission to the European Parliament, the Council, the European Economic and Social Committee and the Committee of the Regions. 20 20 by 2020 Europe's Climate Change Opportunity. Available online: https://eur-lex.europa.eu/legal-content/EN/TXT/?uri=COM:2008:0030:FIN (accessed on 12 December 2020).
11. PE. Decision No 406/2009/EC of the European Parliament and of the Council of 23 April 2009 on the Effort of Member States to Reduce Their Greenhouse Gas Emissions to Meet the Community's Greenhouse Gas Emission Reduction Commitments Up to 2020. Available online: https://eur-lex.europa.eu/legal-content/EN/TXT/?uri=CELEX:32009D0406 (accessed on 12 December 2020).
12. COM. Communication from the Commission to the European Parliament, the Council, the European Economic and Social Committee and the Committee of the Regions. A Policy Framework for Climate and Energy in the Period from 2020 to 2030. Available online: https://eur-lex.europa.eu/legal-content/EN/ALL/?uri=celex%3A52014DC0015 (accessed on 12 December 2020).
13. EEA. European Environment Agency. 2020. Available online: https://www.eea.europa.eu/data-and-maps/data/data-viewers/greenhouse-gases-viewer (accessed on 12 December 2020).
14. IPCC. IPCC Guidelines for National Greenhouse Gas Inventories. 2006. Available online: https://www.ipcc-nggip.iges.or.jp/public/2006gl (accessed on 12 December 2020).
15. Syp, A. Greenhouse Gas Emissions from Agriculture in 1990–2014. Zeszyty Naukowe Szkoły Głównej Gospodarstwa Wiejskiego w Warszawie. *Problemy Rolnictwa Światowego* **2017**, *17*, 244–255. [CrossRef]
16. Kiefer, L.; Menzel, F.; Bahrs, E. The effect of feed demand on greenhouse gas emissions and farm profitability for organic and conventional dairy farms. *J. Dairy Sci.* **2014**, *97*, 7564–7574. [CrossRef]
17. Rotz, A.C. Modeling greenhouse gas emissions from dairy farms. *J. Dairy Sci.* **2017**, *101*, 6675–6690. [CrossRef] [PubMed]
18. Capper, J.L.; Cady, R.A. The effects of improved performance in the U.S. dairy cattle industry on environmental impacts between 2007 and 2017. *J. Anim. Sci.* **2020**, *98*, 1–14. [CrossRef]
19. Rendon-Huerta, J.A.; Pinos-Rodriguez, J.; Kebreab, E. Animal nutrition strategies to reduce greenhouse gas emissions in dairy cattle. *Acta Univ.* **2018**, *28*, 34–41. [CrossRef]
20. Hammond, K.J.; Humphries, D.J.; Crompton, L.A.; Green, C.; Reynolds, C.K. Methane emissions from cattle: Estimates from short-term measurements using a GreenFeed system compared with measurements obtained using respiration chambers or sulphur hexafluoride tracer. *Anim. Feed. Sci. Technol.* **2015**, *203*, 41–52. [CrossRef]
21. O'Brien, D.; Geoghegan, A.; McNamara, K.; Shalloo, L. How can grass-based dairy farmers reduce the carbon footprint of milk? *Anim. Prod. Sci.* **2016**, *56*, 495–500. [CrossRef]
22. Van der Weerden, T.; Beukes, P.; De Klein, C.; Hutchinson, K.; Farrell, L.; Stormink, T.; Romera, A.; Dalley, D.; Monaghan, R.; Chapman, D.; et al. The Effects of System Changes in Grazed Dairy Farmlet Trials on Greenhouse Gas Emissions. *Animals* **2018**, *8*, 234. [CrossRef] [PubMed]
23. Fournel, S.; Pelletier, F.; Godbout, S.; Lagacé, R.; Feddes, J. Greenhouse Gas Emissions from Three Cage Layer Housing Systems. *Animals* **2012**, *2*, 1–15. [CrossRef] [PubMed]
24. Philippe, F.X.; Nicks, B. Review on greenhouse gas emissions from pig houses: Production of carbon dioxide, methane and nitrous oxide by animals and manure. *Agric. Ecosyst. Environ.* **2015**, *199*, 10–25. [CrossRef]
25. Samsonstuen, S.; Åby, B.A.; Crosson, P.; Beauchemin, K.A.; Aass, L. Mitigation of greenhouse gas emissions from beef cattle production systems. *Acta Agric. Scand. Sect. A—Anim. Sci.* **2020**, *69*, 220–232. [CrossRef]
26. Witkowska, D.; Korczyński, M.; Koziel, J.; Sowińska, J.; Chojnowski, B. The Effect of Dairy Cattle Housing Systems on the Concentrations and Emissions of Gaseous Mixtures in Barns Determined by Fourier-Transform Infrared Spectroscopy. *Ann. Anim. Sci.* **2020**, *20*, 1487–1507. [CrossRef]
27. Kupper, T.; Häni, C.; Neftel, A.; Kincaid, C.; Bühler, M.; Amon, B.; Van der Zaag, A. Ammonia and greenhouse gas emissions from slurry storage—A review. *Agric. Ecosyst. Environ.* **2020**, *300*, 106963. [CrossRef]
28. Amon, B.; Kryvoruchko, V.; Amon, T. Influence of different methods of covering slurry stores on greenhouse gas and ammonia emissions. *Int. Congr. Ser.* **2006**, *1293*, 315–318. [CrossRef]
29. Shimizu, M.; Satoru, M.; Desyatkin, A.R.; Jin, T.; Hata, H.; Hatano, R. The effect of manure application on carbon dynamics and budgets in a managed grassland of Southern Hokkaido, Japan. *Agric. Ecosyst. Environ.* **2009**, *130*, 31–40. [CrossRef]

30. Smith, K.; Watts, D.; Way, T.; Torbert, H.; Prior, S. Impact of Tillage and Fertilizer Application Method on Gas Emissions in a Corn Cropping System. *Pedosphere* **2012**, *22*, 604–615. [CrossRef]
31. Duncan, E.W.; Dell, C.J.; Kleinman, P.J.A.; Beegle, D.B. Nitrous Oxide and Ammonia Emissions from Injected and Broadcast—Applied Dairy Slurry. *J. Environ. Qual.* **2017**, *46*, 36–44. [CrossRef] [PubMed]
32. Statuto, D.; Picuno, P. Improving the greenhouse energy efficiency through the reuse of agricultural residues. *Acta Hortic.* **2017**, *1170*, 501–508. [CrossRef]
33. Algieri, A.; Andiloro, S.; Tamburino, V.; Zema, D.A. The potential of agricultural residues for energy production in Calabria (Southern Italy). *Renew. Sustain. Energy Rev.* **2019**, *104*, 1–14. [CrossRef]
34. Viaene, J.; Van Lancker, J.; Vandecasteele, B.; Willekens, K.; Bijttebier, J.; Ruysschaert, G.; De Neve, S.; Reubens, B. Opportunities and barriers to on-farm composting and compost application: A case study from northwestern Europe. *Waste Manag.* **2016**, *48*, 181–192. [CrossRef]
35. Bedoić, R.; Ćosić, B.; Duić, N. Technical potential and geographic distribution of agricultural residues, co-products and by-products in the European Union. *Sci. Total Environ.* **2019**, *686*, 56–579. [CrossRef] [PubMed]
36. Page, K.L.; Allen, D.E.; Dalal, R.C.; Slattery, W. Processes and magnitude of CO_2, CH_4, and N_2O fluxes from liming of Australian acidic soils: A review. *Aust. J. Soil Res.* **2019**, *47*, 747–762. [CrossRef]
37. Barton, L.; Gleeson, D.B.; Maccarone, L.D.; Zúñiga, L.P.; Murphy, D.V. Is Liming Soil a Strategy for Mitigating Nitrous Oxide Emissions from Semi-arid Soils? *Soil Biol. Biochem.* **2013**, *62*, 28–35. [CrossRef]
38. Hénault, C.; Bourennane, H.; Ayzac, A.; Ratié, C.; Saby, N.P.A.; Cohan, J.-P.; Eglin, T.; Le Gall, C. Management of soil pH promotes nitrous oxide reduction and thus mitigates soil emissions of this greenhouse gas. *Sci. Rep.* **2019**, *9*, 20182. [CrossRef]
39. Watson, C.J. Urease activity and inhibition—Principles and practice. In Proceedings of the International Fertiliser Society, York, UK, 1 April 2000.
40. Krasuska, E.; Pudełko, R.; Faber, A.; Jarosz, Z.; Borzęcka-Walker, M.; Syp, A.; Kozyra, J. Optimization and risk analysis of greenhouse gas emissions depending on yield and nitrogen rates in winter wheat cultivation. *J. Food Agric. Environ.* **2013**, *11*, 2217–2219.
41. Byrne, M.P.; Tobin, J.T.; Forrestal, P.J.; Danaher, M.; Nkwonta, C.G.; Richards, K.; Cummins, E.; Hogan, S.A.; O'Callaghan, T.F. Urease and Nitrification Inhibitors—As Mitigation Tools for Greenhouse Gas Emissions in Sustainable Dairy Systems: A Review. *Sustainability* **2020**, *12*, 6018. [CrossRef]
42. Xie, Y.; Tang, L.; Han, Y.L.; Yang, L.; Xie, G.X.; Peng, J.W.; Tian, C.; Zhou, X.; Liu, Q.; Rong, X.M.; et al. Reduction in nitrogen fertilizer applications by the use of polymer-coated urea: Effect on maize yields and environmental impacts of nitrogen losses. *J. Sci. Food Agric.* **2019**, *99*, 2259–2266. [CrossRef] [PubMed]

 MDPI

Article

Nongrowing Season CO_2 Emissions Determine the Distinct Carbon Budgets of Two Alpine Wetlands on the Northeastern Qinghai—Tibet Plateau

Chenggang Song [1,2,3,†], Fanglin Luo [4,5,†], Lele Zhang [1], Lubei Yi [6], Chunyu Wang [4,5], Yongsheng Yang [4,5], Jiexia Li [4,5], Kelong Chen [1,*], Wenying Wang [2,5,*], Yingnian Li [4,5] and Fawei Zhang [4,5,*]

1 School of Geographic Sciences, Qinghai Normal University, Xining 810008, China; scg8088@163.com (C.S.); zhang1986lele@163.com (L.Z.)
2 School of Life Sciences, Qinghai Normal University, Xining 810008, China
3 Qinghai Engineering Consulting Center, Xining 810008, China
4 Key Laboratory of Adaptation and Evolution of Plateau Biota, Northwest Institute of Plateau Biology, Chinese Academy of Sciences, Xinning Streets, Xining 810008, China; luofanglin@nwipb.cas.cn (F.L.); wangchunyu@nwipb.cas.cn (C.W.); ysyang@nwipb.cas.cn (Y.Y.); jxli@nwipb.cas.cn (J.L.); ynli@nwipb.cas.cn (Y.L.)
5 Institute of Sanjiangyuan National Park, Chinese Academy of Sciences, Xinning Streets, Xining 810008, China
6 Forestry Carbon Sequestration Service Center, Qinghai Forestry and Grassland Administration, Xining 810008, China; celtylb@163.com
* Correspondence: ckl7813@163.com (K.C.); wangwy0106@163.com (W.W.); fwzhang@nwipb.cas.cn (F.Z.); Tel.: +86-971-6133353 (F.Z.); Fax: +86-971-6143282 (F.Z.)
† These authors contribute equally.

Citation: Song, C.; Luo, F.; Zhang, L.; Yi, L.; Wang, C.; Yang, Y.; Li, J.; Chen, K.; Wang, W.; Li, Y.; et al. Nongrowing Season CO_2 Emissions Determine the Distinct Carbon Budgets of Two Alpine Wetlands on the Northeastern Qinghai—Tibet Plateau. *Atmosphere* **2021**, *12*, 1695. https://doi.org/10.3390/atmos12121695

Academic Editors: Baojie He, Ayyoob Sharifi, Chi Feng and Jun Yang

Received: 24 November 2021
Accepted: 15 December 2021
Published: 17 December 2021

Publisher's Note: MDPI stays neutral with regard to jurisdictional claims in published maps and institutional affiliations.

Abstract: Alpine wetlands sequester large amounts of soil carbon, so it is vital to gain a full understanding of their land-atmospheric CO_2 exchanges and how they contribute to regional carbon neutrality; such an understanding is currently lacking for the Qinghai—Tibet Plateau (QTP), which is undergoing unprecedented climate warming. We analyzed two-year (2018–2019) continuous CO_2 flux data, measured by eddy covariance techniques, to quantify the carbon budgets of two alpine wetlands (Luanhaizi peatland (LHZ) and Xiaobohu swamp (XBH)) on the northeastern QTP. At an 8-day scale, boosted regression tree model-based analysis showed that variations in growing season CO_2 fluxes were predominantly determined by atmospheric water vapor, having a relative contribution of more than 65%. Variations in nongrowing season CO_2 fluxes were mainly controlled by site (categorical variable) and topsoil temperature (T_s), with cumulative relative contributions of 81.8%. At a monthly scale, structural equation models revealed that net ecosystem CO_2 exchange (NEE) at both sites was regulated more by gross primary productivity (GPP), than by ecosystem respiration (RES), which were both in turn directly controlled by atmospheric water vapor. The general linear model showed that variations in nongrowing season CO_2 fluxes were significantly ($p < 0.001$) driven by the main effect of site and T_s. Annually, LHZ acted as a net carbon source, and NEE, GPP, and RES were 41.5 ± 17.8, 631.5 ± 19.4, and 673.0 ± 37.2 g C/(m^2 year), respectively. XBH behaved as a net carbon sink, and NEE, GPP, and RES were −40.9 ± 7.5, 595.1 ± 15.4, and 554.2 ± 7.9 g C/(m^2 year), respectively. These distinctly different carbon budgets were primarily caused by the nongrowing season RES being approximately twice as large at LHZ ($p < 0.001$), rather than by other equivalent growing season CO_2 fluxes ($p > 0.10$). Overall, variations in growing season CO_2 fluxes were mainly controlled by atmospheric water vapor, while those of the nongrowing season were jointly determined by site attributes and soil temperatures. Our results highlight the different carbon functions of alpine peatland and alpine swampland, and show that nongrowing season CO_2 emissions should be taken into full consideration when upscaling regional carbon budgets. Current and predicted marked winter warming will directly stimulate increased CO_2 emissions from alpine wetlands, which will positively feedback to climate change.

Keywords: CO_2 fluxes; boosted regression trees; structural equation models; alpine wetlands; Qinghai—Tibet Plateau

1. Introduction

Due to their water-logged and relatively low-temperature conditions, wetlands comprise a large component of global terrestrial carbon reserves; they store ~30% of the global soil carbon pool, despite only constituting ~5% of the land surface [1–3]. Although pristine alpine wetlands could be potential future carbon sources because of the projected warming and drying climate [1,4,5], carbon accumulation in mid- and high-latitude wetlands has increased slightly over recent decades [2,6]. Thus, increased knowledge of carbon dynamics and their responses to environmental controls in alpine wetlands is essential to address this discrepancy and to quantify the contributions of wetlands in mitigating atmospheric greenhouse gas concentrations [4,7,8].

Alpine wetlands can be either carbon sources or carbon sinks depending on local hydrothermal conditions, consequent vegetation types, soil physical and chemical properties, and microorganism compositions [9–12]. Gross primary productivity (GPP) and ecosystem respiration (RES) are two contrasting processes that determine carbon budgets, and those are closely related to ecohydrological factors, usually with nonlinear relationships [13–15]. Water availability has been shown to be an important control of the seasonal variability in CO_2 fluxes of alpine wetlands, through its simultaneous effects on vegetation growth and organic matter decomposition [4,9,16]. Vascular plant coverage has an asymptotic relationship with GPP, and an incremental relationship with RES, but their sensitivity varies by leaf area index and vegetation type [17,18]. Under the context of climate warming scenarios, recent studies have shown that plant productivity outweighs ecosystem respiration, and alpine peatlands have become more efficient carbon sinks because of a higher sensitivity of GPP compared to RES, longer growing season, and more plant carbon input [19–21]. Contrarily, other studies have suggested that decreased moisture availability caused by warming could stimulate more soil respiration and potentially reverse the carbon sink function [6,22,23]. Quantifying the relative contributions of biotic and abiotic controls on carbon budgets has the potential to understand this debate but has been poorly addressed because of their intertwined relationships [22]. Moreover, nongrowing season CO_2 emissions are believed to play a considerable role in carbon budgets, but there is still much uncertainty due to limited long-term field observations [4,18]. More detailed knowledge is greatly needed to quantify and predict the fate of the carbon function of alpine wetlands under remarkable winter warming and increasing precipitation [9,13]. Year-round continuous measurements, rather than merely from the growing season, are required in alpine wetlands to address this issue [23,24].

Carbon budgets of alpine wetlands on the Qinghai—Tibet Plateau (QTP) have drawn increased attention recently [11,13,25,26]; the QTP wetlands house relatively large soil carbon reservoirs because of lower decomposition rates and higher vegetation photosynthetic production compared with other surrounding wetland ecosystems [4,17]. Moreover, wetlands on the QTP have been expanding due to the warming climate and decreased human activity [27]. However, the carbon sink/source functions in alpine wetlands are highly variable. For instance, the wetlands of Zoige, Qinghai Lake (Xiaobohu: XBH), and Hehei have recently behaved as carbon sinks, with annual carbon accumulations of approximately 170, 250, and 510 g C/m^2, respectively [12,13,28]. In contrast, Luanhaizi (LHZ) wetland has been a carbon source and released approximately 100 g $C/(m^2$ year) to the atmosphere from 2005 to 2014 [14,17,25]. Therefore, quantifying the carbon dynamics of contrasting alpine wetlands will be helpful in understanding the confounding mechanisms in their carbon budgets. In this study, we obtained CO_2 fluxes and auxiliary environmental variables, measured by eddy covariance systems, over a two-year period from 2018 to 2019, from both LHZ and XBH on the northeastern QTP (Figure S1). This enables us to quantify inter-seasonal and inter-annual CO_2 budgets of the two contrasting alpine wetlands and to explore the relative contributions of each of the main environmental controls and their underlying ecological processes. We hypothesize that RES will contribute more than GPP to the distinct carbon functions of the two wetlands, because prior studies have reported that annual GPP was nearly equivalent in these two wetlands (630 g C/m^2 in LHZ [17] and 640 g C/m^2 in XHB [29]).

2. Materials and Methods

2.1. Site Descriptions

The two alpine wetlands, namely Luanhaizi (LHZ, 37°35′ N, 101°20′ E, 3250 m altitude) and Xiaobohu (XBH, 36°42′ N, 100°47′ E, 3210 m altitude), are both located on the northeastern Qinghai—Tibet Plateau (Figure S1); these can be classified as peatland and swampland, respectively, according to the differences in vegetation community, hydrological processes, and soil history [27]. The climate is cold and humid, belonging to the plateau monsoon climate of the alpine, frigid temperate zone.

LHZ lies adjacent to the Haibei National Field Research Station for Alpine Grasslands, which is a long-standing member of the Chinese Flux Observation and Research Network (ChinaFLUX). XBH lies on the eastern Qinghai Lake and belongs to the Qinghai Normal University. The mean annual air temperature and precipitation were –1.1 °C and 490.0 mm in LHZ [17], respectively, and 1.2 °C and 357.0 mm in XBH [29]. The 0–20 cm soil organic carbon content averaged 15.7% at LHZ and 6.2% at XBH. In LHZ, the dominant plant species were *Kobresis tibetica, Carex pamirensis, C. alrofusca, Blysmus sinocompressus, Hippuris vulgaris*, and *Triglochin palustre* [14]. The relative vegetation coverage, aboveground biomass, and belowground biomass during the flourishing growth stage (July-August) were approximately 95%, 340 g/m^2, and 4200 g/m^2, respectively [25]. LHZ was constantly submerged during the growing season, and the water depth was about 20 cm, with remarkable spatial and temporal variations [9]. In XBH, the dominant plant species were *C. limosa, B. sinocompressusm, K. tibetica, Schoenoplectus tabernaemontani*, and *Phragmites australis*. The relative vegetation coverage, aboveground biomass, and belowground biomass from July to August averaged 85%, 225 g/m^2, and 3200 g/m^2, respectively. The maximum volumetric topsoil water content was about 60% [29].

2.2. Measurements

CO_2 fluxes were measured by eddy covariance techniques at both sites. An eddy flux tower, including a three-dimensional ultrasound anemometer (CSAT3, the offset error < ± 8.0 cm/s, Campbell Scientific, Logan, UT, USA) and an open-path infrared CO_2/H_2O analyzer (Li-7500 for LHZ and Li-7500A for XBH, the typical zero drift was ± 0.1 ppm in CO_2 and ± 0.03 mmol/mol in H_2O, Li-Cor, Lincoln, NE, USA), was installed at a height of approximately 2.0 m above the surface; these have been in place since 2003 in LHZ and since 2010 in XBH. The raw data frequency was 10 Hz. A micro-meteorology station was also constructed adjacent to the flux tower, to monitor air temperature, atmospheric water vapor, wind speed and direction, four-component radiation, precipitation, and topsoil temperature. More detailed measurement information can be found in prior papers [17,29]. The growing season was simply defined as May–October for both sites and was robust because the enhanced vegetation index was above 0.2 during these periods (Figure 1f).

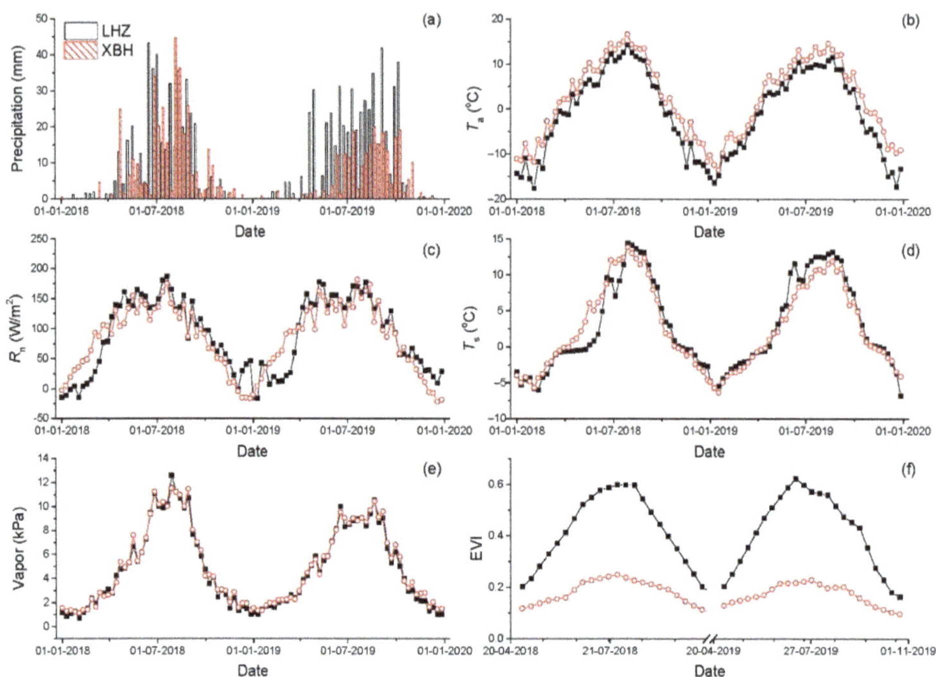

Figure 1. The seasonal variations in precipitation (**a**), air temperature (T_a), (**b**), net radiation (R_n), (**c**), topsoil temperature (T_s), (**d**), atmospheric water vapor (Vapor), (**e**) and enhanced vegetation index (EVI), (**f**) from 2018 to 2019 at the two alpine wetlands.

2.3. Data Collection and Processing

Subsequent to raw high-frequency data collection, half-hour CO_2 fluxes were computed from spike removal, two-dimensional coordinate rotation, time-lag compensation, and the Webb-Pearman-Leuning (WPL) 'Burba correction' for density fluctuations because of the self-heating effect [30] using Eddypro 7.0.6 (Li-Cor, Lincoln, NE, USA). The calculated CO_2 fluxes were screened to improve data quality by removal of a quality flag with "2" [31] and outliers (beyond 4.5 standard deviations) in a 10-day window, and by filtering against lower turbulence (threshold nighttime friction velocity was 0.15 m s^{-1}). The annual valid data coverage was ~40%, with ~70% and ~20% in the daytime and nighttime, respectively, comparable to data reported for ChinaFLUX [32] and other alpine sites [22]. The CO_2 flux data were gap-filled by a machine learning algorithm of boosted regression trees (BRT) [33], which was conducted using a dataset with a valid flux subset and corresponding routine meteorological subset, including air temperature, atmospheric water vapor, net radiation, wind velocity, and topsoil temperature. Machine learning algorithms were found to outperform other conventional gap-filling techniques by incorporating all of the main environmental controls simultaneously [34,35]. Because of the distinctly different processes governing CO_2 exchanges between growing season daytime (Growing: the simultaneous occurrence of plant photosynthesis and ecosystem respiration) and nongrowing season all-time and growing season nighttime (Nongrowing: only ecosystem respiration occurred), we filled data gaps separately. Firstly, boosted regression trees were fitted using the Growing valid dataset (BRT-growing) and the Nongrowing valid dataset (BRT-nongrowing), respectively. Secondly, the fitted BRT-growing and BRT-nongrowing, respectively, were used to fill data gaps with corresponding meteorological variables. After several comparisons, tree complexity, learning rate, and number of trees in BRT were fixed at 5, 0.005, and 5000, respectively.

The growing season daytime ecosystem respiration was estimated from the nongrowing-BRT and extrapolated based on the daytime meteorological variables. Therefore, daily ecosystem respiration (RES) is the sum of daytime respiration and nocturnal respiration. Daily GPP equals the values that daily RES minus daily NEE.

The enhanced vegetation index (EVI, MYD13Q1) was adopted as a metric for capturing the relationship between CO_2 fluxes and biotic variables in alpine grasslands [35–37]. The 8-day EVI was flux tower-centric at 500 × 500 m spatial resolution and was obtained from the Oak Ridge National Laboratory Distributed Active Archive Center (ORNL DAAC, https://modis.ornl.gov/globalsubset/). Half-hour values of CO_2 fluxes and meteorological variables were aggregated into an 8-day scale to match the temporal resolution of the EVI.

2.4. Statistical Analysis

BRT is well documented for ecological studies [33–35] and was here adopted to explore the relative contributions to 8-day CO_2 fluxes of the main environmental controls, which included continuous variables (air temperature, atmospheric water vapor, wind speed, net radiation, topsoil temperature, and enhanced area index) and categorical variables (site name, indicative site attributes). Vapor pressure deficit (VPD) can also represent atmospheric moisture status but was excluded from the key environmental controls because of its limited contribution to variations in CO_2 fluxes (Figure S2) in alpine grasslands [35]. More importantly, this algorithm can tolerate collinearity and nonnormality among environmental controls, which are rather common in environmental studies. Piecewise structural equation models (SEM) were used to identify the total effects and pathways of environmental controls on monthly CO_2 fluxes. BRT and SEM were performed by the package of "Dismo" [33] and "piecewiseSEM" [38] in R 4.0.3 [39]. Other conventional statistical analysis, such as linear regression, correlative analysis, and general linear models were also conducted in R.

3. Results

3.1. Information Regarding Abiotic and Biotic Controls

The two wetlands experienced small differences in climate during the two study years, with LHZ being relatively wetter and colder (Figure 1). Annual precipitation averaged 487.8 mm at LHZ and 291.8 mm at XBH. Annual mean air temperature (T_a) was –0.9 °C at LHZ, which was lower than that of XBH by 3.3 °C. Mean daily net radiation (R_n) was almost equivalent, being approximately 85.0 W/m^2 at both locations. Annual mean topsoil temperature (T_s) and atmospheric water vapor were also similar at the two wetlands, and averaged 2.7 °C and 4.9 kPa, respectively. Vapor pressure deficit (VPD) was 0.26 kPa in LHZ and 0.39 kPa in XBH. However, there was a distinct difference in EVI, with an annual value of 0.40 at LHZ and 0.17 at XBH. The maximum 8-day EVI was approximately 0.62 at the end of July in LHZ, which was over twice more than that of XBH (0.25; Figure 1).

3.2. Eight-Day Variations in CO_2 Fluxes

The trends of seasonal CO_2 fluxes were similar at the two sites, although the magnitudes were to some extent different (Figure 2). Daily GPP averaged 3.23 ± 1.88 g C/(m^2 d) (Mean ± S.D., the same below) at LHZ and 3.43 ± 2.25 g C/(m^2 d) at XBH. The paired-sample t-test showed that the difference was marginally significant (p = 0.07). Daily RES was 1.83 ± 0.90 g C/(m^2 d) at LHZ, which was significantly ($p < 0.001$) more (by 17.6%) than that of XBH. Therefore, daily NEE was 0.11 ± 1.57 g C/(m^2 d) at LHZ, and this was much higher than that of XBH (–0.11 ± 0.96 g C/(m^2 d)).

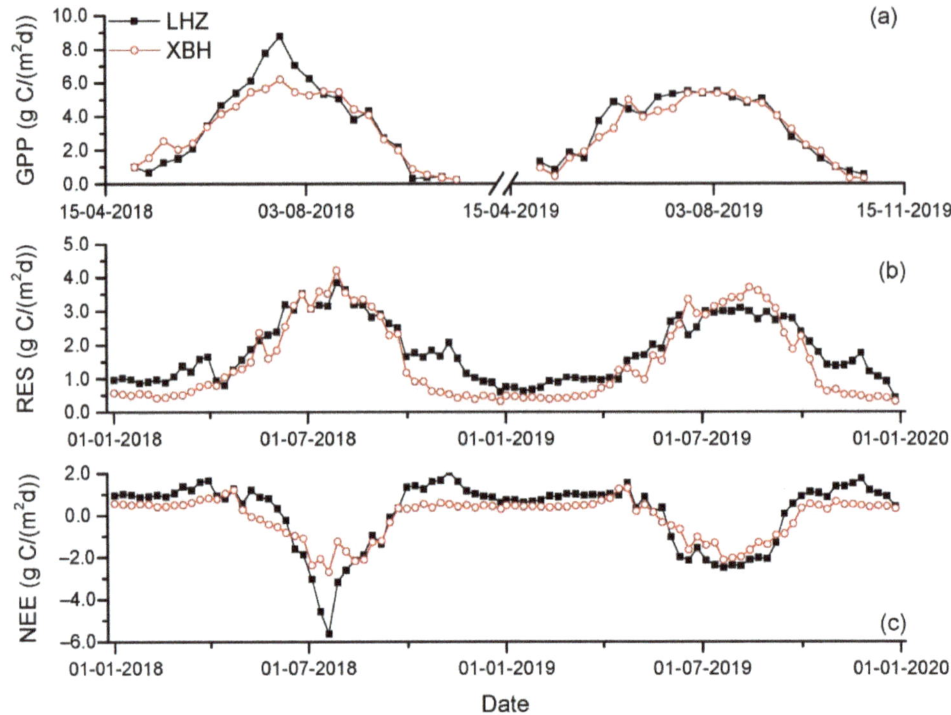

Figure 2. The 8-day variations in CO_2 fluxes of the two alpine wetlands (gross primary productivity GPP, (**a**); ecosystem respiration RES, (**b**); net ecosystem CO_2 exchange NEE, (**c**)).

During the growing season, the mean total deviance and 10-fold cross-validation deviance for 8-day GPP, RES, and NEE were 108.57 and 9.67, 72.14 and 3.47, and 269.57 and 17.64, respectively, which suggested the good performance of the BRT models ($R^2 > 0.90$, Figure S3). The results showed that the variations of 8-day GPP and 8-day NEE were mostly determined by atmospheric water vapor, with relative contributions of more than 65% (Figure 3a). GPP showed a positive correlation with atmospheric water vapor, while NEE showed a negative correlation with atmospheric water vapor (Figure 4a,d). Variations in 8-day RES were jointly controlled by atmospheric water vapor and T_s, which explained 76.5% of the total variations. RES showed a positive correlation with both atmospheric water vapor and T_s (Figure 4b,c). It should be noted that the relative contribution of site (categorical variables) was less than 2%, which indicated that the growing season patterns of 8-day CO_2 fluxes at the two sites should be attributed to variations in environmental controls rather than site attributes. Furthermore, the contribution of EVI to NEE was more than 10%, although it was only 5.0% for GPP and 9.8% for RES (Figure 3a).

During the nongrowing season, the mean total deviance and 10-fold cross-validation deviance for 8-day NEE were 9.57 and 2.59 ($R^2 = 0.73$; Figure S3), respectively. The variability in NEE was jointly determined by site attributes and T_s, with cumulative relative contributions of 81.8% (Figure 3b). Mean daily NEE was approximately 1.07 ± 1.57 g C/(m^2 d) at LHZ, which was about twice that of XBH (0.56 ± 0.96 g C/(m^2 d), Figure 4e). NEE exponentially increased with T_s at both sites (Figure 4f).

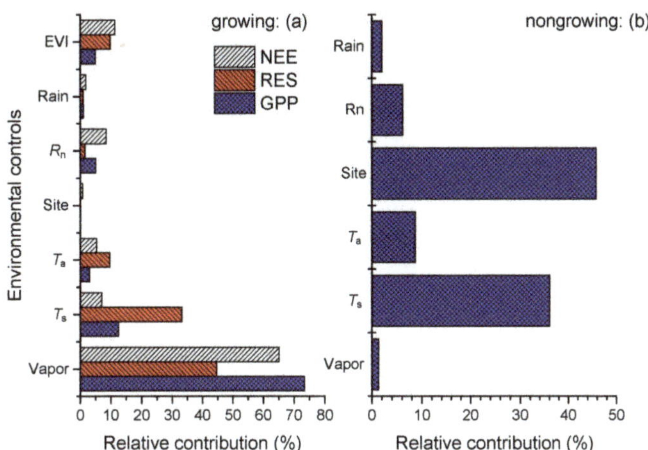

Figure 3. The relative contributions of environmental controls on variations in the growing season (**a**) and the nongrowing season (**b**) 8-day CO_2 fluxes of the two alpine wetlands.

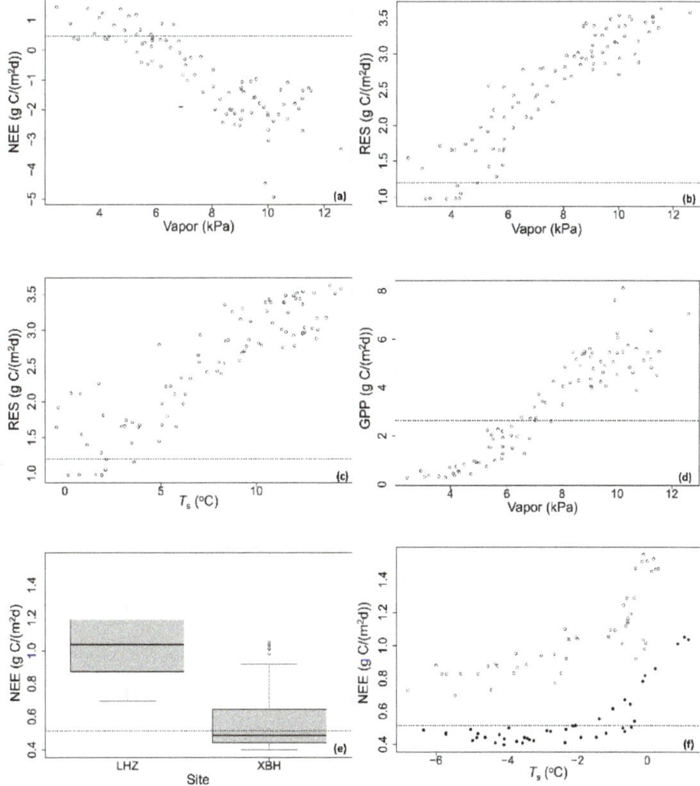

Figure 4. The fitted 8-day CO_2 fluxes (net ecosystem exchange NEE, (**a**,**e**,**f**) ecosystem respiration RES, (**b**,**c**) gross primary productivity GPP, (**d**) in relation to each of the important predictors (Vapor: atmospheric water vapor; T_S: topsoil temperature) used in the model during the growing season (**a**,**d**) and nongrowing season (**e**,**f**).

3.3. Monthly Variations in CO_2 Fluxes

Maximum monthly GPP occurred in July and August at both sites (Figure 5), and averaged 179.6 ± 28.0 g C/(m^2 month) at LHZ and 161.6 ± 11.0 g C/(m^2 month) at XBH. Similarly, maximum monthly RES averaged 97.2 ± 6.0 g C/(m^2 month) at LHZ and 105.2 ± 5.8 g C/(m^2 month) at XBH during July and August. The minimum monthly NEE averaged –82.4 ± 24.3 g C/(m^2 month) at LHZ and –56.4 ± 7.3 g C/(m^2 month) at XBH during these two months. Interestingly, the maximum monthly NEE occurred in November (47.5 ± 3.2 g C/(m^2 month)) at LHZ, while it occurred in April (30.3 ± 1.4 g C/(m^2 month)) at XBH, reflecting different hydrothermal conditions. The monthly RES at LHZ (56.1 ± 27.2 g C/(m^2 month)) was significantly ($p < 0.001$) more than that at XBH (46.2 ± 35.7 g C/(m^2 month)), in contrast to the undetectable difference in monthly GPP ($p = 0.23$) and monthly NEE ($p = 0.10$) at the two sites (Figure 6). Further analysis showed that the difference of RES mainly stemmed from an approximately two-fold larger non-growing season RES at LHZ (Figure 6b). Moreover, the nongrowing season RES comprised as much as 29.5% of the annual RES in LHZ, but only comprised 18.5% in XHB.

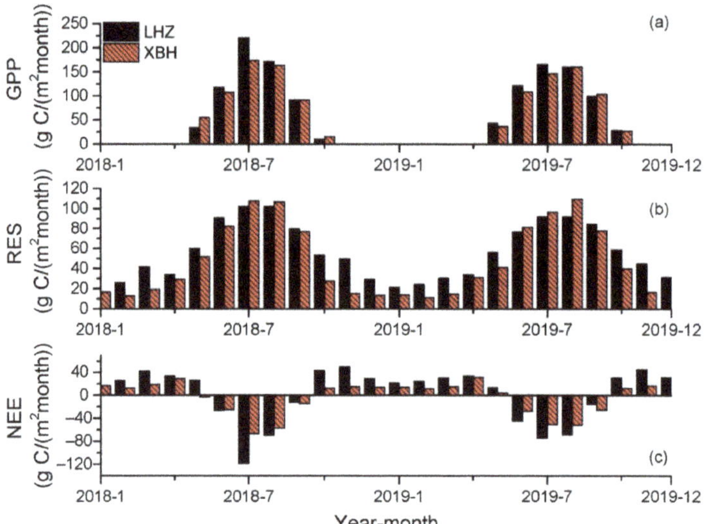

Figure 5. The monthly variations in CO_2 fluxes of the two alpine wetlands (gross primary productivity GPP, (**a**); ecosystem respiration RES, (**b**); net ecosystem CO_2 exchange NEE, (**c**)).

During the growing season, both NEE and RES were significantly correlated with GPP, with 69.0% of the per-unit GPP contributed to RES and 31.0% to NEE. Monthly NEE at both sites was controlled much more by monthly GPP than by monthly RES, indicated by stronger standardized coefficients for the former (Figure 7). GPP and RES were both controlled by atmospheric water vapor and the direct effect was similar for both LHZ and XBH (0.78 versus 0.79 for GPP, and 1.01 versus 0.96 for RES) for the two wetlands. It should be noted that the indirect effect of atmospheric water vapor through EVI on CO_2 fluxes was significant ($p < 0.05$) at XBH alone, and the total effects of atmospheric water vapor on GPP and RES were 1.19 (0.79 + 1.09 × 037) and 1.07 (0.79 + 1.09 × 0.26), respectively (Figure 7b). This might be caused by XBH being a much lower EVI (Figure 1f). Therefore, the variations in monthly CO_2 fluxes were predominantly determined by monthly atmospheric water vapor during the growing season, and the effect of vegetation growth was important at XBH with lower plant coverage. GPP/RES, which evaluates the relative contribution of CO_2 exchange processes to the total CO_2 exchange, averaged 1.21 ± 0.61 at LHZ and

1.20 ± 0.35 at XBH, respectively. The paired-sample t-test showed that the difference in GPP/RES between the two sites was non-significant (p = 0.91, N = 12).

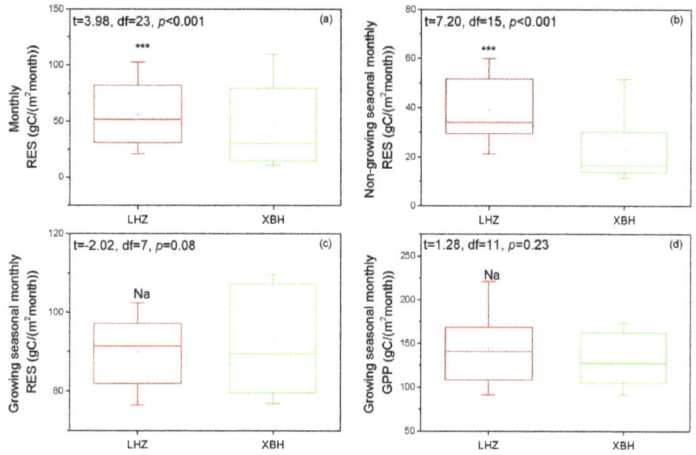

Figure 6. The paired-sample t-test on the monthly ecosystem respiration (RES) at the whole year (**a**), non-growing season (**b**), growing season (**c**) and the growing season monthly gross primary productivity (GPP), (**d**) of the two alpine wetlands (Na: non-significant; ***: p < 0.001).

During the nongrowing season, monthly NEE was controlled by monthly T_s, and the relationship could be depicted by a Q_{10}-based model ($NEE = a \times e^{\ln(Q_{10})(T_s-10)/10}$, Q_{10} describes the proportional increase in the rate of respiration per 10 °C rise in temperature [5]; Figure 8). The reference respiration in LHZ was 29.8% more than that of XBH, while Q_{10} was 27.3% lower at the former site. Furthermore, the general linear model of monthly ln(NEE) with T_s and site (categorical variable) revealed that the main effects of T_s and site attributes were both significant (p < 0.001), but the interaction was insignificant (p = 0.91; Table S1). Thus, the nongrowing season monthly NEE should be jointly determined by T_s and site attributes.

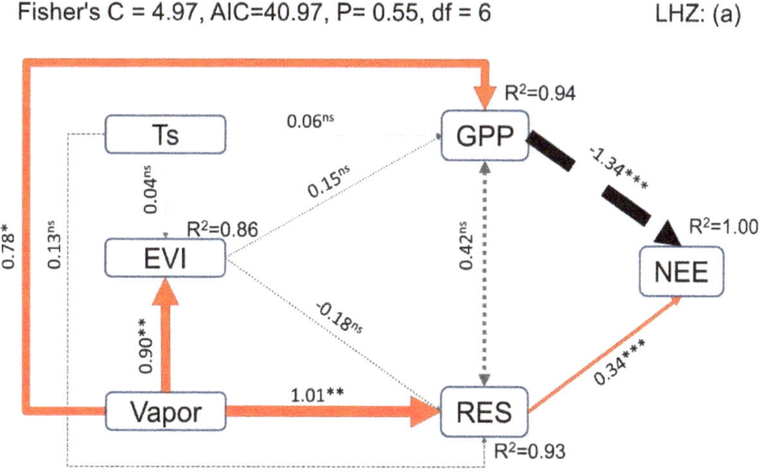

Figure 7. *Cont.*

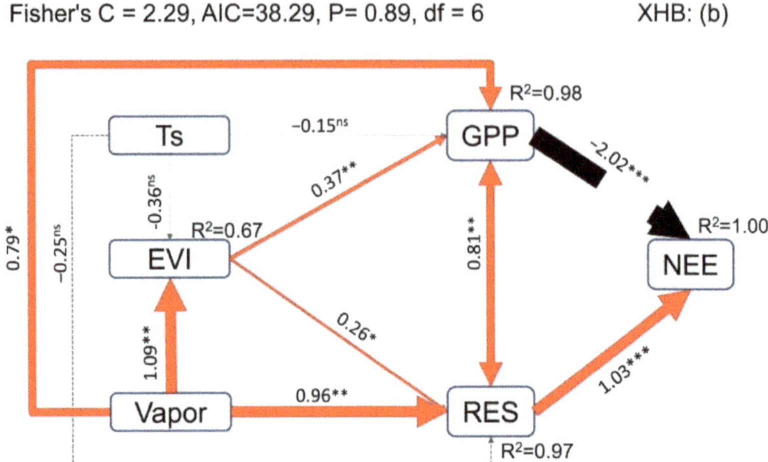

Figure 7. The piecewise SEM models of monthly CO_2 fluxes for LHZ (**a**) and for XBH (**b**) during the growing season. Numbers on arrows are standardized path coefficients, black dashed arrows are negative and red solid are negative, grey thin arrows show non-significant standardized path coefficients (significance levels are ^{ns} $p > 0.05$, * $p < 0.05$, ** $p < 0.01$ and *** $p < 0.001$). R^2 value indicates the variance explained by the model. The line weight is ten times the standardized coefficients. Abbreviations: T_s: topsoil temperature, Vapor: atmospheric water vapor; EVI: enhanced area index; GPP: gross primary productivity; RES: ecosystem respiration; NEE: net ecosystem CO_2 exchange.

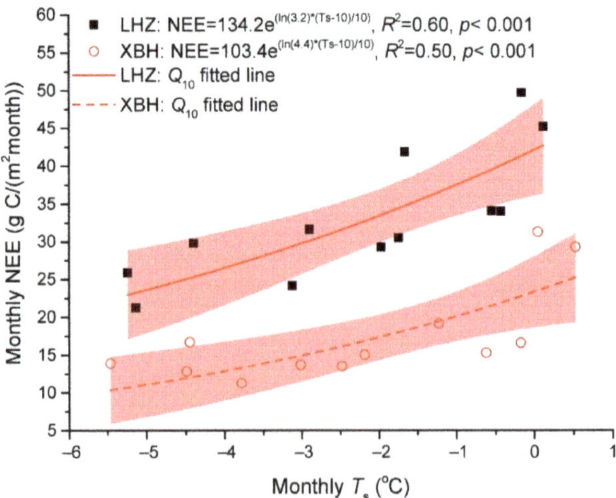

Figure 8. The relationship between monthly NEE and monthly topsoil temperature (T_s) of the two alpine wetlands. Shaded areas are the 95% confidence intervals of the fitted regression.

Overall, LHZ acted as a carbon source at an annual rate of 41.5 ± 17.8 g C/(m^2 year) while XBH was a carbon sink at a rate of -40.9 ± 7.5 g C/(m^2 year). Annual GPP and RES at HLZ were 631.5 ± 19.4 and 673.0 ± 37.2 g C/(m^2 year), respectively, and those at XBH were 595.1 ± 15.4 and 554.2 ± 7.9 g C/(m^2 year) (Figure 5). Bivariate correlation analysis showed that the site-to-site variations in annual NEE were more significantly determined by annual RES ($p = 0.02$, N = 4), than by annual GPP ($p = 0.13$, N = 4). Forward stepwise linear regression showed that the variations in annual NEE and annual RES were both related to annual accumulative EVI ($p < 0.03$, N = 4, $R^2 > 0.94$; Table S2). Annually, the carbon use efficiency (-NEE/GPP) averaged -0.065 ± 0.026 at LHZ and 0.068 ± 0.011 at XBH, respectively.

4. Discussion

4.1. Environmental Controls on CO_2 Fluxes

The variations in growing season CO_2 fluxes were mainly determined by atmospheric water vapor in the two alpine wetlands (Figures 3 and 7), which is inconsistent with the long-standing view that thermal conditions predominantly control CO_2 fluxes in these temperature-limited wetlands [14,15,40]. There are two potential explanations for this finding from the perspectives of the atmospheric water vapor itself and plant physiology. First, atmospheric water vapor is an integral indicator of hydrothermal status and can exert a substantial influence on land carbon uptake variability through its indirect effects from soil moisture-atmosphere feedbacks [41,42]; this is indeed exponentially related to T_a (atmospheric water vapor = $3.9e^{0.08Ta}$, $R^2 = 0.97$, $p < 0.001$) and T_s (atmospheric water vapor = $2.6e^{0.11Ts}$, $R^2 = 0.83$, $p < 0.001$) and would, thus, be a reasonable proxy of thermal conditions. Second, the optimum temperature of plant physiological activity is closely related to the growing season air temperature due to long-term plant evolutionary adaptation [43]. In other words, current thermal conditions should be comparable for analogous plant photosynthesis and growth within a certain cold site [44,45]. Additionally, the effects of thermal conditions on CO_2 fluxes are markedly moisture-dependent, based on reports from warming experiments in alpine regions [45,46]. Thus, atmospheric water vapor should be a reliable metric of hydrothermal conditions controlling growing season CO_2 fluxes in alpine wetlands [12,35]. The nongrowing season CO_2 fluxes were jointly controlled by site attributes and T_s (Figures 3 and 8). This stems from the fact that ecosystem respiration consists of root autotrophic respiration and microbial heterotrophic respiration, which are closely related to substrate availability and soil temperatures in alpine regions [9,44]. Therefore, atmospheric water vapor, more than air/soil temperatures, regulated the variability in growing season CO_2 fluxes, while site attributes and soil temperatures jointly controlled nongrowing season CO_2 emissions in our wetlands. These findings are consistent with previous studies [4,12,18]. Given that the nongrowing season temperature is projected to further increase across the QTP [47], these alpine wetlands will likely release more carbon and positively feedback to climate change [5]; this prediction is different from the model simulations of a single site where little attention was paid to nongrowing season CO_2 dynamics [13,26].

4.2. Carbon Budgets of the Two Wetlands

The distinct carbon function of the two wetlands is explained more by RES, than by GPP, specifically by nongrowing RES (Figures 4 and 6). The smaller annual carbon use efficiency (about 6%) and little difference in growing GPP/RES (1.20) also confirmed the important role of nongrowing RES in carbon budgets. The nongrowing season daily CO_2 effluxes were 1.07 g C/(m^2 d) in LHZ and 0.56 g C/(m^2 d) in XBH, which lie within the reported values of other alpine wetlands [4,40]. The higher nongrowing season RES at LHZ compared to XBH thus contributed to a net carbon source at LHZ. Soil temperature was similar at both sites (Figure 1) and was still high enough to support organic matter decomposition and root respiration due to the snow insulation effect, despite the air temperature being below −10 °C [4,16]; hence, greater root biomass and more soil organic carbon at LHZ, rather than thermal conditions, probably led to more ecosystem respiration at LHZ compared to XBH during the nongrowing season [1,4]. Meanwhile, the easily degradable labile carbon from plant litter, induced by higher vegetation coverage (e.g., two-fold higher EVI at LHZ), will stimulate enhanced microbial growth, carbon mineralization, and soil respiration at LHZ [11]. However, the temperature sensitivity of RES was much higher at XBH (Figure 8), which also indicated lower stability of soil organic matter at this site [21].

The nonsignificant difference in GPP possibly stems from the higher productivity of swampland plants compared to peatland plants [9]. This appears to be confirmed by a three-fold higher sensitivity of GPP to EVI (slope = 40.5 g C/(m^2 d)) at XBH compared to that at LHZ (slope = 14.5 g C/(m^2 d)). The growing season RES was closely correlated with GPP ($R^2 > 0.90$, $p < 0.001$; Figure S4) at both sites, through substrate supply [13]. Therefore, nonsignificant GPP results in little difference in growing season plant autotrophic respiration and the consequent RES, specifically, nongrowing season RES contributes to the distinct carbon functions in the two alpine wetlands.

Higher carbon sink capacity appears to coincide with higher precipitation, which is in agreement with previous reports from alpine wetlands [14,19,40]. However, the underlying responses of the two sites are clearly different. Greater precipitation inhibited carbon dynamics at LHZ while stimulated carbon exchanges at XBH (Figure S5). Specifically, at LHZ, the higher precipitation (536.1 mm versus 439.6 mm) decreased annual GPP less by 27.5 g C/m^2 (645.2 g C/(m^2 year) versus 617.7 g C/(m^2 year)), compared to a decrease in RES of 52.6 g C/m^2 (699.3 g C/(m^2 year) versus 646.7 g C/(m^2 year)). Decreased carbon fluxes might be caused by more waterlogged patches and lower oxygen availability for soil respiration under higher precipitation [14,17,23]. At XBH, the greater precipitation (238.8 mm versus 344.7 mm) stimulated increased GPP by 21.8 g C/m^2 (584.2 g C/(m^2 year) versus 606.0 g C/(m^2 year)), compared to an increase in RES of 11.1 g C/m^2 (548.7 g C/(m^2 year) versus 559.8 g C/(m^2 year)). Such a phenomenon indicates that GPP and RES at XBH should be to some extent water-stressed, which agrees well with previously reported positive relationships between CO_2 fluxes and water availability at the same site from 2012 to 2013 [29]. Indeed, recent studies have suggested that the response of soil carbon decomposition to water-table manipulation in LHZ is controlled by ferrous iron, in contrast to the classic "enzyme latch" theory seen in other wetlands [21,48]. Therefore, the different responses of CO_2 fluxes to promoted precipitation between peatlands and swamplands should be taken into full consideration in predicting the feedback of alpine wetlands to climate change on the QTP. Further, more extensive and long-term studies, including those on another potent greenhouse gas, CH_4, are required to quantify the contribution of alpine wetlands to the regional carbon neutrality of alpine wetlands [8].

5. Conclusions

Using two-year continuous measurements from the northeastern Qinghai—Tibet Plateau, this study revealed that an alpine peatland site acted as a net carbon source, while an alpine swampland site acted as a net carbon sink. Growing season CO_2 fluxes were predominantly controlled by atmospheric water vapor, while nongrowing season ecosystem respiration was regulated by site attributes and topsoil temperatures. Ecosystem respiration, specifically nongrowing season CO_2 emissions, potentially accounted for the distinctly different carbon budgets of the two alpine wetlands. These findings highlight the crucial role of nongrowing season CO_2 dynamics in alpine wetlands, which will positively feedback to climate change under the context of marked winter warming.

Supplementary Materials: The following are available online at https://www.mdpi.com/article/10.3390/atmos12121695/s1, Table S1: The analysis of variance from the general linear model of monthly net ecosystem exchange (ln(*NEE*) with soil temperature (T_s, continuous variable) and site attributes (Site, categorical variable) during the nongrowing season, Table S2: The Pearson correlation analysis between annual CO_2 fluxes (NEE, RES, GPP) with environmental controls (T_a: air temperature; Vapor: atmospheric water vapor; Rain: precipitation; R_n: net radiation; T_s: topsoil temperature; EVI_{sum}: accumulative enhanced vegetation index). Figure S1: The geographic location of the two alpine wetlands, Figure S2: The relative contributions of environmental controls to variations in the growing season (a) and the nongrowing season (b) 8-day CO_2 fluxes of the two alpine wetlands, Figure S3. The fitted 8-day CO_2 fluxes (NEE: (a, d); RES (b); GPP (c)) in relation to each of the predictors (Rain: precipitation; T_a: air temperature; R_n: net radiation; Vapor: atmospheric water vapor; T_s: topsoil temperature; EVI: enhanced vegetation index; Site: categorical variable, LHZ and XBH) used in the model during the growing season (a, b, c) and nongrowing season (d), Figure S4: The relationship between monthly ecosystem respiration (RES) and monthly gross primary productivity (GPP) during the growing season of the two alpine wetlands (LHZ: Luanhaizi and XBH: Xiaobohu). The shading areas are 95% confidence intervals, Figure S5: The relationships of annual carbon fluxes and annual precipitation of the two alpine wetland sites (Xiaobohu: XBH and Luanhaizi: LHZ).

Author Contributions: Conceptualization, F.Z. and W.W.; methodology, C.S., F.L., L.Z., C.W. and Y.L.; software, C.S. and F.Z.; validation, C.S., L.Y. and J.L.; investigation, C.S., F.L., Y.Y. and L.Z.; resources, K.C. and Y.L.; supervision, F.Z.; funding acquisition, F.Z., K.C., W.W. and Y.L. All authors have read and agreed to the published version of the manuscript.

Funding: This work is funded by the Chinese Academy of Sciences-People's Government of Qinghai Province Joint Grant on Sanjiangyuan National Park Research (LHZX-2020-07, LHZX-2020-08), the project of Economic Development Research Center of National Forestry and Grassland Administration (JYFL-2021-20); the National Key R&D Program (2017YFA0604802), the National Natural Science Foundation of China (41877547), the 2021 first funds for central government to guide local science and technology development in Qinghai Province (2021ZY002), and the Second Tibetan Plateau Scientific Expedition and Research (STEP) Program (2019QZKK0302).

Institutional Review Board Statement: Not applicable.

Informed Consent Statement: Not applicable.

Data Availability Statement: The data are available on reasonable request to the corresponding author.

Conflicts of Interest: The authors declare no conflict of interest.

Abbreviations

BRT: boosted regression trees; SEM: structural equation model; EVI: enhanced area index; Rain: precipitation; R_n: net radiation; T_a: air temperature; T_s: topsoil temperature; Vapor: atmospheric water vapor; VPD: vapor pressure deficit; NEE: net ecosystem CO_2 exchange; GPP: gross primary productivity; RES: ecosystem respiration; LHZ: Luanhaizi peatland; XBH Xiaobohu swampland; ChinaFLUX: Chinese Flux Observation and Research Network; QTP: Qinghai—Tibet Plateau.

References

1. Limpens, J.; Berendse, F.; Blodau, C.; Canadell, J.G.; Freeman, C.; Holden, J.; Roulet, N.; Rydin, H.; Schaepman-Strub, G. Peatlands and the carbon cycle: From local processes to global implications—A synthesis. *Biogeosciences* **2008**, *5*, 1475–1491. [CrossRef]
2. Mitsch, W.J.; Bernal, B.; Nahlik, A.M.; Mander, Ü.; Zhang, L.; Anderson, C.J.; Jørgensen, S.E.; Brix, H. Wetlands, carbon, and climate change. *Landsc. Ecol.* **2013**, *28*, 583–597. [CrossRef]
3. Yu, Z.; Loisel, J.; Brosseau, D.P.; Beilman, D.W.; Hunt, S.J. Global peatland dynamics since the Last Glacial Maximum. *Geophys. Res. Lett.* **2010**, *37*, L13402. [CrossRef]
4. Pullens, J.W.M.; Sottocornola, M.; Kiely, G.; Toscano, P.; Gianelle, D. Carbon fluxes of an alpine peatland in Northern Italy. *Agric. For. Meteorol.* **2016**, *220*, 69–82. [CrossRef]
5. Dise, N.B. Peatland Response to Global Change. *Science* **2009**, *326*, 810. [CrossRef] [PubMed]
6. Gallego-Sala, A.V.; Charman, D.J.; Brewer, S.; Page, S.E.; Prentice, I.C.; Friedlingstein, P.; Moreton, S.; Amesbury, M.J.; Beilman, D.W.; Björck, S.; et al. Latitudinal limits to the predicted increase of the peatland carbon sink with warming. *Nat. Clim. Change* **2018**, *8*, 907–913. [CrossRef]
7. Lund, M.; Lafleur, P.M.; Roulet, N.T.; Lindroth, A.; Christensen, T.R.; Aurela, M.; Chojnicki, B.H.; Flanagan, L.B.; Humphreys, E.R.; Laurila, T.; et al. Variability in exchange of CO_2 across 12 northern peatland and tundra sites. *Glob. Chang. Biol.* **2010**, *16*, 2436–2448. [CrossRef]
8. Wei, D.; Zhao, H.; Huang, L.; Qi, Y.; Wang, X. Feedbacks of Alpine Wetlands on the Tibetan Plateau to the Atmosphere. *Wetlands* **2020**, *40*, 787–797. [CrossRef]
9. Hirota, M.; Tang, Y.; Hu, Q.; Hirata, S.; Kato, T.; Mo, W.; Cao, G.; Mariko, S. Carbon Dioxide Dynamics and Controls in a Deep-water Wetland on the Qinghai-Tibetan Plateau. *Ecosystems* **2006**, *9*, 673–688. [CrossRef]
10. Xiao, D.; Deng, L.; Kim, D.; Huang, C.; Tian, K. Carbon budgets of wetland ecosystems in China. *Glob. Chang. Biol.* **2019**, *25*, 2061–2076. [CrossRef]
11. Gao, J.; Feng, J.; Zhang, X.; Yu, F.; Xu, X.; Kuzyakov, Y. Drying-rewetting cycles alter carbon and nitrogen mineralization in litter-amended alpine wetland soil. *Catena* **2016**, *145*, 285–290. [CrossRef]

12. Wang, H.; Li, X.; Xiao, J.; Ma, M.; Tan, J.; Wang, X.; Geng, L. Carbon fluxes across alpine, oasis, and desert ecosystems in northwestern China: The importance of water availability. *Sci. Total Environ.* **2019**, *697*, 133978. [CrossRef]
13. Kang, X.; Li, Y.; Wang, J.; Yan, L.; Zhang, X.; Wu, H.; Yan, Z.; Zhang, K.; Hao, Y. Precipitation and temperature regulate the carbon allocation process in alpine wetlands: Quantitative simulation. *J. Soil. Sediment.* **2020**, *20*, 3300–3315. [CrossRef]
14. Zhu, J.; Zhang, F.; Li, H.; He, H.; Li, Y.; Yang, Y.; Zhang, G.; Wang, C.; Luo, F. Seasonal and interannual variations of CO_2 fluxes over 10 years in an alpine wetland on the Qinghai-Tibetan Plateau. *J. Geophys. Res. Biogeosciences* **2020**, *125*, e2020J–e6011J. [CrossRef]
15. He, G.; Li, K.; Liu, X.; Gong, Y.; Hu, Y. Fluxes of methane, carbon dioxide and nitrous oxide in an alpine wetland and an alpine grassland of the Tianshan Mountains, China. *J. Arid Land* **2014**, *6*, 717–724. [CrossRef]
16. Heikkinen, J.E.P.; Elsakov, V.; Martikainen, P.J. Carbon dioxide and methane dynamics and annual carbon balance in tundra wetland in NE Europe, Russia. *Glob. Biogeochem. Cycles* **2002**, *16*, 61–62. [CrossRef]
17. Zhao, L.; Li, J.; Xu, S.; Zhou, H.; Li, Y.; Gu, S.; Zhao, X. Seasonal variations in carbon dioxide exchange in an alpine wetland meadow on the Qinghai-Tibetan Plateau. *Biogeosciences* **2010**, *7*, 1207–1221. [CrossRef]
18. McFadden, J.P.; Eugster, W.; Chapin, F.S. A regional study of the controls on water vapor and CO_2 exchange in arctic tundra. *Ecology* **2003**, *84*, 2762–2776. [CrossRef]
19. Kang, X.; Hao, Y.; Cui, X.; Chen, H.; Huang, S.; Du, Y.; Li, W.; Kardol, P.; Xiao, X.; Cui, L. Variability and Changes in Climate, Phenology, and Gross Primary Production of an Alpine Wetland Ecosystem. *Remote Sens.* **2016**, *8*, 391. [CrossRef]
20. Yuan, X.; Chen, Y.; Qin, W.; Xu, T.; Mao, Y.; Wang, Q.; Chen, K.; Zhu, B. Plant and microbial regulations of soil carbon dynamics under warming in two alpine swamp meadow ecosystems on the Tibetan Plateau. *Sci. Total Environ.* **2021**, *790*, 148072. [CrossRef]
21. Park, S.; Knohl, A.; Migliavacca, M.; Thum, T.; Vesala, T.; Peltola, O.; Mammarella, I.; Prokushkin, A.; Kolle, O.; Lavrič, J.; et al. Temperature Control of Spring CO_2 Fluxes at a Coniferous Forest and a Peat Bog in Central Siberia. *Atmosphere* **2021**, *12*, 984. [CrossRef]
22. Bridgham, S.D.; Megonigal, J.P.; Keller, J.K.; Bliss, N.B.; Trettin, C. The carbon balance of North American wetlands. *Wetlands* **2006**, *26*, 889–916. [CrossRef]
23. Yu, L.; Wang, H.; Wang, Y.; Zhang, Z.; Chen, L.; Liang, N.; He, J. Temporal variation in soil respiration and its sensitivity to temperature along a hydrological gradient in an alpine wetland of the Tibetan Plateau. *Agric. For. Meteorol.* **2020**, *282–283*, 107854. [CrossRef]
24. Miao, Y.; Liu, M.; Xuan, J.; Xu, W.; Wang, S.; Miao, R.; Wang, D.; Wu, W.; Liu, Y.; Han, S. Effects of warming on soil respiration during the non-growing seasons in a semiarid temperate steppe. *J. Plant Ecol.* **2020**, *13*, 288–294. [CrossRef]
25. Zhang, F.; Liu, A.; Li, Y.; Zhao, L.; Wang, Q.; Du, M. CO_2 flux in alpine wetland ecosystem on the Qinghai-Tibetan Plateau, China. *Acta Ecol. Sin.* **2008**, *28*, 453–462.
26. Jin, Z.; Zhuang, Q.; He, J.; Zhu, X.; Song, W. Net exchanges of methane and carbon dioxide on the Qinghai-Tibetan Plateau from 1979 to 2100. *Environ. Res. Lett.* **2015**, *10*, 85007. [CrossRef]
27. Niu, Z.; Zhang, H.; Wang, X.; Yao, W.; Zhou, D.; Zhao, K.; Zhao, H.; Li, N.; Huang, H.; Li, C.; et al. Mapping wetland changes in China between 1978 and 2008. *Chin. Sci. Bull.* **2012**, *57*, 2813–2823. [CrossRef]
28. Cao, S.; Cao, G.; Chen, K.; Han, G.; Liu, Y.; Yang, Y.; Li, X. Characteristics of CO_2, water vapor, and energy exchanges at a headwater wetland ecosystem of the Qinghai Lake. *Can. J. Soil Sci.* **2019**, *99*, 227–243. [CrossRef]
29. Cao, S.; Cao, G.; Feng, Q.; Han, G.; Lin, Y.; Yuan, J.; Wu, F.; Cheng, S. Alpine wetland ecosystem carbon sink and its controls at the Qinghai Lake. *Environ. Earth Sci.* **2017**, *76*, 210. [CrossRef]
30. Burba, G. *Eddy Covariance Method: For Scientific, Industrial, Agricultural and Regulatory Applications*; LI-COR Biosciences: Lincoln, NE, USA, 2013.
31. Foken, T.; Göockede, M.; Mauder, M.; Mahrt, L.; Amiro, B.; Munger, W. Post-field data quality control. In *Handbook of Micrometeorology*; Lee, X., Massman, W., Law, B., Eds.; Springer: Dordrecht, The Netherlands, 2005; pp. 181–208.
32. Zhang, L.; Luo, Y.; Liu, M.; Chen, Z.; Su, W.; He, H.; Zhu, Z.; Sun, X.; Wang, Y.; Zhou, G.; et al. Carbon and water fluxes observed by the Chinese Flux Observation and Research Network (2003–2005). *Sci. Data Bank* **2019**, *4*, 1–17.
33. Elith, J.; Leathwick, J.R.; Hastie, T. A working guide to boosted regression trees. *J. Anim. Ecol.* **2008**, *77*, 802–813. [CrossRef]
34. Sun, S.; Che, T.; Li, H.; Wang, T.; Ma, C.; Liu, B.; Wu, Y.; Song, Z. Water and carbon dioxide exchange of an alpine meadow ecosystem in the northeastern Tibetan Plateau is energy-limited. *Agric. For. Meteorol.* **2019**, *275*, 283–295. [CrossRef]
35. Li, H.; Wang, C.; Zhang, F.; He, Y.; Shi, P.; Guo, X.; Wang, J.; Zhang, L.; Li, Y.; Cao, G.; et al. Atmospheric water vapor and soil moisture jointly determine the spatiotemporal variations of CO_2 fluxes and evapotranspiration across the Qinghai-Tibetan Plateau grasslands. *Sci. Total Environ.* **2021**, *791*, 148379. [CrossRef] [PubMed]
36. Zhang, F.; Li, H.; Wang, W.; Li, Y.; Lin, L.; Guo, X.; Du, Y.; Li, Q.; Yang, Y.; Cao, G.; et al. Net radiation rather than moisture supply governs the seasonal variations of evapotranspiration over an alpine meadow on the northeastern Qinghai-Tibetan Plateau. *Ecohydrology* **2018**, *11*, e1925. [CrossRef]
37. Li, H.; Zhu, J.; Zhang, F.; He, H.; Yang, Y.; Li, Y.; Cao, G.; Zhou, H. Growth stage-dependant variability in water vapor and CO_2 exchanges over a humid alpine shrubland on the northeastern Qinghai-Tibetan Plateau. *Agric. For. Meteorol.* **2019**, *268*, 55–62. [CrossRef]
38. Lefcheck, J.S. piecewiseSEM: Piecewise structural equation modelling in r for ecology, evolution, and systematics. *Methods Ecol. Evol.* **2016**, *7*, 573–579. [CrossRef]

39. R Development Core Team. *R: A Language and Environment for Statistical Computing;* R Foundation for Statistical Computing: Vienna, Austria, 2006.
40. Hao, Y.B.; Cui, X.Y.; Wang, Y.F.; Mei, X.R.; Kang, X.M.; Wu, N.; Luo, P.; Zhu, D. Predominance of Precipitation and Temperature Controls on Ecosystem CO2 Exchange in Zoige Alpine Wetlands of Southwest China. *Wetlands* **2011**, *31*, 413–422. [CrossRef]
41. Chapin, F.S.; Matson, P.A.; Mooney, H.A. *Principles of Terrestrial Ecosystem Ecology*, 2nd ed.; Springer: New York, NY, USA, 2011.
42. Humphrey, V.; Berg, A.; Ciais, P.; Gentine, P.; Jung, M.; Reichstein, M.; Seneviratne, S.I.; Frankenberg, C. Soil moisture-atmosphere feedback dominates land carbon uptake variability. *Nature* **2021**, *592*, 65–69. [CrossRef] [PubMed]
43. Huang, M.; Piao, S.; Ciais, P.; Peñuelas, J.; Wang, X.; Keenan, T.F.; Peng, S.; Berry, J.A.; Wang, K.; Mao, J.; et al. Air temperature optima of vegetation productivity across global biomes. *Nat. Ecol. Evol.* **2019**, *3*, 772–779. [CrossRef]
44. Körner, C. *Alpine Plant Life: Functional Plant Ecology of High Mountain Ecosystems*, 2nd ed.; Springer: Berlin/Heidelberg, Germany, 2003.
45. Zhang, T.; Zhang, Y.; Xu, M.; Zhu, J.; Chen, N.; Jiang, Y.; Huang, K.; Zu, J.; Liu, Y.; Yu, G. Water availability is more important than temperature in driving the carbon fluxes of an alpine meadow on the Tibetan Plateau. *Agric. For. Meteorol.* **2018**, *256–257*, 22–31. [CrossRef]
46. Quan, Q.; Tian, D.; Luo, Y.; Zhang, F.; Crowther, T.W.; Zhu, K.; Chen, H.Y.H.; Zhou, Q.; Niu, S. Water scaling of ecosystem carbon cycle feedback to climate warming. *Sci. Adv.* **2019**, *5*, v1131. [CrossRef] [PubMed]
47. Ding, J.; Yang, T.; Zhao, Y.; Liu, D.; Wang, X.; Yao, Y.; Peng, S.; Wang, T.; Piao, S. Increasingly Important Role of Atmospheric Aridity on Tibetan Alpine Grasslands. *Geophys. Res. Lett.* **2018**, *45*, 2852–2859. [CrossRef]
48. Wang, Y.; Wang, H.; He, J.; Feng, X. Iron-mediated soil carbon response to water-table decline in an alpine wetland. *Nat. Commun.* **2017**, *8*, 15972. [CrossRef] [PubMed]

 atmosphere

MDPI

Article

Composting and Methane Emissions of Coffee By-Products

Macarena San Martin Ruiz *, Martin Reiser and Martin Kranert

Institute for Sanitary Engineering, Water Quality and Solid Waste Management (ISWA), University of Stuttgart, Bandtäle 2, 70569 Stuttgart, Germany; martin.reiser@iswa.uni-stuttgart.de (M.R.); martin.kranert@iswa.uni-stuttgart.de (M.K.)

* Correspondence: macarena.sanmartin@iswa.uni-stuttgart.de

Abstract: In the last 20 years, the demand for coffee production has increased detrimentally, heightening the need for production, which is currently driving the increase in land cultivation for coffee. However, this increase in production ultimately leads to the amplification of waste produced. This study aims to develop an experimental methodology for sustainable coffee by-products (Pulp (CP)) in Costa Rica for nutrient-rich compost. The performance of the experiments is to explore and optimize composting processes following its key parameters. This will allow quantifying the emissions rate to obtain an emission factor for CP during the open composting process and optimizing the conditions to minimize CH_4 emissions using P and green waste (GW) materials. Five CP and GW mixtures were analyzed for the composting process for ten weeks, acting P as primary input material as a by-product. Quantification of the methane emissions was performed in two areas: composting area and open field deposition. Peak temperatures of compost appeared at twenty-five days for control and five days for GW added treatments. CP emission factors provide a similar result with the standard values recommended by the literature, accomplishing the emission reductions. Thus, this study designed and validated a sustainable protocol for transforming coffee by-products into compost.

Keywords: coffee pulp; coffee by-products; composting; methane; emissions rate; emissions factor

Citation: San Martin Ruiz, M.; Reiser, M.; Kranert, M. Composting and Methane Emissions of Coffee By-Products. *Atmosphere* **2021**, *12*, 1153. https://doi.org/10.3390/atmos12091153

Academic Editors: Baojie He, Ayyoob Sharifi, Chi Feng and Jun Yang

Received: 10 August 2021
Accepted: 3 September 2021
Published: 7 September 2021

1. Introduction

Agriculture is responsible for an essential portion of global emissions, contributing to 45% of their methane (CH_4) emissions globally, impacting climate change [1]. In addition, gas concentrations act as indicators that show biological degradation, and this guides optimization possibilities of developing new strategies for emissions reduction [2]. Therefore, it is required to estimate greenhouse gas (GHG) emission rates as they are the flow of a pollutant expressed in weight per unit of time [3]. Emissions rates are necessary to calculate an emission factor, a representative value that attempts to relate the amount of a pollutant released into the atmosphere with an activity associated with the release of that pollutant [4]. The detection of gaseous emissions during composting coffee by-products is one of the most critical tools to meet the challenge of reducing CH_4 emissions from the waste residues generated in the coffee processing industry [5,6]. Coffee is a worldwide used product and of the most valued commodities in trade, being one of the most important agricultural exports in Costa Rica [7]. In coffee processing, the production chain comprehends several steps. Firstly, the berries from the coffee plants are transported to be washed and peeled (de-pulping), separating the green beans from the pulp or husk [8]. The outer membrane that envelops the coffee bean is called the pulp (CP) (mesocarp), which contains 43% w/w of the morphology of the coffee fruit [9–11]. CP is one of the main by-products generated during the process [11]. It contributes to pollution, environmental and health problems of the surrounding waters, soil, and atmosphere when the coffee berries are ripe and processed during the wet method and mishandled [12,13]. Some researchers and the Costa Rican Coffee Institute (ICAFE) indicate CP management has been one of the challenged coffee by-products with the most significant volume of waste [9,14]. In addition, it accumulates for long periods, and it leads to the generation of foul odors, being a favorable environment for reproducing flies and other pests responsible for multiple diseases [15,16]. Currently, the country has a top priority: reducing GHG emissions in the coffee industry, together with a National Decarbonization Plan 2050 [17]. Hence, studying methane emissions during the composting of the coffee by-products and finding new approaches will be crucial to achieving future goals and mitigating the current challenges with coffee by-products each harvest.

Composting has been a promising technique for waste treatment in converting organic matter and agricultural residues into compost, even using minor technologies and operational expenses [18,19]. Aerobic composting involves the changes in the properties and degradation of the substrates [20]. In addition, the existence of aeration in the system gives biological products from the metabolism of the process, such as carbon dioxide, water, and [21,22] heat. During the composting process, three phases are observed in the aerobic decomposition: initial and degradation, conversion, and maturation. In addition, possess two types of microbial activity during this decomposition: thermophilic stage (45–70 °C) and mesophilic stage (15–45 °C) [21,23–26] The first stage is linked with the microbial activity of the material and is followed by a second stage, where the conversion of the organic material occurs [27]. The final stage is the maturation process, which occurs at ambient temperature and mesophilic micro-organisms play a role in finishing the product (mainly bacteria and fungi) [28]. Even if a composting technique is beneficial to the environment, GHG is present during the process, enhancing global warming [29]. The emissions will depend on the waste type and composition, key composting parameters such as C/N, temperature, moisture, pH [27,30], and the final use of the compost [31].

This study aims to develop an experimental methodology for sustainable coffee by-products for nutrient-rich compost following key composting parameters and quantify CH_4 emissions rate to obtain an emission factor for CP from open composting processes to optimize operating conditions CH_4 emissions. Furthermore, with the usage of improved management of coffee by-products, it will grand new approaches of the residue in the coffee industry. This will allow the communities and the coffee sector to receive a positive environmental impact where the aim is to reduce GHG emissions, odors, and pathogens generated. On the other hand, the farmers could make sustainable use of the compost in the future, returning soil amendment into their coffee plantations.

2. Materials and Methods

2.1. Compost Pile Construction and Mixing in the Study Area

The study was conducted in the biggest Costa Rican coffee mill in the country and located about 70 km south of the Capital San Jose, between 9°39′25.41″ N and 84°01′32.08″ W. The performance of the experiment was during the harvest and dry coffee season of 2019–2020 at an altitude of 1200–1900 m.a.s.l. [32]. The status quo of the current mill of study generates about 37,000 tons/ harvest of coffee by-products (1 harvest per year) and, 80–90% of CP is used for composting purposes. The so-called open field depositions are fields where CP is buried every harvest without composting treatment due to the lack of space for composting.

Currently, in the mill, the composting process is carried out just with CP as an input material; therefore, this current technique is taken as the control (TC). Open Windrow composting trials were formed using five different percentages of two types of feedstock: CP as a coffee by-product and green waste (GW) (Table 1). CP was collected directly from the mill after the industrial de-pulping process of the coffee cherries and transported into the composting area by trucks. The collection of GW was from the surroundings of the mill and shredded for the trials and windrow formations. GW was a mixture of wood sticks, Elephant grass (*Pennisetum purpureum*), African Stargrass (*Cynodon plectostachyus*), and pruning of trees from the surroundings, including coffee plants. The feedstock was in a triangular cross-section formed (approximately 1.2 m high, 2.6 m wide, and between 15–25 m long for all the treatments). The windrows were operated without forced aeration and turned weekly using a mechanical turner (Backhus® windrow turner) for the entire process. Previous studies [10,33] have shown that the optimal turning frequency for CP during the composting process was once per week to avoid anaerobic conditions inside the windrow and methane emissions. Therefore, these considerations were also taken into account for this study.

Table 1. Overview of the treatments based on volume and amount of pulp used for each treatment for 23 windrows.

Treatment	Windrow Percentage Based on Volume	Total CP in Mg per Treatment	Total Number of Windrows	Total of Fanegas per Treatment **
T1	80% CP–20% GW	98.54	7	937
T2	75% CP–25% GW	36.95	3	351
T3	70% CP–30% GW	49.27	4	468
T4	60% CP–40 % GW	18.48	2	176
T5	50% CP–50% GW	9.24	1	88
TC	100% CP (Control)	105.58	6	1004

CP: coffee pulp, GW: green waste (grass clippings and weeds + structural materials), ** Fanega: typical Costa Rican coffee measure, where one Fanega corresponds to 105.2 kg of P and 253 kg of coffee fruit [34,35].

The primary input material was CP as a coffee by-product (>50% of input). A total of 318 Mg of CP and 74 Mg of GW was used to form 23 windrows for ten weeks (Figure 1). The treatments (percentage based on volume) were decided according to the feasibility of the mill in collecting the input materials of GW. Since the interest of the mill is the implementation of the proposed treatments, this information was considered for the windrow formation. Table 1 shows that the total of windrows was different for each treatment. The study was performed during the summer season in an open space in the mill. Hence, water loss and depletion to avoid a decrease in the activity of the microorganisms were important factors to consider. Irrigation starting from week three until week six before each turning was needed to maintain the proper moisture content in the process for T1 to T5. The irrigation was based on the WC monitored weekly and temperature profiles for all the treatments.

Figure 1. Aerial photo of the proposed treatments T1–5 (**right** section) and aerial photo of the current treatment at the mill TC (**left** section).

2.2. Compost Sample Analysis

Fresh Compost samples were used to quantify the gravimetric moisture content (MC), pH, bulk density (p_t), volatile solids (VS), and carbon to nitrogen (C/N) ratio during the process shown in Table 2. These results were analyzed in an accredited laboratory in Costa Rica for compliance with INTE ISO/IEC 17025 standards. The analysis was performed to obtain the properties of the raw materials and microbial population counts by a serial dilution technique. The procedure was made for pH and EC in water 10:25; Acidity, P, and K with Olsen Modified pH 8.5 ($NaHCO_3$ 0.5 N, EDTA 0.01M, Superfloc 127) 1:10. Acidity is determined by titration with NaOH; P by Colorimetry with Flow Injection Analyzer (FIA), and the rest of the elements by Atomic Absorption Spectrophotometry (AAS). Total %C and %N were determined with the C/N auto analyzer by dry combustion.

Table 2. Properties for the raw material used for the composting process.

Parameters	MC (%)	pH	CE (mscm [1])	C/N	VS (%) DM *	C (%)	N (%)	P (%)	K (%)	p_t g/L
SM [1]	55	6.2	2.7	59.3	77.2	47	0.8	0.08	0.7	95.4
CP [2]	82	4.9	3.9	16.1	94.76	44	2.3	0.11	3.17	432
DGW [3]	7.9	7	5.8	35	91.27	38	1.1	0.39	2.5	31.4

DM *: dry matter; SM [1]: structural material; CP [2]: coffee pulp; DGW [3]: dry grass clippings and weeds.

Once the windrow piles were formed, the MC, pH, and C/N ratio were measured weekly, and the temperature was taken manually daily by triplicate in all the windrows. The weekly samples were collected from five different locations and depths along the windrow. Thereafter, a representative sampling over the entire windrows to analyze MC and pH was obtained. All the sample analyses followed the German Quality Assurance Organization standards for Compost (BKG) [36]. Water content was calculated from field moist and oven-dry (105 °C for 48–72 h) mass of compost according to the DIN EN 1304. The pH was extracted from 20 g (wet weight) of compost with 180 mL of $CaCl_2$ and assessed by potentiometric measurements. VS were performed and calculated according to the Federal Compost Quality Assurance Organization (FCQAO) and the DIN 18128. C/N ratios were analyzed using a Vario Max CN element analyzer elementar Analysensysteme GmbH® following the DIN ISO 10694. Three replicates of 10 g were inserted into a porcelain crucible into a muffle furnace at 550 °C. The samples were burned until constant weight according to the DIN 18128 and determined as sample weight loss.

After ten weeks, once the compost was finished, samples of each treatment were shipped to Germany to analyze each nutrient content and chemical parameters. A certified laboratory (PLANCO-TEC, Neu-Eichenberg) performed the mature compost analysis in Germany, following the BKG standards. The following methods were used for each parameter enlisted in Table 3: total nitrogen, MB BGK: 2013-05; total phosphate, potassium and magnesium DIN EN ISO 1 1885: 2009-09; soluble nitrate, ammonia, phosphate and potassium, VDLUFA I A 6.2.4.1: 2012; organic substances, DIN EN 13039: 2000-02; alkaline active ingredients MB BGK, 2006-09, soluble magnesium VDLUFA I A 6.2.4.1: 1997; Rotting degree, Methodenbuch (MB) BGK; Salinity, DIN EN 13038; Bulk Density, Plant tolerance (25% and 50% Substrate), MB BGK, 2006 and C/N ratio, QMP_BIK_C3808: 2018-09.

Table 3. Final parameters of plant nutrients, soil amendment, and physical parameters of the compost treatments.

Parameters	Units	TC	T1	T2	T3	T4	T5
		Plant Nutrients					
Total Nitrogen (N)	% *	1.61	1.14	1.06	1.37	1.36	1.31
Total Phosphate (P_2O_5)	% *	0.59	0.44	0.46	0.46	0.44	0.47
Total Potassium (K_2O)	% *	4.28	2.95	2.87	3.19	2.96	2.93
Total Magnesium (MgO)	% *	0.36	0.29	0.36	0.35	0.33	0.33
Nitrate $CaCl_2$-soluble (NO_3-N)	mg/L **	8	114	128	103	276	162
Ammonia $CaCl_2$-soluble (NH_4-N)	mg/L **	45	40	22	23	66	108
Phosphate $CaCl_2$-soluble (P_2O_5) 1	mg/L **	298	133	150	122	65	51
Potassium $CaCl_2$-soluble (K_2O)	mg/L **	10,700	10,900	11,400	9750	9920	10,400
		Soil Amendment					
Organic Substances	% *	60.4	43.9	37.5	47.3	45.1	40.4
Alkaline Active Ingredients (CaO)	% *	2.4	2.31	2.25	2.69	2.28	2.18
Magnesium $CaCl_2$-soluble	mg/L **	19	27	44	30	48	43
		Physical Parameters					
Degree of Rotting	-	3	4	4	5	5	5
Salinity	g KCl/L **	5.97	5.14	3.96	3.76	4.53	3.68
Bulk density	g/L **	345	502	520	414	464	504
PC [1] 50% Substrate	% *	75	97	95	103	100	102
PC 25% Substrate	% *	52	82	90	87	88	95
C/N Ratio	-	22	22	21	20	19	18

[1] PC: plant compatibility (relative yield); * Fresh mass. ** Dry Mass.

2.3. *Methane Gas Sampling System*

The measurements were carried out in open windrows and open field depositions. First, the focus was on whether the different windrow mixtures with coffee by-products could be comparable or compatible. Second, the differences in terms of gaseous emissions and when the volume mixture is distinct. Third, the results were compared with the mill (TC) current treatment where the composting occurs using CP as a raw material. Moreover, the determination of methane concentration was measured in the windrows and the field depositions. In this last one, the main focus occurred on determining the emissions over time (years), with the purpose to estimate how long the coffee by-products emit methane gaseous emissions when the CP is not treated.

Field deposition and TC are common practices in the country; therefore, it is a quantification essential to consider. A flux chamber was placed on top of the windrow piles and inserted approximately 5–10 cm deep into the windrow. This was made to seal the chamber against atmospheric influences to quantify the emissions. The upper part of the flux chamber is designed with two ports (Figure 2), one inlet connected to a hose, allowing the ambient air to enter and be mixed inside the chamber, producing a constant airflow. The second port is an outlet to connect the gas analyzer in the chamber to collect the inner gas. Then, with a gas detector device, the methane concentrations are determined.

Before the gas detection, an estimation of time for the measurements was made. The estimation was considered until the emissions remained permanently constant; therefore, no variation during the measurements could occur. The measurements were conducted weekly for 15 min by quadruplicate before turning the piles in two different windrows per treatment. The methane emissions sampling took place during the first six weeks of the composting process. Measurements were halted during summer without rain events. According to the manufacturer, the sensitivity of the gas detector device was from 0 ± 1 to 60 ± 3, represented in volume percent [37]. The calculation of the methane concentrations is crucial to determine the emission rates and the emission factors.

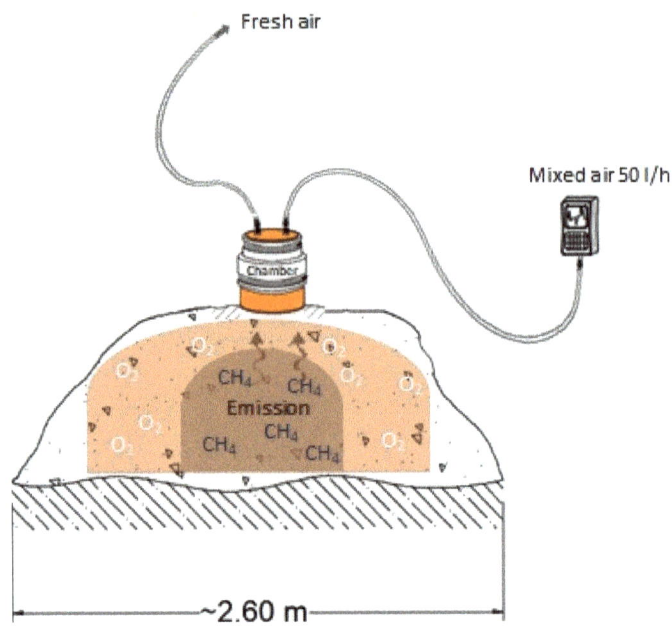

Figure 2. Diagram flow principle of gas sampling on a passive source area in a windrow.

2.4. Emission Rates and Emission Factors

The flow quantifies emission rates through the flux chamber related to the treatment and quality of composting. After identifying a spot, the area is encapsulated to form an aerated chamber. A defined amount of air is extracted, covering the entire area required for sampling to function the constant flow of emissions and supply of ambient air. The flux chamber allows the flux gas to diffuse, which can measure the methane concentration over time and estimate the flux gas emission [38,39] The result is the volumetric flow rate extracted per unit of time [40]. Several factors were considered for the emissions rate calculation: the sampling chamber and chamber area, the volume, and flow volume. Methane emissions rates were calculated in $g \times m^{-2} \times h^{-1}$ with following the equation [10,33,41,42]:

$$q_{CH_4} = \frac{C_{CH_4} * V_{gas}}{A} \quad (1)$$

q_{CH4}—the emission rate of methane ($g \times m^{-2} \times h^{-1}$); C_{CH4}—methane concentration in ($mg \times m^{-3}$); A—flux chamber area in (m^2); V_{gas}—gas flow volume, ($L \times h^{-1}$).

The emission factor from a given source can be calculated as the mass ratio of gas emitted to initial fresh matter mass ($kg \times Mg^{-1}$). However, sometimes the feedstock is reported in units of dry mass. Emission factors related to the mass of CP treated in each treatment and calculated as:

$$EF_{CH_4} = \frac{qCH_4 \times t_{treat} \times A_{treat}}{m_{treat}} \quad (2)$$

EF_{CH4}—the emission factor of methane related to the mass of CP treated ($g \times kg^{-1}$); q_{CH4}—the emission rate of methane ($g \times m^{-2} \times h^{-1}$); t_{treat}—duration (time) of treatment (h); A_{treat}—area of treatment (surface area of the emission) (m^2); m_{treat}—the mass of treated material (mass of CP at the pile) (kg).

2.5. Statistical Analysis

A one-way analysis of variance (ANOVA) test was carried out to investigate the correlations between emissions rates and the period in weeks of composting in the windrows. A significance level of $p \leq 8.9 \times 10^{-8}$ for the composting treatments was used for all mean values. In addition, the Tukey HSD test ($\alpha = 0.05$) was used to assess significant differences between treatment means at a 5% probability level.

3. Results and Discussions

3.1. Environmental Conditions: Temperature, Moisture, and pH

At first, all the material is nearly identical, but heat is generated by increasing the temperature as the micro-organisms grow [43]. One of the indicators of microbial activity is the increase in temperature inside the windrow, where the temperature has traditionally been considered a fundamental variable in the control of composting [44]. Figure 3 shows the temperature profiles for all the windrows. TA refers to the ambient temperature. Additionally, during those 70 days of the composting process, the windrows experienced no rain events. Figure 3 shows a typical development for composting processes containing self-heating, cooling, and stabilization phases. The turning of the material caused low peaks, reflected in the graphs. Sanitation or hygiene is crucial to destroy pathogenic micro-organisms, seeds, and plant components for future sprouting [30,45]. The sanitation process in a windrow should experience at least 14 days of high temperatures (above 55 °C) to enhance the sanitation process [24,46]. The sanitation process was achieved in the temperatures profile for all the treatments maintaining the high temperature at least for the recommended period. The addition of green waste produced different behavior regarding temperature profiles and thermophilic stages within the windrows.

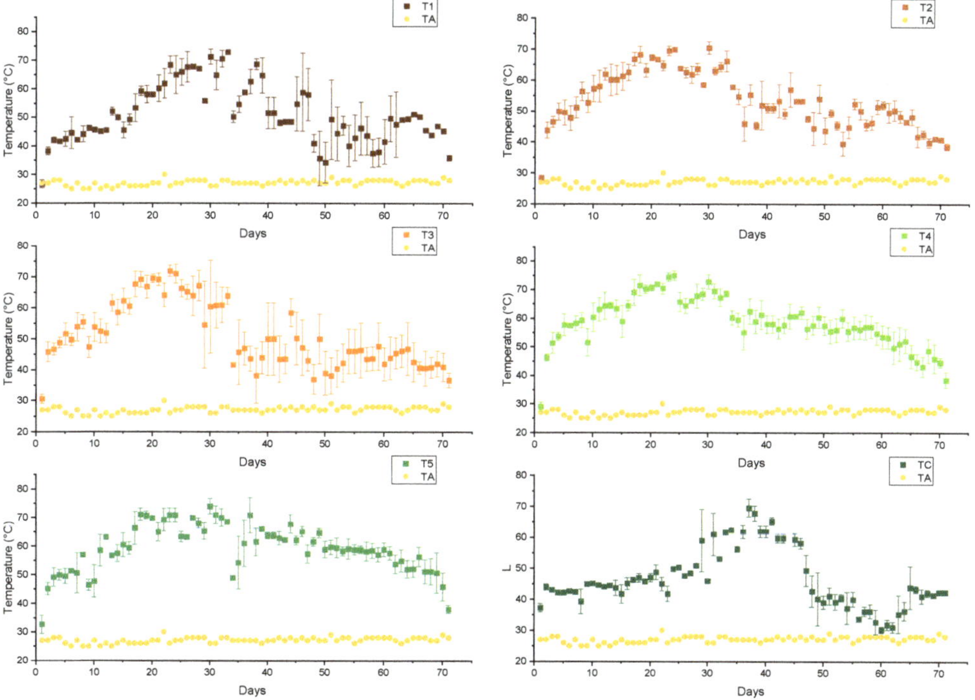

Figure 3. Temperature profiles in all the types of windrows. TA: ambient temperature. Bars represent the standard error of the mean (n = 4).

Comparing the temperature increased in the treatments, T1 accelerated the temperature profile by 38%, T2, T3, and T5 by 72%, and T4 is the most notorious by 83% compared with TC. During the maturation phase, the temperature in the windrows began to decrease until it reached temperatures of 35 °C among the windrows. TC experienced the increase in temperature and the thermophilic stage after 29 days, T1 after 18 days, T2, T3, and T5 after eight days, and T4 after five days of the composting process. The addition of green waste into the windrow has accelerated the degradation, microbial activity, and temperature profile within the windrows. T1 experienced the most extended period for reaching the thermophilic stage, attributed to the CP percentage in the piles (80% P). The observed variation in periods of high temperature could be attributed to the variation in the percentages of GW in each treatment [47]. For example, the behavior of T1 is the closest to the profile of the control TC. This is attributed toT1 containing the highest amount of CP in the composting treatment.

During aerobic windrow systems, moisture and aeration are important parameters to consider to enhance the biological activity inside the windrow [48]. It is recommended during the composting process to maintain a moisture content between 50–60% for favorable results [23] and the avoidance that water fully occupies the pores of the composting mass [44]. This parameter is essential for CH4 emissions control during composting and affects degradation and end-product features [5]. Furthermore, evaporation is linked to the lack of porosity and aeration in the system and the relationship between air to water ratio [30,49].

Microbial activity, including bacteria, fungi, and yeast, depends on temperature and moisture content [27]. The aim of this study was not the microbial counts weekly during the composting process. Instead, a general overview of the microbial activity was considered important to control the microbes in at least two treatments to ensure that the activity is higher than the control TC with the proposed technique. Quantitative analysis was based on colony counts and subsequent calculation of colony-forming units per gram (CFU/g) to estimate the viable number of bacteria, fungal cells in the samples during the third week of composting process for TC, T1, and T3 (Figure 4).

Results show that the microbial activity is higher for T1 and T5 than for TC. During the third week of composting, T1 and T2 reached thermophilic temperatures, and the amount of yeast from Figure 4 compared to TC can be linked to this event. Some studies show that higher temperatures for T1 and T5 induced an earlier microbial activity [50]. For this case, Figure 4 shows that temperature during the third week was below 50 °C, whereas T1 and T5 had temperatures above 50 °C. Low moisture content (below 40%) limits microbial activity [21].

On the other hand, very high moisture enhance anaerobic conditions because the pore spaces are filled with water rather than air [51]. Among those variables, the moisture content is considered one of the key parameters affecting the biodegradation process. Furthermore, some studies have suggested an important influence on microbial activity linked with MC than with the temperature [50]. In addition, the number of input materials can change the microbial communities, where T5 possesses the highest amount of green waste material among the piles. This ensures that adding materials to P improves the activity of the micro-organisms, enhancing the composting process.

Each windrow treatment varied regarding the response to water consumption and evaporation. The monitoring was important during the composting process to obtain sufficient conditions for T1–T5 and compare with the current treatment at mill TC. Over ten weeks, moisture content was measured among the different windrows represented in Figure 5. During the first week, all the treatments showed a high moisture content due to the high levels of moisture that CP possesses by itself (Table 2). After the second week, all the windrows except TC reached the recommended moisture content levels for the composting process. Thus, TC has reached the levels recommended after six weeks of the process. In TC occurred no variation during four weeks of pile age. The high value is attributed to the high MC of the coffee by-product. T1–T5 had rapid absorption with the percentages of the mixture of GW and CP. It was observed that the highest amount GW added in the windrow, the fastest was the reduction of the MC in the system after the second week of the composting process.Additionally, pH directly influences composting due to its action on the dynamics of microbial processes [52]. For this study, pH in all windrows rapidly increases within three weeks of all the treatments (Figure 5), showing a tendency. Subsequently, an alkaline state was seen for the rest of the composting process. Some authors have seen a link between the loss of organic acids and the generation of ammonia from the decomposition of proteins and this alkaline state [10,53–55]. In all the windrows, at the end of the process, the pH of the treatments was basic (8.8–9.8), indicating maturity since lower pH values would indicate anaerobic processes [56].

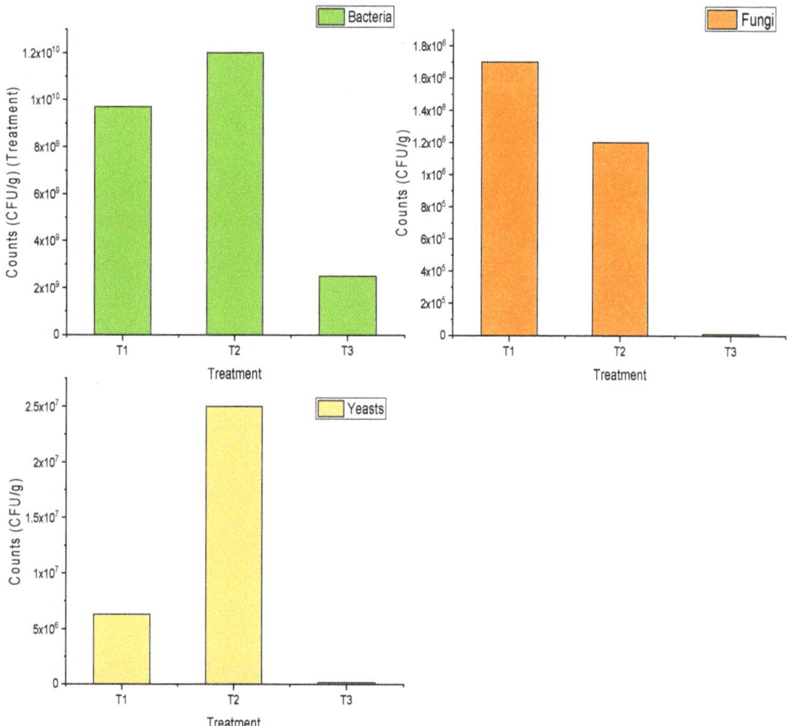

Figure 4. Comparison of mean colony-forming unit (CFU/g) for samples T1, T5, and TC during the third week of the composting process.

Some studies have shown that the alkalinity can be attributed to the high potassium (K) content of the input materials [57], such as using the CP as the primary raw input material for the process (Table 2) [58]. This observation was consistent with the results of this study, showing that the highest pH values were obtained in TC for the last week of the process. The windrows T1–T5 show results as a mature compost, whereas TC indicates an immaturity. The turning and aeration are also important to factor during the pH behavior in the window systems. Having good aeration and oxygen concentration is obtained low concentrations of organic acids enhancing the decomposition of these acids and giving a faster rise in the pH [59].

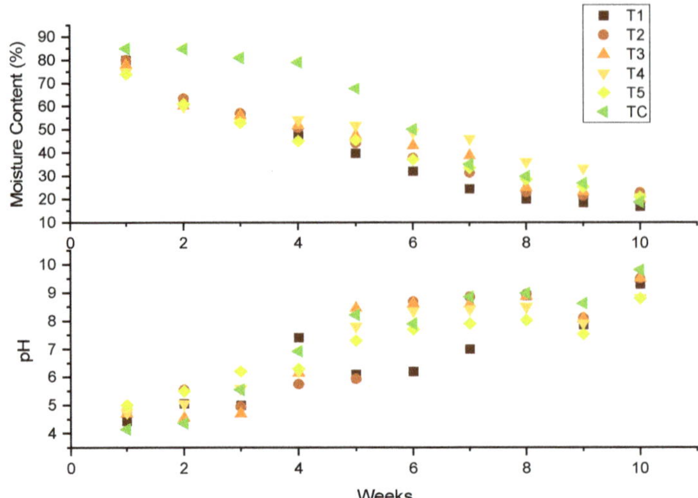

Figure 5. Moisture content and pH profiles in all the types of windrows throughout CP composting. Values are the means ($n = 3$).

Once the ten-week composting period was achieved, and the low temperatures remained constant, the samples were shipped to Germany. They were delivered to an accredited testing laboratory for the BGK to analyze plant nutrients, soil amendment, and final physical parameters for the windrows. These results are summarized in Table 3. The elements include nitrogen (N), phosphorous (P), and K, which are fundamental macronutrients for microbial growth. [60]. The total of nutrients is essential since it can vary and allows the mill to determine an appropriate end-use for the compost. N, P, and K are the macronutrients that make the biggest uptake [61]. N is assimilated by the plants in the forms of nitrate (NO_3^-) and ammonium (NH_4^+), P in forms of orthophosphate ($H_2PO_4^-$), and K in the forms of potassium oxide (K_2O) [62]. Nevertheless, attention must be focused on the availability of these nutrients, which is known as a limiting growth and uptake factor [62]. In this case, T4 presents high levels of N and NO_3^-, which the plant can assimilate immediately. It also contains high NH_4^+ levels, which need a process in the soil for the plant to absorb [60], giving a good combination of NO_3^- and NH_4^+. Finally, it contains the highest value of soluble magnesium (Mg), meaning practical use in crops in the future since the soils of coffee production possess a common Mg deficiency [63]. Furthermore, Mg gives color to the plant as it is the central atom of the chlorophyll molecule [64]. To summarize, T4 shows the best performance in the nutrient content, and it is suggested for future usage in the coffee plantations as an additional amendment. It contains high N, NO_3^-, and Mg, which could be used as a plant stimulant in some phenological parts of the crop.

High salt levels can be unfavorable for seeds and plants when compost is for a nursery medium [65].TC showed the lowest plant compatibility test associated with the degree of rotting and the salt levels; meanwhile, the rest of the treatments proposed showed a >95% for plant compatibility and low salt content. Bulk density among the proposed treatments fulfills the BKG limit values (between 400–900 g/L); meanwhile, the control obtained the lowest bulk density value. This could carry consequences in compost application since the material can suffer high pore space and common water retention values, causing difficulties in future compost applications. The lowest C/N ratio was obtained in T5, whereas TC possesses the highest value. Even though CP owns low C/N, this increase is linked to the mineralization of organic N, promoting the ammonia emissions, leading to a high N loss when the initial C:N is low. [66].

On the other hand, the results show high total K values. They are attributed, as previously mentioned, to the high potassium content of the CP since it contains more nitrogen and K than other common finished materials as compost [9]. Regarding the organic substances, BKG recommends values higher of 30% based on dry mass for finished compost. Higher levels indicate that the compost is not finished, as is the case of TC. The rotting degree or self-heating test is considered an important parameter indicating heat in the window, showing signs of immaturity. The five categories of the interpretation scale are grouped, often made by professionals and European agencies, into three main classes. The lowest grade (I) is called "fresh compost", the two intermediate grades (II, III) are called "active compost," and the top two grades (IV, V) are called "finished compost" [36]. For all the treatments, the grade was IV and V except for the control.

3.2. Methane Gas Emissions

The CH_4 is influenced by different factors such as temperature, moisture, and pH directly [67]. When the aeration is not proper and the moisture content increases in the window, this can result in high CH_4 emissions affecting the oxygen restrictions in the microbiological metabolism in the windrow [5,68,69]. Methane emissions rates shown in Figure 6A were measured during the first six weeks of the composting process and compared if the addition of GW influences their emissions in the windrows. The highest value is TC with 38 g $\times$ m^{-2} $\times$ h^{-1} in the first week of composting. Comparing Figure 5 of MC and Figure 6, the first week of composting process possesses the highest value for moisture content among the study. Thus, the production of CH_4 is also increased exponentially with the moisture level, allowing the formation of undesirables anaerobic zones enhancing the methanogenesis and anaerobic metabolism in the windrow [30,33]. Even if the composting treatment is under aerobic conditions, the diversity of the input materials, moisture content, temperature, biological microbial activity, and redox requirements could be developed in the window [2].

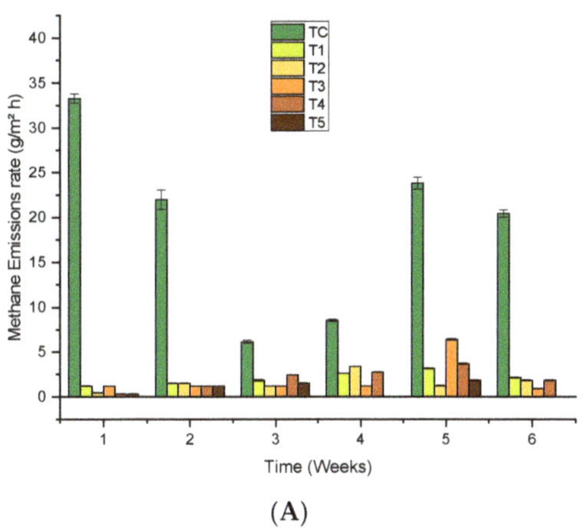

(A)

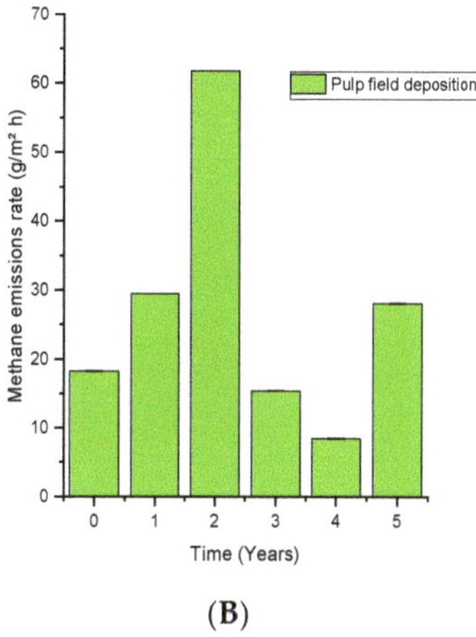

(B)

Figure 6. (**A**) Methane emissions rate for all the treatments using coffee pulp for composting and (**B**) methane emissions rate in field deposition. Bars represent the standard error of the mean (n = 4).

In this study, the reduction of CH_4 during composting is accomplished if additional material is used, such as GW. The weekly highest values among the treatments were found for T1 and T2 with 3.1 g $\times$ m^{-2} $\times$ h^{-1} and 3.3 g $\times$ m^{-2} $\times$ h^{-1}, respectively. Each value was found during the fourth week and for T3, T4, T5 with 6.4 g $\times$ m^{-2} $\times$ h^{-1}, 3.6 g $\times$ m^{-2} $\times$ h^{-1}, and 1.8 g $\times$ m^{-2} $\times$ h^{-1}, respectively in the fifth week. Emissions formation increases faster when temperatures are over 65 °C [3]. Furthermore, a correlation between the temperature profiles and the emissions peaks is shown since the highest peaks of temperatures and emissions were found for the windrows during the same weeks of the composting process.

The presence of low pH, which illustrates the presence latency of organic acids, and the present of also CH_4 emissions, indicates that anaerobic conditions have been since the initial formation of the windrow pile [68]. The reduction of total emissions CH_4 was between 89–95% compared to the control treatment, which shows the difference of aeration and proper management within the treatments compared to the control. Researchers recommend that increasing the oxygen level available in the system is necessary during the first week [3]. In the fifth week, methane emissions were higher than in the first week for T1–T5. This could be attributed to the moisture suppressing the airflow since the material at that week is more compact. The pores in the feedstock are filled with water, favoring the formation of anaerobic conditions and methane emissions [31]. The emissions in the fifth week were the highest among the treatment compared to the control TC. The reduction of emissions with GW is added in the window is seen. In other studies, it has been found that CH_4 emissions are present during the initial stage of the thermophilic phase since there is an oxygen solubility reduction, enhancing anaerobic zones in the windrow [70]. During CH_4 emission exists other microbial factors affecting the gas transport and gas diffusion, including the presence of methanotrophic bacteria (these bacteria are colonizing the area nearby anaerobic zones being able to oxidize up to 98% of the CH_4 formed in the windrow) [71].

A previous study [10] shows the magnitude of the CH_4 emissions in open field depositions when the CP is not composted or pre-treated before field deposition, generating serious environmental concerns. Figure 6B shows the behavior of the methane emissions over the years when the CP is buried for a lifetime. The highest emission was in the second year with 53.7 g $\times$ m^{-2} $\times$ h^{-1}. Equivalent emissions are seen in the first and the fifth year of field deposition with 25.6 and 24.4 g $\times$ m^{-2} $\times$ h^{-1}. These results imply that the material is not degraded over time, producing continuous emissions of great magnitude when not treated correctly. This behavior of high CH_4 emissions over this period is due to the properties of the CP. When it is buried with high moisture content, high material density, and a lack of aeration, anaerobic zones are created, enhancing the methane emissions.

Given this problem and its emissions, the mill aims for successive harvests to avoid this practice. For example, during the 2019–2020 harvest, open field depositions were made with about 630 Mg (6000 fanegas). In previous years, a minimum of 1500 Mg of CP was transported in the fields. It is clear that the depositions are a risk and a focus of emissions since they are large spaces of at least 925 m^2 of surface area; therefore, the high levels of emissions over the years can be attributed to the surface area, amount of material, and the management in that area.

For the estimation of EF among all the treatments, it is shown in two different options. Firstly, a calculation of EF regarding the amount of CP added in each treatment. It is crucial to establish since there is no EF associated directly with CP in the literature, where the results show that the more CP added into the system, the higher EF is obtained. After this, an EF calculation was input to represent the treatments better and was compared with literature regarding green waste from some regulations in composting plants.

The government is currently developing a new national composting plan accomplishing the strategic guidelines of the National Decarbonization Plan 2050. The inventories of GHGs in Costa Rica regarding waste management are made by the National Meteorological Institute (IMN for its abbreviation in Spanish). Therefore, the results in this research are compared with the National Inventory's values; nevertheless, it is necessary to be analyzed. These inventories are for municipal solid waste; therefore, these results cannot be thoroughly compared since the country does not explicitly relate to agricultural waste. On the other hand, there is no emission factor directly linked to CP; a comparison in the literature shows 4 g CH_4/kg solid waste [72], 4 g CH_4/kg waste treated [73], 2.2 g CH_4/Mg GW FM [5], 4.7 and 7.6 g CH_4/kg GW [74].

The emissions factors shown in Table 4 show the decrease of emissions comparing TC with the rest of the treatments. The utmost emission values found were in T3 and T4 with 14 g CH_4/kg CP. The closest value compared to the literature is T5. One of the reasons could be the amount of GW added since it had the most significant amount of material in a pile. Compared with the literature values of biological treatments, these results present elevated values. However, with the proposed methodology for all the five treatments, these values are approaching the values recommended by the literature for composting. For the open field depositions, even if there is no management involved, the emissions factors represented the highest values among all the treatments, including the current treatment of the mill.

Table 4. Emissions factors from all the treatments proposed.

Treatment	EF (g CH_4/kg CP)	EF (g CH_4/kg Input)	SD *
T1	43.9	35.1	0.23
T2	14.4	10.8	0.17
T3	20	14	0.23
T4	23.4	14	0.24
T5	11.6	5.8	0.14
TC	129	129	2.96

D *: standard deviation of the mean values (n = 4).

4. Conclusions

The study achieved the development of an experimental methodology using coffee by-products and GW. An improvement in the key parameters of composting was observed, such as temperature, pH, and WC profiles, when coffee by-products were mixed with GW for composting treatment. Therefore, waste valorization within the process is concluded together with the reduction of methane emissions. The proposed treatments experienced fewer methane emissions rates than the control; hence, implementing this technique suggests a good practice in the future for the coffee sector and the mill in Costa Rica. Results show that T2–T5 are strongly recommended treatments involving methane emissions, physical parameters during the process, plant nutrient content, and finished compost classified (Grade IV and V) following the BKG standards. Overall, this study promotes a better understanding of the performance of CP when the material is composted and their methane emissions during the process. This approach might be necessary for the future to guide a national mitigation plan in the agricultural and coffee sector of the country. In addition, it will be a helpful tool for the future calculations of the global emissions using a technology already studied in another place or another treatment plant. Suppose it is considered the agronomic and environmental aspects in an integrated manner. It is recommended to investigate further the benefits of using the compost and the relationship between GHG emitted during the process. The compost utilization can also compensate for this reduction in the long term in the coffee plantations. Continuous but robust research is suggested to develop emissions and factors that adequately cover national conditions to establish new inventories, especially for the coffee sector, including coffee by-products and management.

Author Contributions: Conceptualization: M.S.M.R., M.R.; methodology: M.R., M.S.M.R., and M.K.; validation: M.R., M.K., and M.S.M.R.; formal analysis: M.S.M.R., M.R.; investigation: M.S.M.R.; M.R., and M.K.; data curation: M.S.M.R., M.R.; writing—original draft preparation: M.S.M.R., M.R.; supervision: M.R., M.K., and M.S.M.R.; project administration: M.R., M.K. All authors have read and agreed to the published version of the manuscript.

Funding: This work has been financed by the Coffee Mill Coopetarrazú R.L in Costa Rica.

Acknowledgments: We gladly thank the personnel, laboratory assistances, students, and colleagues of the Institute for Sanitary Engineering, Water Quality and Solid Waste Management (ISWA) at the University of Stuttgart.

Conflicts of Interest: The authors declare no conflict of interest.

References

1. Smith, P.; Bustamante, M.; Ahammad, H.; Clark, H.; Dong, H.; Elsiddig, E.A.; Haberl, H.; Harper, R.; House, J.; Jafari, M.; et al. Agriculture, Forestry and Other Land Use (AFOLU). In *Climate Change 2014 Mitigation of Climate Change Mitigation of Climate Change. Contribution of Working Group III to the Fifth Assessment Report of the Intergovern-Mental Panel on Climate Change*; Edenhofer, O., Pichs-Madruga, R., Sokona, Y., Farahani, E., Kadner, S., Seyboth, K., Adler, A., Baum, I., Brunner, S., Eickemeier, P., Eds.; Cambridge University Press: Cambridge, UK, 2014; pp. 811–922.
2. Bridgham, S.D.; Richardson, C.J. Mechanisms controlling soil respiration (CO_2 and CH_4) in southern peatlands. *Soil Biol. Biochem.* **1992**, *24*, 1089–1099. [CrossRef]
3. Vergara, S.E.; Silver, W.L. Greenhouse gas emissions from windrow composting of organic wastes: Patterns and emissions factors. *Environ. Res. Lett.* **2019**, *14*, 124027. [CrossRef]
4. US EPA. *AP 42, Fifth Edition Compilation of Air Pollutant Emission Factors, Volume 1: Stationary and Point Sources*; US EPA: Research Triangle Park, NC, USA, 1995; pp. 1–10.
5. Amlinger, F.; Peyr, S.; Cuhls, C. Green house gas emissions from composting and mechanical biological treatment. *Waste Manag. Res.* **2008**, *26*, 47–60. [CrossRef] [PubMed]
6. Sánchez, A.; Gabarrell, X.; Artola, A.; Barrena, R.; Colón, J.; Font, X.; Komilis, D.; Taherzadeh, M.J.; Richards, T. *Composting of Wastes. Resource Recovery to Approach Zero Municipal Waste*, 1st. ed.; Taherzadeh, M.J., Richards, T., Eds.; Green Chemistry and Chemical Engineering: Boca Raton, FL, USA, 2015; pp. 77–106.
7. Blinová, L.; Sirotiak, M.; Bartošová, A.; Soldán, M. Review: Utilization of Waste from Coffee Production. *Res. Pap. Fac. Mater. Sci. Technol. Slovak Univ. Technol.* **2017**, *25*, 91–101. [CrossRef]
8. Chong, J.A.; Dumas, J.A. Coffee pulp compost: Chemical properties and distribution of humic substances. *J. Agric. Univ. Puerto Rico* **2012**, *96*, 77–87. [CrossRef]
9. Braham, J.; Bressani, R. *Coffee Pulp: Composition, Technology, and Utilization*; IDRC: Ottawa, ON, Canada, 1979.
10. San Martin Ruiz, M.; Reiser, M.; Hafner, G.; Kranert, M. A Study about Methane Emissions from Different Composting Systems for Coffee By-products on Costa Rica. *Environ. Ecol. Res.* **2018**, *6*, 461–470. [CrossRef]
11. Chala, B.; Oechsner, H.; Latif, S.; Müller, J. Biogas Potential of Coffee Processing Waste in Ethiopia. *Sustainability* **2018**, *10*, 2678. [CrossRef]
12. Beyene, A.; Kassahun, Y.; Addis, T.; Assefa, F.; Amsalu, A.; Legesse, W.; Kloos, H.; Triest, L. The impact of traditional coffee processing on river water quality in Ethiopia and the urgency of adopting sound environmental practices. *Environ. Monit. Assess.* **2011**, *184*, 7053–7063. [CrossRef]
13. Desai, N.M.; Varun, E.; Patil, S.; Pimpley, V.; Murthy, P.S. Environment Pollutants During Coffee Processing and Its Valorization BT. In *Handbook of Environmental Materials Management*; Hussain, C.M., Ed.; Springer International Publishing: Cham, Switzerland, 2020; pp. 1–13.
14. Misra, R.V.; Roy, R.N.; Hiraoka, H.; Food and Agriculture Organisation of the United Nations. FAO On-farm com-posting methods, Composting Methods and Techniques, Rome. 2003. Available online: http://www.fao.org/docrep/007/y5104e/y5104e05.htm (accessed on 17 February 2021).
15. Sánchez, A.; Artola, A.; Font, X.; Gea, T.; Barrena, R.; Gabriel, D.; Sánchez-Monedero, M.Á.; Roig, A.; Cayuela, M.L.; Mondini, C. Greenhouse Gas from Organic Waste Composting: Emissions and Measurement. In *CO_2 Sequestration, Biofuels and Depollution. Environmental Chemistry for a Sustainable World*; Lichtfouse, E., Schwarzbauer, J., Robert, D., Eds.; Springer: Cham, Switzerland, 2015; Volume 5. [CrossRef]
16. Diaz, L.F.; De Bertoldi, M.; Bidlingmaier, W. *Compost Science and Technology*; Elsevier: Amsterdam, The Netherlands, 2007.
17. Haug, R.T. *The Practical Handbook of Compost Engineering*, 1st ed.; Lewis Publishers: Atascadero, CA, USA, 1993.
18. Shemekite, F.; Gómez-Brandón, M.; Franke-Whittle, I.H.; Praehauser, B.; Insam, H.; Assefa, F. Coffee husk composting: An investigation of the process using molecular and non-molecular tools. *Waste Manag.* **2014**, *34*, 642–652. [CrossRef]
19. *VDI Guideline: Emission Control Mechnical-Biological Treatment Facilities for Municipal Solid Waste*; VDI 3475 Part 3; Beuth Verlag GmbH: Berlin, Germany, 2006.
20. Kranert, M. *Einführung in die Kreislaufwirtschaft: Planung-Recht-Verfahren*; Springer: Vieweg, Wiesbaden, 2017.
21. Shilev, S.; Naydenov, M.; Vancheva, V.; Aladjadjiyan, A. Composting of Food and Agricultural Wastes. In *Utilization of By-Products and Treatment of Waste in the Food Industry*; Springer: Berlin, Germany, 2006; pp. 283–301.
22. Ghazifard, A.; Kasra-Kermanshahi, R.; Far, Z.E. Identification of thermophilic and mesophilic bacteria and fungi in Esfahan (Iran) municipal solid waste compost. *Waste Manag. Res.* **2001**, *19*, 257–261. [CrossRef]
23. Azim, K.; Soudi, B.; Boukhari, S.; Perissol, C.; Roussos, S.; Alami, I.T. Composting parameters and compost quality: A literature review. *Org. Agric.* **2018**, *8*, 141–158. [CrossRef]
24. IVillar, I.; Alves, D.; Garrido, J.; Mato, S. Evolution of microbial dynamics during the maturation phase of the composting of different types of waste. *Waste Manag.* **2016**, *54*, 83–92. [CrossRef]

25. Zhu-Barker, X.; Bailey, S.K.; Paw, U.K.T.; Burger, M.; Horwath, W.R. Greenhouse gas emissions from green waste composting windrow. *Waste Manag.* **2017**, *59*, 70–79. [CrossRef]
26. *VDI Guideline: Emission Control Biological Waste Treatment Facilities Composting and Anaerobic Digestion Plant Capacities More Than Approx. 6*; VDI 3575 Part 1; Beuth Verlag GmbH: Berlin, Germany, 2003.
27. Sánchez, A.; Artola, A.; Font, X.; Gea, T.; Barrena, R.; Gabriel, D.; Sanchez-Monedero, M.A.; Roig, A.; Cayuela, M.L.; Mondini, C. Greenhouse gas emissions from organic waste composting. *Environ. Chem. Lett.* **2015**, *13*, 223–238. [CrossRef]
28. Díaz, M.V.; Prada, P.; Mondragon, M. Optimización del proceso de compostaje de productos post-cosecha (cereza) del café con la aplicación de microorganismos nativos. *Nova* **2010**, *8*, 214. [CrossRef]
29. Muzaifa, M.; Rahmi, F.; Syarifudin. Utilization of Coffee By-Products as Profitable Foods—A Mini Review. *IOP Conf. Ser. Earth Environ. Sci.* **2021**, *672*, 012077. [CrossRef]
30. Ijanu, E.M.; Kamaruddin, M.A.; Norashiddin, F.A. Coffee processing wastewater treatment: A critical review on current treatment technologies with a proposed alternative. *Appl. Water Sci.* **2019**, *10*, 11. [CrossRef]
31. Duong, B.; Marraccini, P.; Maeght, J.-L.; Vaast, P.; Lebrun, M.; Duponnois, R. Coffee Microbiota and Its Potential Use in Sustainable Crop Management. A Review. *Front. Sustain. Food Syst.* **2020**, *4*, 237. [CrossRef]
32. Government of Costa Rica. National Decarbonization Plan, Costa Rica. 2019. Available online: https://unfccc.int/sites/default/files/resource/NationalDecarbonizationPlan.pdf (accessed on 16 February 2021).
33. Moldvaer, A. *Coffee Obsession*, 1st ed.; Dorling Kindersley Limited: London, UK, 2014.
34. San Martin Ruiz, M.; Reiser, M.; Kranert, M. Enhanced composting as a way to a climate-friendly management of coffee by-products. *Environ. Sci. Pollut. Res.* **2020**, *27*, 24312–24319. [CrossRef]
35. SEPSA. Costa Rica. Equivalencias, Rendimientos, Pesos y Factores de Conversión Utilizados en Algunos Productos Agropecuarios. SEPSA, 2017. Available online: http://www.infoagro.go.cr/BEA/BEA24/BEA24/img_superficie_produccion/superficie_produccion_17.pdf (accessed on 16 February 2021).
36. Ramírez, S.M.; Ballestero, Y.Q. Boletín Estadístico Agropecuario №29 | Serie Cronológica 2015–2018, Costa Rica. 2019. Available online: http://www.mag.go.cr/bibliotecavirtual/BEA-0029.PDF (accessed on 16 February 2021).
37. BGK. Qualitätsmanagement-Handbuch (QMH), RAL-Gütesicherung Kompost (RAL-GZ 251). In *QM-Handbuch Kompost*; BGK-Bundesgütegemeinschaft Kompost e.V: Köln, Germany, 2017; p. 45.
38. *Hermann Sewering: Technisches Datenblatt · Multitec 540*; Hermann Sewerin GmbH: Gütersloh, Germany, 2012.
39. Bidlingmaier, W. *Methods Book for the Analysis of Compost*, 3rd ed.; Federal Compost Quality Assurance Organisation (FCQAO) Bundesgütegemeinschaft Kompost e.V. (BGK): Köln, Germany, 2003.
40. *VDI Guideline: Emission Control Facilities for Biological Waste Composting an Anaerobic (Co-)Digestion Plant Capacities up to Approx 6000 Mg/a A*; VDI 3475 Part 2; Beuth Verlag GmbH: Berlin, Germany, 2005.
41. *VDI Guideline: Olfactometry Static Sampling*; VDI 3880; Beuth Verlag GmbH: Berlin, Germany, 2011.
42. Clauß, T.; Reinelt, T.; Vesenmaier, A.; Flandorfer, C.; Reiser, M.; Ottner, R.; Huber-Humer, M.; Flandorfer, C.; Stenzel, C.; Piringer, M.; et al. *Recommendations for Reliable Methane Emission Rate Quantification at Biogas Plants*; Deutsches Bio-masseforschungszentrum gemeinnützige GmbH: Leipzig, Germany, 2019.
43. Liebetrau, J.; Reinelt, T.; Clemens, J.; Hafermann, C.; Friehe, J.; Weiland, P. Analysis of greenhouse gas emissions from 10 biogas plants within the agricultural sector. *Water Sci. Technol.* **2013**, *67*, 1370–1379. [CrossRef]
44. Taiwo, L.B.; Oso, B.A. Influence of composting techniques on microbial succession, temperature and pH in a composting municipal solid waste. *Afr. J. Biotechnol.* **2004**, *3*, 239–243.
45. Miyatake, F.; Iwabuchi, K. Effect of compost temperature on oxygen uptake rate, specific growth rate and enzymatic activity of microorganisms in dairy cattle manure. *Bioresour. Technol.* **2006**, *97*, 961–965. [CrossRef]
46. Inserra, R.N.; Hampton, M.O.; Schubert, T.S.; Stanley, J.D.; Brodie, M.W.; Bannon, J.H.O. Guidelines for compost sanitation. *Proc. Soil Crop Sci. Soc. Florida* **2006**, *65*, 31–37.
47. Jenkins, J. Sanitation by Composting, United States. 2011. Available online: https://humanurehandbook.com/downloads/Compost_Sanitation_Paper_9_2011.pdf (accessed on 22 April 2019).
48. MHimanen, M.; Hänninen, K. Composting of Bio-Waste, Aerobic and Anaerobic Sludges–Effect of Feedstock on the Process and Quality of Compost. *Bioresour. Technol.* **2010**, *102*, 2842–2852. [CrossRef]
49. Richard, T.; Hamelers, H.; Veeken, A.; Silva, T. Moisture Relationships in Composting Processes. *Compos. Sci. Util.* **2002**, *10*, 286–302. [CrossRef]
50. Richard, T.L.; Veeken, A.H.M.; De Wilde, V.; Hamelers, H. Air-Filled Porosity and Permeability Relationships during Solid-State Fermentation. *Biotechnol. Prog.* **2004**, *20*, 1372–1381. [CrossRef]
51. Liang, C.; Das, K.; McClendon, R. The influence of temperature and moisture contents regimes on the aerobic microbial activity of a biosolids composting blend. *Bioresour. Technol.* **2003**, *86*, 131–137. [CrossRef]
52. Kim, E.; Lee, D.-H.; Won, S.; Ahn, H. Evaluation of Optimum Moisture Content for Composting of Beef Manure and Bedding Material Mixtures Using Oxygen Uptake Measurement. *Asian-Australas. J. Anim. Sci.* **2016**, *29*, 753–758. [CrossRef] [PubMed]
53. Sundberg, C. Low pH as an inhibiting factor in the transition from mesophilic to thermophilic phase in composting. *Bioresour. Technol.* **2004**, *95*, 145–150. [CrossRef] [PubMed]

54. Burg, P.; Zemánek, P.; Michálek, M. Evaluating of selected parameters of composting process by composting of grape pomace. *Acta Univ. Agric. Silvic. Mendel. Brun.* **2014**, *59*, 75–80. Available online: https://acta.mendelu.cz/media/pdf/actaun_2011059060075.pdf (accessed on 23 April 2019). [CrossRef]
55. Sundberg, C. Improving Compost Process Efficiency by Controlling Aeration, Temperature and pH. Ph.D. Thesis, Swedish University of Agricultural Sciences, Uppsala, Sweden, 2005.
56. Sánchez-Monedero, M.; Roig, A.; Paredes, C.; Bernal, M.P. Nitrogen transformation during organic waste composting by the Rutgers system and its effects on pH, EC and maturity of the composting mixtures. *Bioresour. Technol.* **2001**, *78*, 301–308. [CrossRef]
57. Suler, D.J.; Finstein, M.S. Effect of Temperature, Aeration, and Moisture on CO_2 Formation in Bench-Scale, Continuously Thermophilic Composting of Solid Waste. *Appl. Environ. Microbiol.* **1977**, *33*, 345–350. [CrossRef] [PubMed]
58. Yeo, D.; Dongo, K.; Mertenat, A.; Lüssenhop, P.; Körner, I.; Zurbrügg, C. Material Flows and Greenhouse Gas Emissions Reduction Potential of Decentralized Composting in Sub-Saharan Africa: A Case Study in Tiassalé, Côte d'Ivoire. *Int. J. Environ. Res. Public Heal.* **2020**, *17*, 7229. [CrossRef] [PubMed]
59. Zoca, S.M.; Penn, C.J.; Rosolem, C.A.; Alves, A.R.; Neto, L.O.; Martins, M.M. Coffee processing residues as a soil potassium amendment. *Int. J. Recycl. Org. Waste Agric.* **2014**, *3*, 155–165. [CrossRef]
60. Beck-Friis, B.; Smårs, S.; Jónsson, H.; Eklind, Y.; Kirchmann, H. Composting of Source-Separated Household Organics At Different Oxygen Levels: Gaining an Understanding of the Emission Dynamics. *Compos. Sci. Util.* **2003**, *11*, 41–50. [CrossRef]
61. Roy, R.N.; Finck, A.; Blair, G.J.; Tandon, H.L.S. Plant Nutrition for Food Security. A Guide for Integrated Nutrient Management. In *Fertilizer and Plant Nutrition Bulletin 16*; FAO, Ed.; Food and Agricultural Organization of United Nations: Rome, Italy, 2006; p. 348.
62. Tong, J.; Sun, X.; Li, S.; Qu, B.; Wan, L. Reutilization of Green Waste as Compost for Soil Improvement in the Afforested Land of the Beijing Plain. *Sustainability* **2018**, *10*, 2376. [CrossRef]
63. Gondek, M.; Weindorf, D.C.; Thiel, C.; Kleinheinz, G. Soluble Salts in Compost and Their Effects on Soil and Plants: A Review. *Compos. Sci. Util.* **2020**, *28*, 59–75. [CrossRef]
64. Nzeyimana, I.; Hartemink, A.E.; de Graaff, J. Coffee Farming and Soil Management in Rwanda. *Outlook Agric.* **2013**, *42*, 47–52. [CrossRef]
65. Senbayram, M.; Gransee, A.; Wahle, V.; Thiel, H. Role of magnesium fertilisers in agriculture: Plant–soil continuum. *Crop. Pasture Sci.* **2015**, *66*, 1219–1229. [CrossRef]
66. Paradelo, R.; Devesa-Rey, R.; Cancelo-González, J.; Basanta, R.; Peña, M.; Diaz-Fierros, F.; Barral, M.T. Effect of a compost mulch on seed germination and plant growth in a burnt forest soil from NW Spain. *J. Soil Sci. Plant Nutr.* **2012**, *12*, 73–86. [CrossRef]
67. Hao, X.; Benke, M.B. *Nitrogen Transformation and Losses during Composting and Mitigation Strategies*; Global Science Books: Isleworth, UK, 2008.
68. Taconi, K.A.; Zappi, M.E.; French, W.T.; Brown, L.R. Methanogenesis under acidic pH conditions in a semi-continuous reactor system. *Bioresour. Technol.* **2008**, *99*, 8075–8081. [CrossRef] [PubMed]
69. Ermolaev, E.; Sundberg, C.; Pell, M.; Smårs, S.; Jönsson, H. Effects of moisture on emissions of methane, nitrous oxide and carbon dioxide from food and garden waste composting. *J. Clean. Prod.* **2019**, *240*, 118165. [CrossRef]
70. Jiang, T.; Schuchardt, F.; Li, G.; Guo, R.; Zhao, Y. Effect of C/N ratio, aeration rate and moisture content on ammonia and greenhouse gas emission during the composting. *J. Environ. Sci.* **2011**, *23*, 1754–1760. [CrossRef]
71. Jäckel, U.; Thummes, K.; Kämpfer, P. Thermophilic methane production and oxidation in compost. *FEMS Microbiol. Ecol.* **2005**, *52*, 175–184. [CrossRef] [PubMed]
72. IMN. Factores de Emisión Gases Efecto Invernadero, San José, Costa Rica. 2020. Available online: http://cglobal.imn.ac.cr/index.php/publications/factores-de-emision-gei-decima-edicion-2020/ (accessed on 10 February 2021).
73. IPCC. Volume 5 Waste-Chapter 4 Biological Treatment of Solid Waste, 2006 IPPC Guidel. National Greenhouse Gas Inventories. 2006, pp. 4.1–4.8. Available online: http://www.ipcc-nggip.iges.or.jp/public/2006gl/vol5.html (accessed on 5 February 2021).
74. ARB. Air Resources Board (ARB): Method for Estimating Greenhouse Gas Emission Reduction from Diversion of Organic Waste from Landfills to Compost Facilities. 2017. Available online: https://ww2.arb.ca.gov/sites/default/files/classic/cc/waste/cerffinal.pdf (accessed on 18 February 2021).

atmosphere

MDPI

Article

Radiative Energy Flux Variation from 2001–2020

Hans-Rolf Dübal [1,*] and Fritz Vahrenholt [2]

[1] Am Langenstück 13, 65343 Eltville, Germany
[2] Department of Chemistry, University of Hamburg, Papenkamp 14, 22607 Hamburg, Germany; fritz.vahrenholt@chemie.uni-hamburg.de
* Correspondence: duebal@t-online.de

Abstract: Radiative energy flux data, downloaded from CERES, are evaluated with respect to their variations from 2001 to 2020. We found the declining outgoing shortwave radiation to be the most important contributor for a positive TOA (top of the atmosphere) net flux of 0.8 W/m^2 in this time frame. We compare clear sky with cloudy areas and find that changes in the cloud structure should be the root cause for the shortwave trend. The radiative flux data are compared with ocean heat content data and analyzed in the context of a longer-term climate system enthalpy estimation going back to the year 1750. We also report differences in the trends for the Northern and Southern hemisphere. The radiative data indicate more variability in the North and higher stability in the South. The drop of cloudiness around the millennium by about 1.5% has certainly fostered the positive net radiative flux. The declining TOA SW (out) is the major heating cause (+1.42 W/m^2 from 2001 to 2020). It is almost compensated by the growing chilling TOA LW (out) ($-$1.1 W/m^2). This leads together with a reduced incoming solar of $-$0.17 W/m^2 to a small growth of imbalance of 0.15 W/m^2. We further present surface flux data which support the strong influence of the cloud cover on the radiative budget.

Keywords: radiative energy flux; CERES; shortwave flux; longwave flux; cloud thinning

Citation: Dübal, H.-R.; Vahrenholt, F. Radiative Energy Flux Variation from 2001–2020. *Atmosphere* **2021**, *12*, 1297. https://doi.org/10.3390/atmos12101297

Academic Editors: Baojie He, Ayyoob Sharifi, Chi Feng and Jun Yang

Received: 1 September 2021
Accepted: 1 October 2021
Published: 5 October 2021

1. Introduction

In the big picture, climate variations originate from variations of the radiative balance at the top of atmosphere (TOA). Surpluses of the EEI (Earth energy imbalance) or net radiative energy fluxes, as measured by satellite mounted radiometers, lead to an increase of the climate system enthalpy and vice versa [1–4]. For about two decades, the CERES Energy Balanced and Filled (EBAF) Ed4.1 [5,6] offers datasets for a variety of radiative fluxes, and, thus, provides a basis to scrutinize the radiative climate driving forces and shine light on the cause-and-effect relation between radiation and temperature change.

As an independent but less direct source of information of the climate system enthalpy change, there are several studies and reconstructions [7–12] of the ocean heat content (OHC) which represents the bulk of the climate system enthalpy, estimated to be about 90%. Assuming this fraction of 90% were a longer-term constant, one can trace back the time-development of the climate system enthalpy. Von Schuckmann et al. [13] have combined radiative, ocean heat and other data to reconstruct an enthalpy curve back to 1960 and have found an accelerated heating since 2010. Recently, Loeb et al. [14] found a good agreement between radiative (CERES) and OHC data for the period mid-2005 to mid-2019. These authors have further studied the influencing factors for the shortwave (SW) and longwave (LW) radiative fluxes and concluded that cloud changes have fostered the downwelling shortwave radiation.

Dewitte et al. [15] have analyzed CERES datasets for the period from 2000 to 2018 and found an EEI value of about 0.9 W/m^2 but with a declining trend going in line with a declining time-derivative of the latest OHC data obtained from Cheng et al. [9]. Based upon recent CERES data, Loeb et al. [16], Wong et al. [17] and Ollila [18] reported an increasing

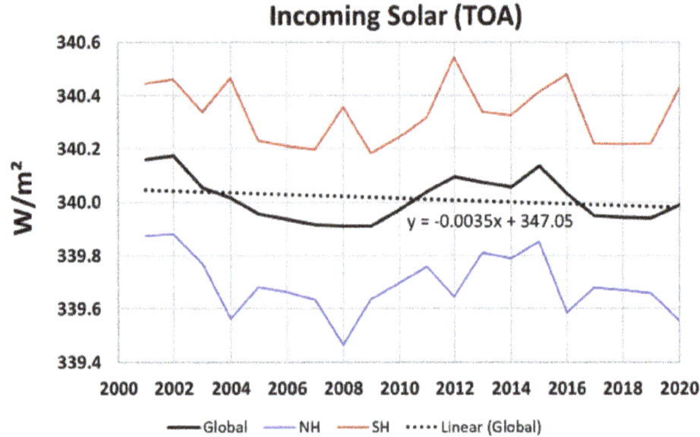

Figure 1. The weakly declining annual mean value (-0.07 ± 0.06 W/m^2) of the incoming solar irradiation for the globe, NH and SH. The Southern hemisphere obtains ca. 0.7 W/m^2 more than the North due to the eccentric orbit and the inclined axis of the Earth.

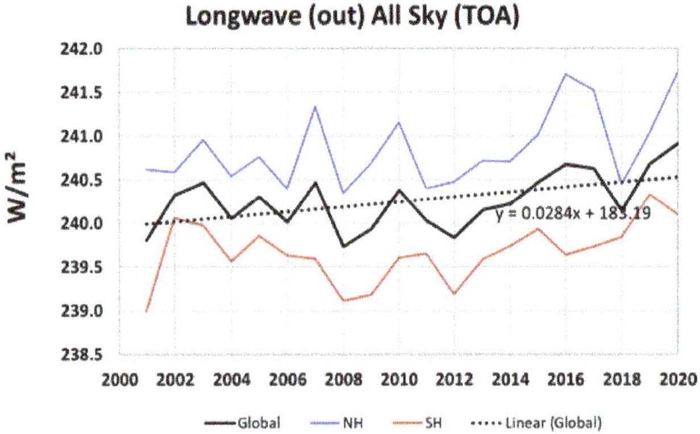

Figure 2. The increasing trend of the LW flux. The average effect is +0.6 $\pm$ 0.2 W/m^2 for two decades. The Northern hemisphere is warmer than the South and therefore shows a higher LW (out) value.

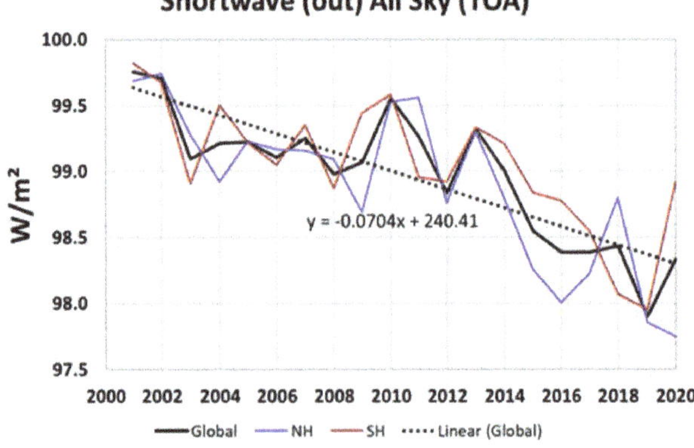

Figure 3. The declining trend of the outgoing SW flux at TOA. The average effect is -1.4 ± 0.2 W/m^2 for two decades, stronger than both effects shown in Figures 1 and 2 combined. Other than above, there are no significant differences between NH and SH.

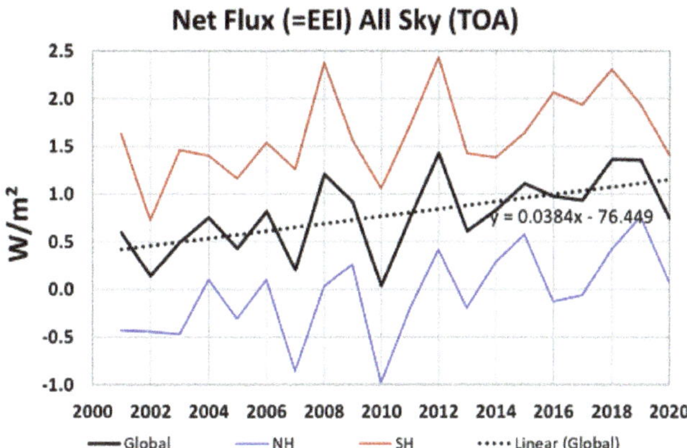

Figure 4. TOA net flux for global, NH, and SH. The Southern hemisphere obtains a higher flux (+ 1.7 W/m^2) than the North which is due to the aphelion/perihelion of the Earth's rotation around the sun, resulting in a stronger radiative flux during the SH summer, not compensated by the weaker SH winter. Nevertheless, the NH has warmed up more strongly during the given period of time.

Some residual uncertainty remains regarding the sign of the TOA net flux. By simply adding the uncertainties of the gross in- and outgoing fluxes, the uncertainty of the difference could easily be in the order of 1 W/m^2 or more, which is larger than the 0.75 W/m^2 reported for the year 2020. However, as shown below, the independently observed OHC data comply well with a positive radiative net flux, and it can be rather safely assumed that there was indeed a positive net flux during the last two decades.

The declining outgoing SW was mainly caused by less reflected solar light which points towards the role of the clouds. Only a minor SW portion may be absorbed by the increasing amount of water vapor by means of their OH stretching vibrations (ca. 3 µm) and overtone, respectively combination band absorption (ca. 1.9 µm, 1.5 µm, 1.0 µm, etc.). Figure 5 compares outgoing SW under "Clear Sky" conditions (left side) with the "Cloudy Areas" (right side) using identical scale ranges. The decline is pronounced in the NH and the absolute effect is greater over the "Cloudy Areas". The "Clear Sky" SH trend is almost flat. Over the clouds, we find a declining trend in NH and SH, still stronger in the North but with a visible decline in the South also. It seems as if the trend accelerated around 2010–2014. In fact, for the global "Cloudy Areas" data the slope is −0.043 W/m^2a for 2001–2010 and −0.16 W/m^2a for 2011–2020, i.e., four times larger.

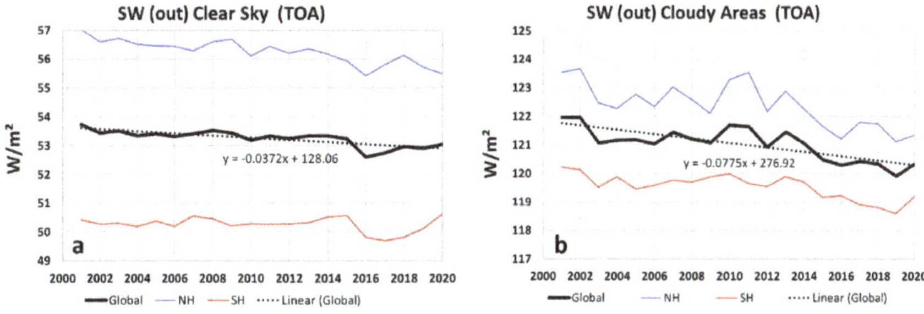

Figure 5. Shortwave outgoing radiative flux (TOA) over "Clear Sky" (**a**) and "Cloudy Areas" (**b**) for global average, NH, and SH. Please note, there is an apparent inconsistency when comparing these plots with Figure 3, in which there is no significant difference between NH and SH. The reason is a different average cloud area fraction of about 70% in the South and 64% in the North. Therefore, the cloud-area-weighted "All Sky" data are very similar.

The changes in the cloud albedo are overwhelming the changes of the land/ocean albedo. If we define surface albedo as the ratio of the TOA outgoing SW and the incoming solar fluxes, we find a reduction from 15.80% to 15.56% (−0.23% absolute) in the "Clear Sky" from 2001 to 2020 stemming from changes in the land and ocean surface. In the "Cloudy Areas" we find a much stronger reduction from 35.85% to 35.28% (−0.58% absolute). The "All Sky" values changed from 29.33% to 28.80% (−0.53% absolute). The area-weighted effect of the albedo change is −0.08% (absolute) for the land/ocean part and −0.39% (absolute) for the clouds. The cloud effect on the albedo change is 5.2 times larger than the land/ocean effect. In other terms, about 15% of the observed albedo change is due to the diminishing land/ocean albedo, and 85% is due to the diminishing cloud albedo. Hartmann and Ceppi [21] as well as Dewitte et al. [15] have studied the influence of sea ice extend (SIE) in the Northern polar region on the reflected SW and found a good correlation of the "Clear Sky" reflected solar light with SIE for that region. We also find such a pronounced regional effect, however, it is limited to a too small area to explain the globally declining "All Sky" outgoing SW presented here.

The outgoing TOA LW flux shows a different trend. Figure 6 reveals the increasing LW flux which is much steeper over the clouds (0.35 ± 0.13 W/m^2 per decade) than over the clear sky (0.04 ± 0.1 W/m^2 per decade) which is essentially constant. Furthermore, there is a striking difference between North and South. The difference between NH and SH is smaller for the "Clear Sky" areas, and the North emits a higher LW flux than the South. For the "Cloudy Areas" the difference between NH and SH is twice as large but the South emits a higher LW flux than the North. The latter goes in line with a higher cloud effective temperature (Figure 7) in the South and the former corresponds to a higher surface temperature on the Northern hemisphere. The smaller absolute value of the flux over the "Cloudy Areas" is due to a lower cloud temperature compared with the surface. Only the much steeper trend for the "Cloudy Areas" is puzzling and points towards a change in the cloud structure.

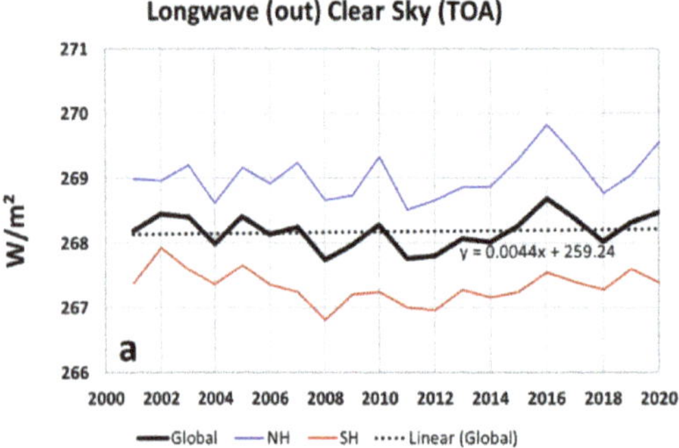

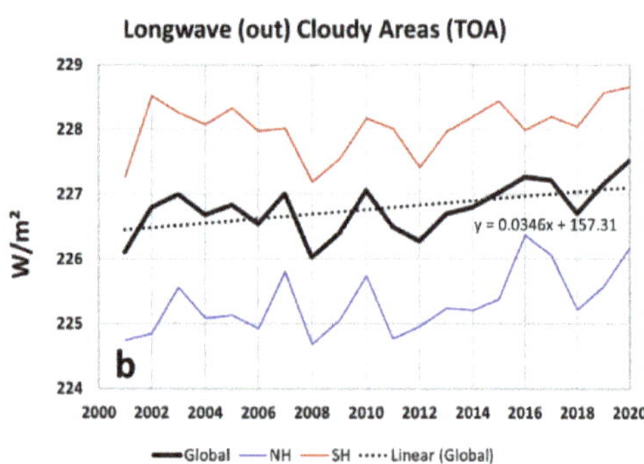

Figure 6. Outgoing longwave flux (TOA) over "Clear Sky" (**a**) and "Cloudy Areas" (**b**). The slope over the "Cloudy Areas" is eight times higher.

The TOA net fluxes over the "Clear Sky" and "Cloudy Areas" exhibit—as expected—a striking difference (Figure 8). The former has a strongly positive ("heating") net flux whereas the latter has a large negative ("cooling") value. About 2/3rds of the sky is covered with clouds of all types, heights, thicknesses and other properties, and the mix of ca. 2/3rd negative flux (Figure 8b) and ca. 1/3rd positive flux (Figure 8a) results in an average net flux shown in Figure 4 above.

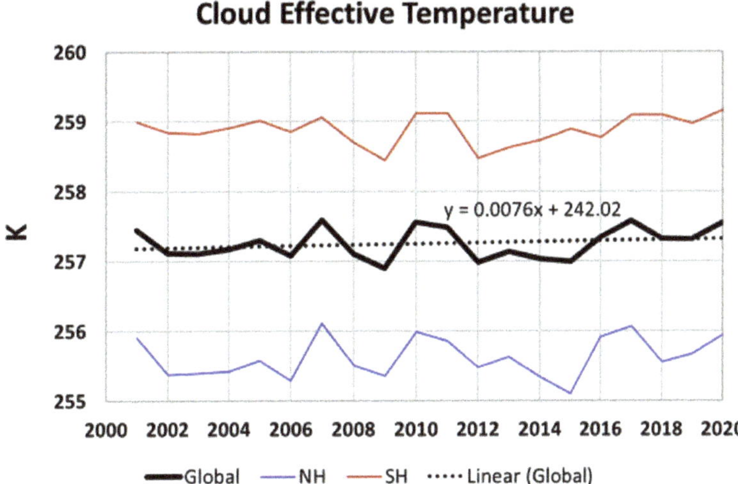

Figure 7. The effective temperature of the clouds is ca. 3.5 K higher in the Southern hemisphere compared with the North. In 20 years, it has increased by ca. 0.1 K in the SH, by ca. 0.2 K in the NH, and by 0.15 K in the global average. The data were downloaded from CERES. Source: https://ceres.larc.nasa.gov/data/ (accessed on 29 July 2021).

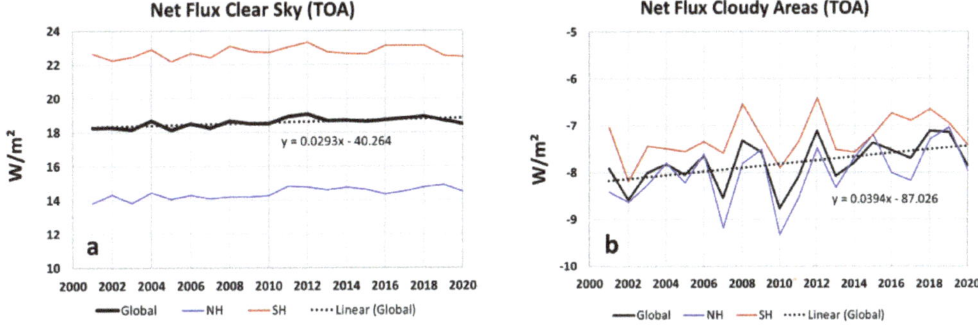

Figure 8. Net flux (TOA) over "Clear Sky" (**a**) and "Cloudy Areas" (**b**) for global average, NH, and SH. For the global values, the difference between "Clear Sky" and "Cloudy Areas" is 26 W/m^2. This 1% change of the cloud cover corresponds to 0.26 W/m^2 flux change.

The difference between NH and SH is much more pronounced over the "Clear Sky", but the rising trend is somewhat steeper over the "Cloudy Areas". The enormous influence of the clouds for the energy balance and, hence, for the climate system enthalpy is obvious. This is also visible from Figure 9 in which the cloudiness and the outgoing LW (here: OLR) are presented for a longer time span. The (inverse) correlation coefficient between OLR and cloudiness is 74%. The cloud window that opened around the year 2000 has certainly fostered the heating. The CERES data agree quite well with the HIRS OLR data, but the comparison also shows that some methodical uncertainties in the order of 1 W/m^2 must be taken into account.

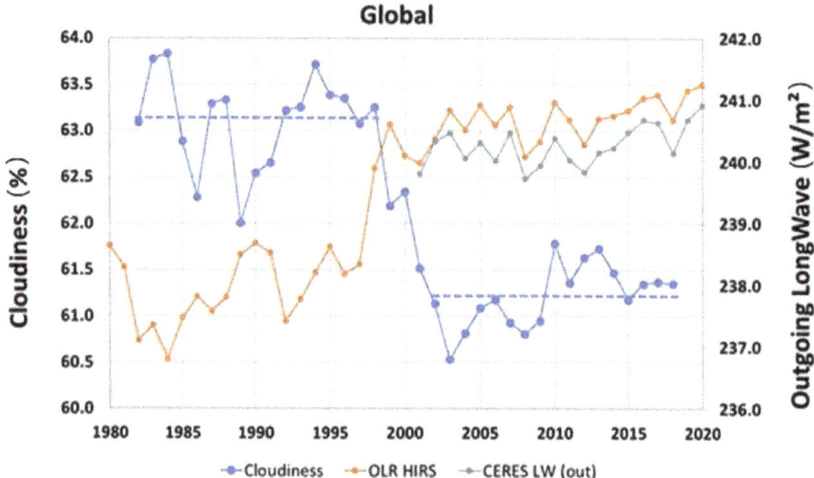

Figure 9. A longer-term plot of the outgoing longwave flux and cloudiness, based upon datasets downloaded from the WMO homepage [20], source: https://climatedata-catalogue.wmo.int/explore accessed on 29 July 2021. The broken lines denote the average of the years before (1982–1997) and after (2003–2018) the drop of the cloud cover. The difference is $-1.86 \pm 0.80\%$.

The terms "cloudiness" and the CERES term "cloud area fraction" have different definition and methodology but both are associated with the cloud cover. The cloudiness is currently ca. 61.3% whereas the cloud area fraction from CERES is ca. 67.5%. Nevertheless, from Figure 8 it follows, that a change of cloud cover by 1% would cause a change of the TOA global net flux of 0.26 W/m^2 so that the drop of 1.86% of cloudiness in Figure 9 could have caused an effect of ca. 0.5 W/m^2, a large portion of the observed average value for 2001–2020 (0.8 W/m^2). Actually, adding the 0.5 W/m^2 to the 20th century average (ca. 0.3 W/m^2) gives the current value.

4. Results and Discussion: Surface Fluxes

CERES data provide an opportunity to differentiate between "Clear Sky" and "Cloudy Areas" not only at TOA, but also for the surface fluxes and this enables us to analyze the influence of clouds on the surface radiation budget as well. At first, under idealized balanced conditions, the TOA radiative budget is exactly zero. At the surface, however, the radiation budget under idealized balanced conditions is far away from zero. It is ca. 110 W/m^2 and this radiative imbalance is compensated by enthalpy changes, mainly by the evaporation of water so that the sum of the overall balance of radiative flux and the time derivative of the enthalpy at the surface strives towards zero.

The surface fluxes show striking differences between "Clear Sky" and "Cloudy Areas" as well. Figure 10 shows SW up- and downwelling surface fluxes for different cloud coverages. While in the "Clear Sky" regions the downwelling SW (Figure 10a) essentially follows the declining incoming solar flux, there is a significantly rising downwelling surface flux (Figure 10c) for the "Cloudy Areas". The upwelling SW fluxes (Figure 10d–f) have similarly declining trends irrespective of the clouds and their absolute values are about 12.5% of the downwelling fluxes (Figure 10a–c), with decreasing trend. The surface SW net flux is rising more pronounced in the "Cloudy Areas" by +0.93 $\pm$ 0.18 W/m^2 per decade compared with the "Clear Sky" regions with only +0.24 $\pm$ 0.10 W/m^2 per decade. We conclude that the clouds have increased their transmittance for the SW radiation. This is possible by means of changes of the cloud type, height, particle size, phase mix, geographical location, interaction, etc. The parameter "Cloud Optical Density", which is available from CERES, however, could neither support nor deny this interpretation because the rather high variance prevents to identify a clear trend.

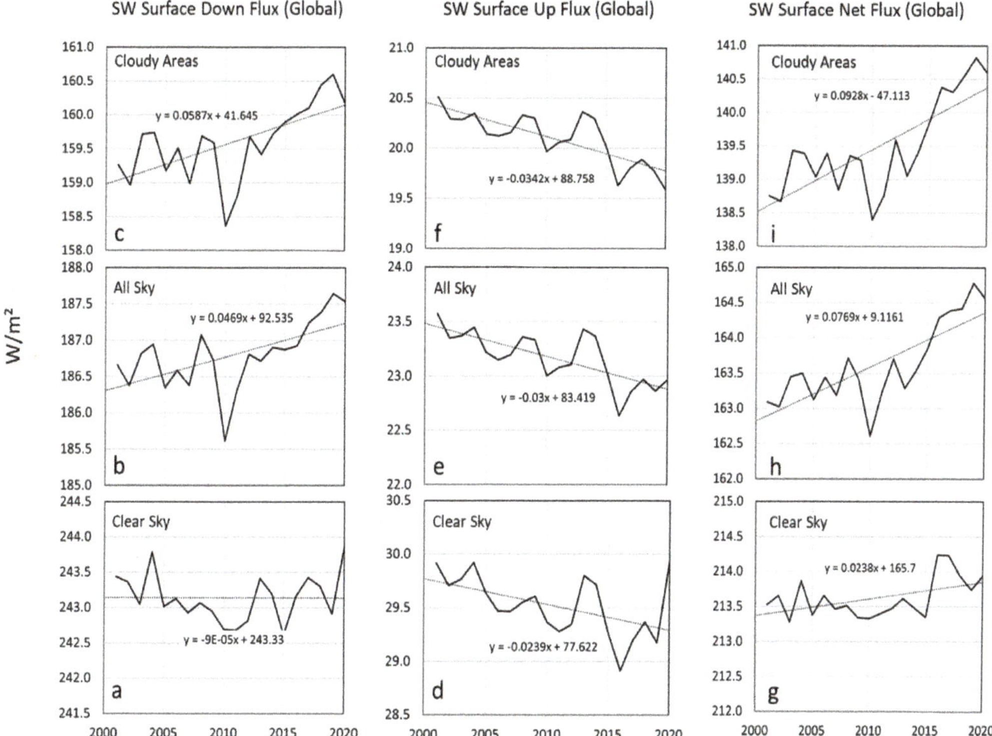

Figure 10. Global average SW downwelling (**a**–**c**), upwelling (**d**–**f**), and net (**g**–**i**) surface fluxes for "Clear Sky", "All Sky", and "Cloudy Areas".

The surface downwelling LW flux (Figure 11) has a different trend than the SW. The "Clear Sky" areas (Figure 11a) show a strongly rising trend, compared with a flat or slightly declining trend in the "Cloudy Areas". In the "All Sky" average, the downwelling LW has a rising trend half as strong as the downwelling SW and with high variance, namely +0.27 ± 0.34 W/m^2 per decade. The LW upwelling fluxes (Figure 11d–f) show strong trends around 1 W/m^2 per decade and are almost independent of the cloud coverage in size and trend. Here, the surface temperature is the main driver, and the radiative flux increases as a function of the surface temperature. The LW net surface fluxes show a strongly declining trend that is accelerating after 2016 with the exception of the "Clear Sky" regions. As a consequence of the rising SW and the falling LW net fluxes, the "All Sky" surface net flux has no clear trend but a significant fluctuation (Figure 12).

At this point, we describe the direct observation of the greenhouse effect, namely by comparing the "Clear Sky" LW upwelling radiation (Figure 11d) and the outgoing TOA LW flux (Figure 6a, left side). The first is rising with +1.22 ± 0.22 W/m^2 per decade, the latter trend is literally flat with +0.04 ± 0.10 W/m^2 per decade. Hence, in the absence of clouds, a large portion of this additional upwelling LW radiation is absorbed by the increasing greenhouse gas concentration (CO_2 has increased from 371 to 414 ppm obtained from https://gml.noaa.gov/webdata/ccgg/trends/co2/co2_annmean_mlo.txt (accessed on 29 July 2021), water from 10.44 to 10.71 g/kg at 1000 mbar from 2001 to 2020 obtained from https://psl.noaa.gov/cgi-bin/data/timeseries/timeseries1.pl, (accessed on 29 July 2021). We have applied the Lambert–Beer's law and correlated the logarithmic ratio of the outgoing TOA LW flux and the "Clear Sky" LW upwelling flux with the rising greenhouse gas concentrations and found a very good linear correlations for CO_2 (R^2 = 0.92) and water vapor (R^2 = 0.72).

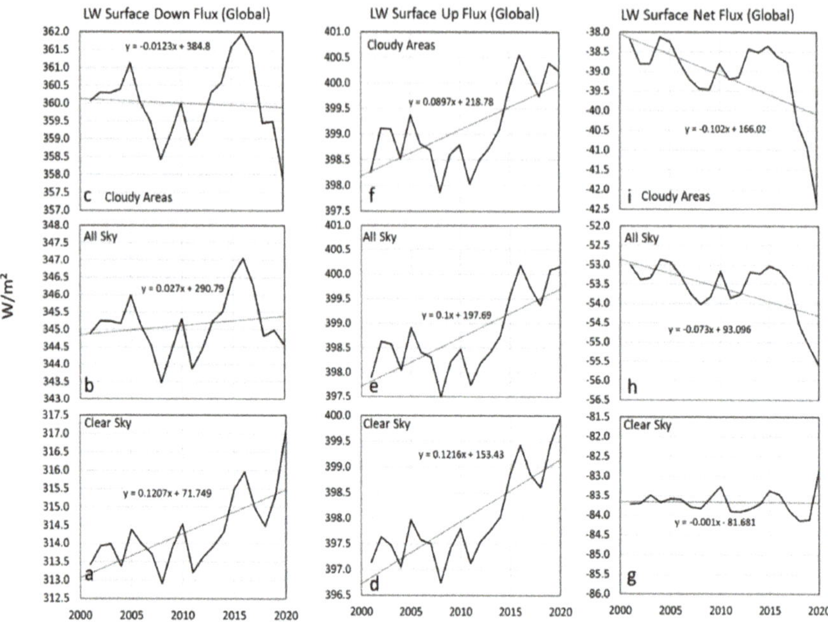

Figure 11. Global average LW downwelling (**a**–**c**), upwelling (**d**–**f**), and net (**g**–**i**) surface fluxes for "Clear Sky", "All Sky", and "Cloudy Areas".

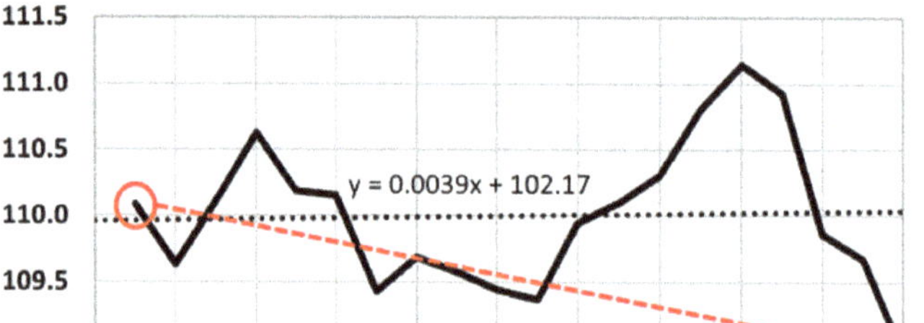

Figure 12. The global "All Sky" surface net flux shows no clear tendency. The 2020 value is below the 2001 value (arrow), and the mean trend is slightly positive but with overwhelming variance.

Another observation is that these correlations vanish for the "Cloudy Areas". The clouds absorb most of the LW flux and are emitting a weaker thermal flux in all directions. The part designated to reach the TOA point has little chance to be absorbed by greenhouse gases because the pathway is shorter, the pressure is lower, the water vapor is sparse, and the initial flux is weaker. All this works towards much lower infrared absorption above the clouds and the TOA outgoing LW rises with 0.35 ± 0.13 W/m^2 per decade.

In absolute terms, the LW surface upwelling flux is about 398 W/m^2 for "clear sky" (Figure 11d) and only ca. 268 W/m^2 at TOA (Figure 6a) which is an attenuation of 130 W/m^2 or 33%, caused by greenhouse gases, by scattering, and by small clouds below the measurement threshold. In the "Cloudy Areas" we start with about 399 W/m^2 LW surface emission (Figure 11f), which is absorbed by the clouds and re-emitted at a lower temperature of about 258 K (Figure 7). This cloud average temperature corresponds to about 260 W/m^2 emission (estimated from Stefan–Boltzmann's law), and ca. 227 W/m^2 of it reaches the TOA (Figure 6b). The attenuation here is only 33 W/m^2 or 12.6%. There again, the corresponding downwelling back-radiation is absorbed by the clouds and contributes to elevating their temperature. Only a minor share of this smaller part reaches the Earth's surface.

Hence, the rise of the greenhouse gas concentration from 2001 to 2020 had a measurable effect on the LW flux in the "Clear Sky", covering about 1/3rd of the Earth surface. In the cloudy part, about 2/3rd, this effect was much smaller, if significant at all.

5. Effect on the Climate System's Enthalpy

With the one and only exception of radiation, the Earth is an almost perfect adiabatic thermodynamical system. Material exchange, convection and heat diffusion is prevented by gravity and all other sources or mechanism of energy exchange with the environment, the space, such as starlight and cosmic particle flux are negligible in terms of enthalpy. The cosmic particle flux could, however, have a significant indirect effect by its influence on the cloud formation which again provides a very strong leverage for the radiation budget. This potentially very important indirect effect was described by Svensmark [22] and by Shaviv [23].

Radiative datasets are available for only a few decades, however, ocean heat content (OHC) data were reconstructed for much longer and, assuming that constantly ca. 90% of the climate system's enthalpy ended up as ocean heat, provide an opportunity to reconstruct the enthalpy development for a longer period of time. OHC reconstructions were reported by Levitus et al. [7], Roemmich et al. [24], Cheng et al. [10] and Gebbie and Huybers [11]. Von Schuckmann et al. [13] have developed an enthalpy summary for 1971–2018. They also reported the distribution over the various degrees of freedom such as the heat of oceans, air and land, evaporation of water and melting of ice and volume work. Changes of the kinetic energy of air and ocean water would have to be added. They also used CERES radiative data and reported a net flux of 0.87 W/m^2 for 2010–2018, which is larger compared with the longer term mean value of only 0.47 W/m^2 for 1971–2018 and the long-term 0.2 W/m^2 reported by Baggenstos et al. [25]. Earlier, Allen et al. [26] reported net radiative fluxes of 0.34 ± 0.67 W m^2 for the period 1985–1999, and 0.62 ± 0.43 W m^2 for 2000–2012. Roemmich et al. [24] reported a value between 0.4 and 0.6 W/m^2 for the period of 2006–2013 for the 0–2000 m OHC. Recently, Loeb et al. [14] reported a trend of 0.43 ± 0.40 W/m^2 per decade from mid-2005 to mid-2019 for the 0–2000 m OHC, compared with the trend for the net CERES TOA energy flux of 0.50 ± 0.47 W/m^2 per decade over that same time-period. The data analyzed in this paper resulted in a net flux of 0.8 W/m^2 for the period 2001–2020 and this corresponds to an enthalpy increase of about 240 ZJ during this period. However, the root cause of this enthalpy gain is in dispute. This becomes clear from the strong effects of the shortwave radiation changes, the cloud cover changes and, also, from a longer-term development of the enthalpy.

Figure 13 presents a longer-term enthalpy curve, constructed from various sources, radiative and oceanic, the latter under the assumption of a constant 90% OHC share. Figure 14 shows a shorter-term view of the same data. In both figures, the distinct zero-level of each of the enthalpy data was shifted to fit. The current enthalpy is still ca. 600 ZJ below the medieval maximum 1000 years ago grounded on the 2000 years OHC reconstruction reported by Gebbie and Huybers [11]. This would be compensated in only

50 years with the presently observed +0.8 W/m^2 net flux and in about 100–200 years for the average rate as during the 20th century. With a possible interception by another phase of a negative radiative net flow, it would take centuries until we reach the medieval OHC maximum again.

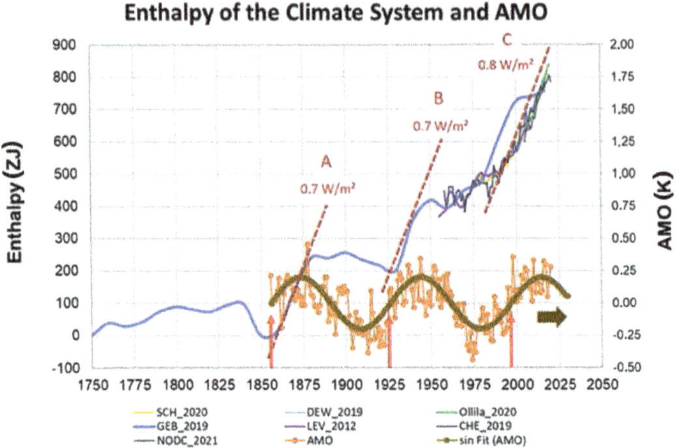

Figure 13. Climate system enthalpy since 1750, reconstructed from radiative and ocean heat data taken from the publications of Schuckmann et al. [13], Ollila [18], Dewitte et al. [15], Levitus et al. [7] and NODC [7], Cheng et al. [10], Gebbie and Huybers [11] and this work. The OHC-Data were divided by 0.9. The two heating impulses A and B had a span of about 25 years, and similarly, high net fluxes as the presently observed phase C. The zero levels of the enthalpy datasets was set to match. The AMO values were downloaded from http://www.psl.noaa.gov/data/timeseries/AMO/ (accessed on 29 July 2021) and transformed into annual averages. A sine-fit is added to the AMO-data (right scale). The arrows at ca. 1860, 1925 and 2000 indicate the sign change of the AMO-Index from negative to positive.

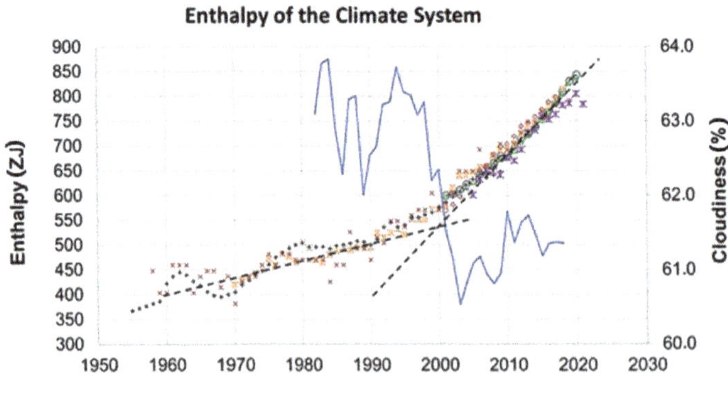

Figure 14. Expanded enthalpy plot. The cloudiness [20] is included (right scale). The rise of the enthalpy after the millennium coincides with the drop of cloudiness. The latest OHC data from NODC exhibit such a flatter tendency for about one year.

Hence, the decisive question is; whether the currently observed strong heating phase C is temporary like the phases A and B in Figure 13 or the beginning of a steadily rising warming period. A third possibility is that phase C is a combination of the natural induced

A/B type and steadily warming. The enthalpy-enhancing effect of the greenhouse gases (including water vapor) is clearly visible in our analysis, however, we found, in line with other authors [14,15,18], that this effect is only one out of several other important factors. In the period 2001–2020, the greenhouse gas effect is prominent in the "Clear Sky" areas but overcompensated by the SW fluxes and cloud effects. This is also evident from the bridge chart in Figure 15 below. As an alternative, the phases A, B and C since 2001 could be interpreted as a kind of "global brightening" phases followed by "global dimming" (for a review of global brightening/dimming see [27]).

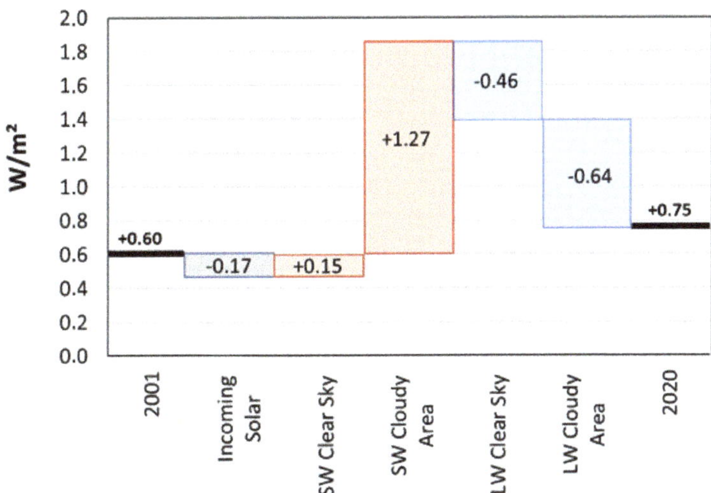

Figure 15. Bridge-chart for the actual changes at TOA from 2001 to 2020 ("start-to-end change"), dominated by the change in the outgoing shortwave radiation over "Cloudy Areas". The declining SW (out) is the major heating cause. It is almost compensated by the chilling LW (out). The incremental effects are weighted by area (see the example in the caption of Figure 16).

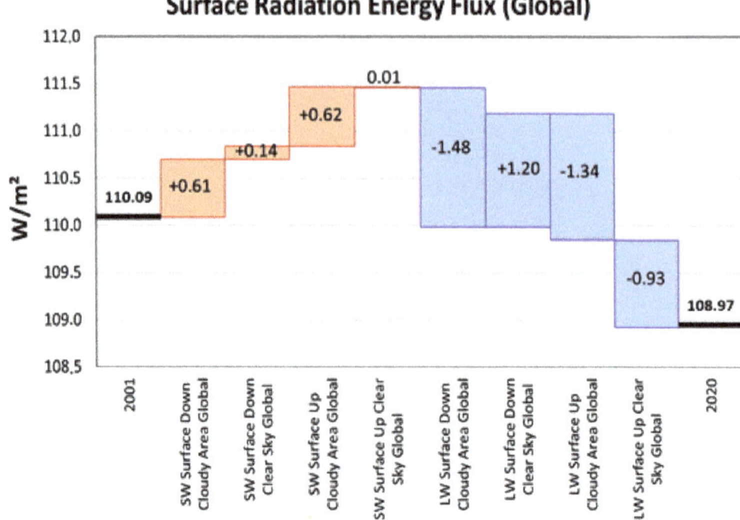

Figure 16. Bridge-chart for the actual changes at the surface from 2001 to 2020 ("start-to-end change").

Please note that the individual effects, shown in Figures 10 and 11, are weighted by the cloud area fraction (ca. 1/3rd for "Clear Sky" and ca. 2/3rd for "Cloudy Areas") and that the sign is positive for increasing downwelling or decreasing upwelling fluxes and vice versa. Reading example: "LW Surface Down Cloudy" in Figure 11c changes from 360.07 to 357.88 W/m^2 = −2.19 W/m^2. Multiplied by 0.674 (67.4% cloud area fraction) gives a contribution of −1.48 W/m^2 from 2001 to 2020.

The rather short period of only 20 years prevents us from answering this key question at this time. Furthermore, there is no validated physical mechanism that led to the declining TOA shortwave emission and to the cloudiness change around the year 2000.

6. Conclusions

Radiative energy flux data from CERES were analyzed and showed in accordance with OHC data a further increasing climate system enthalpy during the period 2001–2020. The total enthalpy rise amounted to about 240 ZJ in these two decades. As Figure 15 shows, the major driving effect was the declining shortwave TOA emission. The TOA outgoing longwave emission has increased and therefore reduced the TOA net flux.

Generating the CERES data is a demanding task and requires sophisticated technology and models which are vulnerable and prone to uncertainties. Liu et al. [28] have pointed out significant uncertainties of satellite datasets and discussed the role of the lateral energy flow. Su et al. [29] have recently pointed out the importance of maintaining the consistency among the components of the measuring system. On the other hand, Loeb et al. [14], Johnson et al. [30] and, also, Dewitte et al. [15] have shown that the CERES net flux agrees well with the independently observed OHC data. This good agreement, also confirmed by our analysis, justifies some confidence in the CERES datasets used.

We could identify the effect of the anthropogenic greenhouse gas emissions from 2001 to 2020 in the "Clear Sky" LW part but not in the "Cloudy Areas" and not in the SW. At the same time, we find, in accordance with the analysis of Loeb et al. [14] and Ollila [18], that the major changes for the TOA energy budget during this period of time stemmed from the clouds for SW and LW, as well as the ground temperature in the LW.

Loeb et al. [14] pointed out, that the direct aerosol effect is rather small, but the indirect effect via the cloud formation may be larger. The shift from a negative to a positive PDO

(Pacific Decadal Oscillation) index as an additional factor for the net TOA flux is mentioned in [14].

Our analysis, which differentiates between clear sky and cloudy areas, support that in view of the historical heating steps shown in Figure 13, that the currently observed high radiative net flux has a large intrinsic component. As shown in Figure 13, the heating phases coincide with the AMO change from negative to positive. It has been shown by several authors that AMO may be an important intrinsic climate factor [31–33]. A similar discussion was held about 20 years ago in papers of Chen et al. [34] and Wielicki et al. [35], both of them emphasized the underestimated decadal natural variabilities in the tropical regions.

The start-to-end bridge charts in Figure 15 (TOA) and Figure 16 (surface) are highlighting the effect of the "Cloudy Areas" and the shortwave radiation in a slightly different view. In these figures, we look at the actual start and end data, their actual differences from 2001 to 2020, and the actual incremental contribution of each of the radiative categories. Please note, that the increments are weighted by area. In Figure 15 (TOA), the biggest changes originate from the "Cloudy Areas". In Figure 16 (surface) the largest heating impulse stemmed from the increasing downwelling and decreasing upwelling shortwave fluxes in the "Cloudy Areas" (together +1.23 W/m^2), the strongest cooling effect was the decreased longwave downwelling flux in the "Cloudy Areas" (-1.48 W/m^2), followed by increasing LW upwelling fluxes in the "Cloudy Areas" (-1.34 W/m^2) and the "Clear Sky" (-0.93 W/m^2). The change of the longwave downwelling radiation can be interpreted in part as the additional effect of the increased greenhouse gas concentration. For the "Clear Sky" it is +1.20 W/m^2. In the "Cloudy Areas", this effect is negative (-1.48 W/m^2) so that the sum of these values is -0.14 W/m^2. The -0.93 W/m^2 of the "Clear Sky" upwelling longwave should be caused by the increased thermal emission due to the higher surface temperature.

There are distinct differences between the Northern and the Southern hemisphere. Generally speaking, the South was more stable than the North in trends and variances of almost all radiative quantities. This could be due to the larger ocean share of the surface in the South.

Finally, the key issue, i.e., whether the current heating phase is a temporary phase or a permanent phenomenon, can be judged only on the basis of a longer observation time. In the latter case, the physical mechanism behind the "shortwave heating" [18] or a possible "cloud thinning", as discussed by several other authors [36–38] should be understood, because it could accelerate the warming trend. In the former case, the strong net flux of +0.8 W/m^2 should decrease naturally.

Author Contributions: Conceptualization, F.V. and H.-R.D.; methodology, F.V. and H.-R.D.; software, H.-R.D.; validation, F.V. and H.-R.D.; formal analysis, H.-R.D.; investigation, F.V. and H.-R.D.; writing—original draft preparation, H.-R.D.; writing—review and editing, F.V. and H.-R.D. All authors have read and agreed to the published version of the manuscript.

Funding: This research received no external funding.

Institutional Review Board Statement: Not applicable.

Informed Consent Statement: Not applicable.

Data Availability Statement: See Section 2 and references.

Acknowledgments: We are grateful to Nic Lewis and Frank Bosse, whose reviews helped to improve the manuscript.

Conflicts of Interest: H.-R.D. is a physical chemist with many years of research experience in the chemical industry. His scientific training enables him to work effectively with interdisciplinary datasets and interpret their respective implications. The project did not receive any funding, nor was it commissioned or supported by industry. FV has been for many years Professor at the University of Hamburg, Department of Chemistry and was employed in the Renewables Energy sector. He has many years of academic experience and a solid publication track record in chemistry and climate science.

Appendix A

Table A1. Statistical Error Analysis of the TOA Fluxes (part1).

Quantity	Intercept	Error of Intercept	Slope	Error of Slope	Trend	Rel. Error
Unit	W/m^2	W/m^2	W/m^2 a	W/m^2 a	W/m^2 dec	%
Inc. Solar Flux in-TOA Global	340.05	0.04	−0.0035	0.0032	−0.03	92
LW Flux out-All Sky-TOA Global	239.99	0.12	0.0284	0.0112	0.28	39
SW Flux out-All Sky-TOA Global	99.64	0.12	−0.0704	0.0107	−0.70	15
Net Flux-All Sky-TOA Global	0.42	0.14	0.0384	0.0130	0.38	34
LW Flux out-Clear Sky-TOA Global	268.14	0.11	0.0044	0.0102	0.04	229
SW Flux out-Clear Sky-TOA Global	53.61	0.07	−0.0372	0.0066	−0.37	18
Net Flux-Clear Sky-TOA Global	18.30	0.09	0.0293	0.0080	0.29	27
LW Flux out-Cloudy Areas-TOA Global	226.45	0.15	0.0346	0.0134	0.35	39
SW Flux out-Cloudy Areas-TOA Global	121.77	0.16	−0.0775	0.0142	−0.78	18
Net Flux-Cloudy Areas-TOA Global	−8.18	0.19	0.0394	0.0167	0.39	42
Inc. Solar Flux in-TOA NH	339.74	0.05	−0.0048	0.0043	−0.05	90
LW Flux out-All Sky-TOA NH	240.53	0.17	0.0344	0.0153	0.34	45
SW Flux out-All Sky-TOA NH	99.66	0.16	−0.0813	0.0145	−0.81	18
Net Flux-All Sky-TOA NH	−0.45	0.17	0.0421	0.0151	0.42	36

Table A1. *Cont.*

Quantity	Intercept	Error of Intercept	Slope	Error of Slope	Trend	Rel. Error
Unit	W/m^2	W/m^2	W/m^2 a	W/m^2 a	W/m^2 dec	%
LW Flux out-Clear Sky-TOA NH	268.87	0.14	0.0174	0.0129	0.17	74
SW Flux out-Clear Sky-TOA NH	56.83	0.10	−0.0616	0.0086	−0.62	14
Net Flux-Clear Sky-TOA NH	14.04	0.09	0.0393	0.0084	0.39	21
LW Flux out-Cloudy Areas-TOA NH	224.88	0.18	0.0469	0.0164	0.47	35
SW Flux out-Cloudy Areas-TOA NH	123.31	0.23	−0.0969	0.0209	−0.97	22
Net Flux-Cloudy Areas-TOA NH	−8.45	0.25	0.0452	0.0221	0.45	49
Inc. Solar Flux in-TOA SH	340.35	0.05	−0.0022	0.0045	−0.02	203
LW Flux out-All Sky-TOA SH	239.45	0.14	0.0224	0.0128	0.22	57
SW Flux out-All Sky-TOA SH	99.61	0.15	−0.0594	0.0134	−0.59	22
Net Flux-All Sky-TOA SH	1.29	0.18	0.0348	0.0158	0.35	45
LW Flux out-Clear Sky-TOA SH	267.41	0.11	−0.0085	0.0100	−0.09	118
SW Flux out-Clear Sky-TOA SH	50.38	0.11	−0.0129	0.0095	−0.13	74
Net Flux-Clear Sky-TOA SH	22.56	0.13	0.0192	0.0118	0.19	61
LW Flux out-Cloudy Areas-TOA SH	227.82	0.17	0.0231	0.0154	0.23	67
SW Flux out-Cloudy Areas-TOA SH	120.09	0.12	−0.0572	0.0111	−0.57	19
Net Flux-Cloudy Areas-TOA SH	−7.56	0.18	0.0319	0.0165	0.32	52

Explanation: "Intercept" is the result of a linear regression, i.e., the estimate for the year 2001. Please note that the first year 2001 was set to zero so that "Intercept" is equal to the linear estimate for the year 2001. "Error Intercept" denotes the statistical standard error. "Slope" is the result of a linear regression. "Error of Slope" denotes the statistical standard error. "Trend" is the slope multiplied by 10 and the "Rel. Error" is the Ratio of "Error of Slope" and "Slope", multiplied by 100.

Table A2. Statistical Error Analysis of the TOA Fluxes (part 2).

Quantity	*p*-Value (Slope)	CI Low (Slope)	CI High (Slope)	Δ CI (Slope)	R^2	Fig.
Unit	1	W/m^2 a	W/m^2 a	W/m^2 a	1	-
Inc. Solar Flux in-TOA Global	0.2937	−0.010	0.003	0.014	0.06	1
LW Flux out-All Sky-TOA Global	0.0204	66 0.005	0.052	0.047	0.26	2
SW Flux out-All Sky-TOA Global	0.0000	−0.093	−0.048	0.045	0.71	3
Net Flux-All Sky-TOA Global	0.0086	0.011	0.066	0.055	0.33	4
LW Flux out-Clear Sky-TOA Global	0.6674	−0.017	0.026	0.043	0.01	6

Table A2. *Cont.*

Quantity	*p*-Value (Slope)	CI Low (Slope)	CI High (Slope)	Δ CI (Slope)	R^2	Fig.
SW Flux out-Cloudy Areas-TOA SH	0.0001	−0.080	−0.034	0.047	0.60	5
Net Flux-Cloudy Areas-TOA SH	0.0693	−0.003	0.067	0.070	0.17	8

Explanation: The "*p*-value" for the slope is the result of a linear regression. A *p*-value below 0.05 rejects the Null hypothesis (i.e., slope = 0) with 95% confidence. *p*-values below 1E-4 are cut off after four digits. "CI low" is the lower boundary and "CI high" is the upper boundary of the 95% confidence interval for the slope. "Δ CI" is the width of the confidence interval. "R^2" is coefficient of determination and indicates the strength of the correlation. An R^2 near zero means that the respective flux is approximately a constant for the time-period considered. For convenience, "Fig." gives the reference to the figures in this paper.

Table A3. Statistical Error Analysis of the Surface Fluxes (part1).

Quantity	Intercept	Error of Intercept	Slope	Error Slope	Trend	Rel. Error
Unit	W/m^2	W/m^2	W/m^2 a	W/m^2 a	W/m^2 dec	%
SW Flux Down-All Sky-Surface Global	186.35	0.17	0.047	0.015	0.47	32
SW Flux Down-Clear Sky-Surface Global	243.14	0.15	0.000	0.014	0.00	14723
SW Flux Up-All Sky-Surface Global	23.45	0.07	−0.030	0.006	−0.30	22
SW Flux Up-Clear Sky-Surface Global	29.75	0.11	−0.024	0.010	−0.24	41
LW Flux Down-All Sky-Surface Global	344.88	0.37	0.027	0.034	0.27	125
LW Flux Down-Clear Sky-Surface Global	313.18	0.32	0.121	0.029	1.21	24
LW Flux Up-All Sky-Surface Global	397.81	0.25	0.100	0.023	1.00	23
LW Flux Up-Clear Sky-Surface Global	396.84	0.24	0.122	0.022	1.22	18
SW Flux Down-Cloudy Areas-Surface Global	159.04	0.19	0.059	0.017	0.59	29
SW Flux Up-Cloudy Areas-Surface Global	20.42	0.07	−0.034	0.006	−0.34	19
LW Flux Down-Cloudy Areas-Surface Global	360.12	0.46	−0.012	0.041	−0.12	335
LW Flux Up-Cloudy Areas-Surface Global	398.27	0.26	0.090	0.024	0.90	27
SW Flux Net-All Sky-Surface Global	162.90	0.16	0.077	0.015	0.77	19
SW Flux Net-Clear Sky-Surface Global	213.39	0.11	0.024	0.010	0.24	42
SW Flux Net-Cloudy Areas-Surface Global	138.61	0.20	0.093	0.018	0.93	19
LW Flux Net-All Sky-Surface Global	−52.93	0.26	−0.073	0.023	−0.73	32
LW Flux Net-Clear Sky-Surface Global	−83.66	0.13	−0.001	0.012	−0.01	1188
LW Flux Net-Cloudy Areas-Surface Global	−38.15	0.37	−0.102	0.033	−1.02	33
LW + SW-All Sky-Surface Global	109.96	0.25	0.004	0.022	0.04	579
LW + SW-Clear Sky-Surface Global	129.73	0.16	0.023	0.015	0.23	65
LW + SW-Cloudy Areas-Surface Global	100.46	0.37	−0.009	0.034	−0.09	364

Explanation: "Intercept" is the result of a linear regression, i.e., the estimate for the year 2001. Please note that the first year 2001 was set to zero so that "Intercept" is equal to the linear estimate for the year 2001. "Error Intercept" denotes the statistical standard error. "Slope" is the result of a linear regression. "Error of Slope" denotes the statistical standard error. "Trend" is the slope multiplied by 10 and the "Rel. Error" is the Ratio of "Error of Slope" and "Slope", multiplied by 100.

Table A4. Statistical Error Analysis of the Surface Fluxes (part2).

Quantity	*p*-Value (SLOPE)	CI Low (slope)	CI High (Slope)	Δ CI (Slope)	R^2	Fig.
Unit	1	W/m^2 a	W/m^2 a	W/m^2 a	1	-
SW Flux Down-All Sky-Surface Global	0.0064	0.015	0.079	0.064	0.35	10
SW Flux Down-Clear Sky-Surface Global	0.9947	−0.029	0.029	0.057	0.00	10
SW Flux Up-All Sky-Surface Global	0.0002	−0.044	−0.016	0.027	0.54	10
SW Flux Up-Clear Sky-Surface Global	0.0246	−0.044	−0.003	0.041	0.25	10
LW Flux Down-All Sky-Surface Global	0.4327	−0.044	0.098	0.142	0.03	11
LW Flux Down-Clear Sky-Surface Global	0.0006	0.059	0.182	0.123	0.49	11
LW Flux Up-All Sky-Surface Global	0.0004	0.052	0.148	0.096	0.52	11
LW Flux Up-Clear Sky-Surface Global	0.0000	0.076	0.167	0.091	0.64	11
SW Flux Down-Cloudy Areas-Surface Global	0.0032	0.022	0.095	0.073	0.39	10
SW Flux Up-Cloudy Areas-Surface Global	0.0000	−0.048	−0.021	0.027	0.61	10
LW Flux Down-Cloudy Areas-Surface Global	0.7685	−0.099	0.074	0.173	0.00	11
LW Flux Up-Cloudy Areas-Surface Global	0.0014	0.040	0.140	0.100	0.44	11

Table A4. *Cont.*

Quantity	p-Value (SLOPE)	CI Low (slope)	CI High (Slope)	Δ CI (Slope)	R^2	Fig.
SW Flux Net-All Sky-Surface Global	0.0001	0.046	0.107	0.061	0.61	10
SW Flux Net-Clear Sky-Surface Global	0.0290	0.003	0.045	0.042	0.24	10
SW Flux Net-Cloudy Areas-Surface Global	0.0001	0.055	0.131	0.076	0.60	10
LW Flux Net-All Sky-Surface Global	0.0059	−0.122	−0.024	0.098	0.35	11
LW Flux Net-Clear Sky-Surface Global	0.9339	−0.026	0.024	0.049	0.00	11
LW Flux Net-Cloudy Areas-Surface Global	0.0067	−0.172	−0.032	0.140	0.34	11
LW + SW -All Sky-Surface Global	0.8647	−0.043	0.051	0.094	0.00	12
LW + SW -Clear Sky-Surface Global	0.1405	−0.008	0.054	0.062	0.12	-
LW + SW -Cloudy Areas-Surface Global	0.7865	−0.080	0.061	0.141	0.00	-

Explanation: The "p-value" for the slope is the result of a linear regression. A p-value below 0.05 rejects the Null hypothesis (i.e., slope = 0) with 95% confidence. p-values below 1E-4 are cut off after four digits. "CI low" is the lower boundary and "CI high" is the upper boundary of the 95% confidence interval for the slope. "Δ CI" is the width of the confidence interval. "R^2" is coefficient of determination and indicates the strength of the correlation. An R^2 near zero means that the respective flux is approximately a constant for the time period considered. For convenience, "Fig." gives the reference to the figures in this paper.

References

1. Hansen, J.; Nazarenko, L.; Ruedy, R.; Sato, M.; Willis, J.; Del Genio, A.; Koch, D.; Lacis, A.; Lo, K.; Menon, S.; et al. Earth's energy imbalance: Confirmation and implications. *Science* **2005**, *308*, 1431–1435. [CrossRef] [PubMed]
2. Hansen, J.; Sato, M.; Kharecha, P.; von Schuckmann, K. Earth's energy imbalance and implications. *Atmos. Chem. Phys.* **2011**, *11*, 13421–13449. [CrossRef]
3. Trenberth, K.E.; Fasullo, J.T.; Balmaseda, M.A. Earth's Energy Imbalance. *J. Clim.* **2014**, *27*, 3129–3144. [CrossRef]
4. Dewitte, S.; Clerbaux, N. Measurement of the Earth Radiation Budget at the Top of the Atmosphere—A Review. *Remote Sens.* **2017**, *9*, 1143. [CrossRef]
5. Kato, S.; Rose, F.G.; Rutan, D.A.; Thorsen, T.; Loeb, N.G.; Doelling, D.R.; Huang, X.; Smith, W.L.; Su, W.; Ham, S.-H. Surface Irradiances of Edition 4.0 Clouds and the Earth's Radiant Energy System (CERES) Energy Balanced and Filled (EBAF) Data Product. *J. Clim.* **2018**, *31*, 4501–4527. [CrossRef]
6. Loeb, N.G.; Doelling, D.R.; Wang, H.; Su, W.; Nguyen, C.; Corbett, J.G.; Liang, L.; Mitrescu, C.; Rose, F.G.; Kato, S. Clouds and the Earth's Radiant Energy System (CERES) Energy Balanced and Filled (EBAF) Top-of-Atmosphere (TOA) Edition-4.0 Data Product. *J. Clim.* **2018**, *31*, 895–918. [CrossRef]
7. Levitus, S.; Antonov, J.I.; Boyer, T.P.; Baranova, O.K.; Garcia, H.E.; Locarnini, R.A.; Mishonov, A.V.; Reagan, J.R.; Seidov, D.; Yarosh, E.S.; et al. World ocean heat content and thermosteric sea level change (0–2000 m), 1955–2010. *Geophys. Res. Lett.* **2012**, *39*. [CrossRef]
8. Abraham, J.P.; Baringer, M.; Bindoff, N.; Boyer, T.; Cheng, L.; Church, J.A.; Conroy, J.L.; Domingues, C.M.; Fasullo, J.; Gilson, J.; et al. A review of global ocean temperature observations: Implications for ocean heat content estimates and climate change. *Rev. Geophys.* **2013**, *51*, 450–483. [CrossRef]
9. Cheng, L.; Trenberth, K.E.; Fasullo, J.; Boyer, T.; Abraham, J.; Zhu, J. Improved estimates of ocean heat content from 1960 to 2015. *Sci. Adv.* **2017**, *3*, e1601545. [CrossRef]
10. Cheng, L.; Zhu, J.; Abraham, J.; Trenberth, K.E.; Fasullo, J.T.; Zhang, B.; Yu, F.; Wan, L.; Chen, X.; Song, X. 2018 Continues Record Global Ocean Warming. *Adv. Atmos. Sci.* **2019**, *36*, 249–252. [CrossRef]
11. Gebbie, G.; Huybers, P. The Little Ice Age and 20th-century deep Pacific cooling. *Science* **2019**, *363*, 70–74. [CrossRef]
12. Meyssignac, B.; Boyer, T.; Zhao, Z.; Hakuba, M.Z.; Landerer, F.W.; Stammer, D.; Köhl, A.; Kato, S.; L'Ecuyer, T.; Ablain, M.; et al. Meas-uring Global Ocean Heat Content to Estimate the Earth Energy Imbalance. *Front. Mar. Sci.* **2019**, *6*, 432. [CrossRef]
13. Von Schuckmann, K.; Cheng, L.; Palmer, M.D.; Hansen, J.; Tassone, C.; Aich, V.; Adusumilli, S.; Beltrami, H.; Boyer, T.; Cuesta-Valero, F.J.; et al. Heat stored in the Earth system: Where does the energy go? *Earth Syst. Sci. Data* **2020**, *12*, 2013–2041. [CrossRef]
14. Loeb, N.G.; Johnson, G.C.; Thorsen, T.J.; Lyman, J.M.; Rose, F.G.; Kato, S. Satellite and Ocean Data Reveal Marked In-crease in Earth's Heating Rate. *Geophys. Res. Lett.* **2021**, *48*, e2021GL093047. [CrossRef]
15. Dewitte, S.; Clerbaux, N.; Cornelis, J. Decadal Changes of the Reflected Solar Radiation and the Earth Energy Imbalance. *Remote Sens.* **2019**, *11*, 663. [CrossRef]
16. Loeb, N.; Thorsen, T.; Norris, J.; Wang, H.; Su, W. Changes in Earth's Energy Budget during and after the "Pause" in Global Warming: An Observational Perspective. *Climate* **2018**, *6*, 62. [CrossRef]
17. Wong, T.; Stackhouse Jr, P.W.; Kratz, D.P.; Sawaengphokhai, P.; Wilber, A.C.; Gupta, S.K.; Loeb, N.G. State of the Climate in 2019. *Bull. Am. Meteorol. Soc.* **2020**, *101*, S66–S69.
18. Ollila, A. The Pause End and Major Temperature Impacts during Super El Niños are Due to Shortwave Radiation Anomalies. *Phys. Sci. Int. J.* **2020**, 1–20. [CrossRef]

19. Stephens, G.L. Cloud Feedbacks in the Climate System: A Critical Review. *J. Clim.* **2005**, *18*, 237–273. [CrossRef]
20. Karlsson, K.-G.; Anttila, K.; Trentmann, J.; Stengel, M.; Solodovnik, I.; Meirink, J.F.; Devasthale, A.; Kothe, S.; Jääskeläinen, E.; Sedlar, J.; et al. *CLARA-A2.1: CM SAF CLoud, Albedo and Surface RAdiation Dataset from AVHRR Data—Edition 2.1*; EUMETSAT: Darmstadt, Germany, 2020. [CrossRef]
21. Hartmann, D.L.; Ceppi, P. Trends in the CERES Dataset, 2000–2013: The Effects of Sea Ice and Jet Shifts and Comparison to Climate Models. *J. Clim.* **2014**, *27*, 2444–2456. [CrossRef]
22. Svensmark, H. *FORCEMAJEURE—The Sun's Role in Climate Change*; The Global Warming Policy Foundation: London, UK, 2019; ISBN 978-0-9931190-9-5.
23. Shaviv, N.J. On climate response to changes in the cosmic ray flux and radiative budget. *J. Geophys. Res. Space Phys.* **2005**, *110*. [CrossRef]
24. Roemmich, D.; A Church, J.; Gilson, J.; Monselesan, D.; Sutton, P.; Wijffels, S. Unabated planetary warming and its ocean structure since 2006. *Nat. Clim. Chang.* **2015**, *5*, 240–245. [CrossRef]
25. Baggenstos, D.; Häberli, M.; Schmitt, J.; Shackleton, S.A.; Birner, B.; Severinghaus, J.P.; Kellerhals, T.; Fischer, H. Earth's radiative imbalance from the Last Glacial Maximum to the present. *Proc. Natl. Acad. Sci. USA* **2019**, *116*, 14881–14886. [CrossRef] [PubMed]
26. Allan, R.P.; Liu, C.; Loeb, N.G.; Palmer, M.D.; Roberts, M.; Smith, D.; Vidale, P.-L. Changes in global net radiative imbalance 1985–2012. *Geophys. Res. Lett.* **2014**, *41*, 5588–5597. [CrossRef]
27. Wild, M. Global dimming and brightening: A review. *J. Geophys. Res. Space Phys.* **2009**, *114*, 114. [CrossRef]
28. Liu, C.; Allan, R.P.; Mayer, M.; Hyder, P.; Loeb, N.G.; Roberts, C.; Valdivieso, M.; Edwards, J.M.; Vidale, P.-L. Evaluation of satellite and reanalysis-based global net surface energy flux and uncertainty estimates. *J. Geophys. Res. Atmos.* **2017**, *122*, 6250–6272. [CrossRef]
29. Su, W.; Liang, L.; Wang, H.; Eitzen, Z.A. Uncertainties in CERES Top-of-Atmosphere Fluxes Caused by Chang-es in Accompanying Imager. *Remote Sens.* **2020**, *12*, 2040. [CrossRef]
30. Johnson, G.C.; Lyman, J.M.; Loeb, N.G. Improving estimates of Earth's energy imbalance. *Nat. Clim. Chang.* **2016**, *6*, 639–640. [CrossRef]
31. Chylek, P.; Klett, J.D.; Lesins, G.; Dubey, M.K.; Hengartner, N. The Atlantic Multidecadal Oscillation as a dominant factor of oceanic influence on climate. *Geophys. Res. Lett.* **2014**, *41*, 1689–1697. [CrossRef]
32. Kravtsov, S.; Grimm, C.; Gu, S. Global-scale multidecadal variability missing in state-of-the-art climate models. *Npj Clim. Atmos. Sci.* **2018**, *1*, 34. [CrossRef]
33. Zhou, J.; Tung, K.-K. Deducing Multidecadal Anthropogenic Global Warming Trends Using Multiple Regression Analysis. *J. Atmos. Sci.* **2013**, *70*, 3–8. [CrossRef]
34. Chen, J.; Carlson, B.E.; Del Genio, A.D. Evidence for Strengthening of the Tropical General Circulation in the 1990s. *Science* **2002**, *295*, 838–841. [CrossRef] [PubMed]
35. Wielicki, B.A.; Wong, T.; Allan, R.P.; Slingo, A.; Kiehl, J.T.; Soden, B.J.; Gordon, C.T.; Miller, A.J.; Yang, S.-K.; Randall, D.A.; et al. Evidence for large decadal variability in the tropical mean radiative en-ergy budget. *Science* **2002**, *295*, 841–844. [CrossRef] [PubMed]
36. Kubar, T.L.; Jiang, J.H. Net Cloud Thinning, Low-Level Cloud Diminishment, and Hadley Circulation Weak-ening of Precipitating Clouds with Tropical West Pacific SST Using MISR and Other Satellite and Reanalysis Data. *Remote Sens.* **2019**, *11*, 1250. [CrossRef]
37. Del Genio, A.D.; Wolf, A.B. The Temperature Dependence of the Liquid Water Path of Low Clouds in the Southern Great Plains. *J. Clim.* **2000**, *13*, 3465–3486. [CrossRef]
38. Gordon, N.D.; Klein, S.A. Low-cloud optical depth feedback in climate models. *J. Geophys. Res. Atmos.* **2014**, *119*, 6052–6065. [CrossRef]

Article

Grassland Phenology's Sensitivity to Extreme Climate Indices in the Sichuan Province, Western China

Benjamin Adu, Gexia Qin, Chunbin Li * and Jing Wu

College of Resources and Environment, Gansu Agricultural University, Lanzhou 730070, China; benjamin@st.gsau.edu.cn (B.A.); qingexia2021@163.com (G.Q.); wujing@gsau.edu.cn (J.W.)
* Correspondence: licb@gsau.edu.cn; Tel.: +86-139-1907-0897

Abstract: Depending on the vegetation type, extreme climate and drought events have a greater impact on the end of the season (EOS) and start of the season (SOS). This study investigated the spatial and temporal distribution characteristics of grassland phenology and its responses to seasonal and extreme climate changes in Sichuan Province from 2001 to 2020. Based on the data from 38 meteorological stations in Sichuan Province, this study calculated the 15 extreme climate indices recommended by the Expert Team on Climate Change Detection and Indices (ETCCDI). The results showed that SOS was concentrated in mid-March to mid-May (80–140 d), and 61.83% of the area showed a significant advancing trend, with a rate of 0–1.5 d/a. The EOS was concentrated between 270–330 d, from late September to late November, and 71.32% showed a delayed trend. SOS was strongly influenced by the diurnal temperature range (DTR), yearly maximum consecutive five-day precipitation (RX5), and the temperature vegetation dryness index (TVDI), while EOS was most influenced by the yearly minimum daily temperature (TNN), yearly mean temperature (TEMP_MEAN), and TVDI. The RX5 day index showed an overall positive sensitivity coefficient for SOS. TNN index showed a positive sensitivity coefficient for EOS. TVDI showed positive and negative sensitivities for SOS and EOS, respectively. This suggests that extreme climate change, if it causes an increase in vegetation SOS, may also cause an increase in vegetation EOS. This research can provide a scientific basis for developing regional vegetation restoration and disaster prediction strategies in Sichuan Province.

Keywords: ridge regression method; Hurst index; NDVI; dynamic thresholds; Sichuan province; SOS; EOS

Citation: Adu, B.; Qin, G.; Li, C.; Wu, J. Grassland Phenology's Sensitivity to Extreme Climate Indices in the Sichuan Province, Western China. *Atmosphere* **2021**, *12*, 1650. https://doi.org/10.3390/atmos12121650

Academic Editors: Baojie He, Ayyoob Sharifi, Chi Feng, Jun Yang and Gianni Bellocchi

Received: 4 October 2021
Accepted: 6 December 2021
Published: 9 December 2021

1. Introduction

Phenology is defined as the study of the timing of recurrent biological events, the causes of their timing in terms of biotic and abiotic forces, and the interrelationship between the phases of the same or different species [1,2]. The Intergovernmental Panel on Climate Change (IPCC) stated that vegetation phenology is the most direct and effective indicator of climate change because it can effectively represent the broad interrelationships among ground-level, pedosphere, biospheric, and atmospheric biogeochemical features [2,3]. The average surface temperature of the world has increased substantially since the twentieth century, especially in the last 20 years (from 1951 to 2012, the global average surface temperature increased by 0.72 °C) [4,5], and vegetation phenology has changed dramatically in response [5]. Climate change encompasses both significance and severity [6,7]. Some studies have shown that extreme weather events have more significant and direct impacts than normal climate events, with extreme climate events occurring with increasing frequency and intensity [5,8]. As a result of global climate change, more and more scientists are studying the effects of climate change on vegetation dynamics [8–11]. The phenological status of vegetation not only regulates the duration of photosynthesis of the vegetation cover, but also has a significant impact on the carbon balance of terrestrial ecosystems [12–14], as it allows researchers to analyze the seasonal variation of vegetation. Climate variability is one of the most important drivers of vegetation development, and extreme climate changes appear to have a greater impact on vegetation dynamics than the average climate state [6,15]. Extreme events can disrupt vegetative activity and cause certain species to sprout earlier than expected or even fail to complete their flowering cycle [16]. In a study of vegetation phenology in the United States, researchers found that [17] an increase in daily maximum temperature was the most important factor in earlier SOS. Xie et al. (2015) studied the effects of drought, frost, heat, and humidity on autumn phenology in Greenland [18] and found that an increasing cold index caused forests to end the growing season earlier, while an increased heat index and drought stress delayed dormancy. According to some researchers, extreme temperature events on the Tibetan Plateau have a greater impact on EOS than extreme precipitation. An increase in the heat index during extreme temperature events causes a delay in vegetation EOS, while an increase in the cold index can lead to an earlier end of vegetation EOS, with warmer days and nights in Inner Mongolia also causing delays in autumn phenology [19]. This suggests that extreme climate events have a greater impact on vegetation phenology in areas affected by climate change than average climate values. Extreme events such as the European heatwave of 2003 and the Australian bushfires of 2019 [20,21] can affect vegetation dynamics and carbon balance, leading to a carbon sink being displaced by carbon sources.

Sichuan province is the most important agricultural province in western China, with rich natural resources that play an important role in the long-term sustainability of economic and environmental development. Considering its large area and geographical features, the relationship between vegetation and climate is spatially heterogeneous, which has led to some controversies about the effects of climatic factors on vegetation growth. It is therefore very important to study the response to climatic indices in this province. Previous scholars have quantified the phenology of the Zoigê Plateau in Sichuan and found that with the deterioration of mean temperatures and total annual precipitation from low to high elevations and from south to north, SOS was gradually delayed and EOS slowly advanced [22]. Recent studies have shown that there is a significant correlation between the normalized difference vegetation index (NDVI), annual mean temperature, and precipitation index in the hilly area of central Sichuan province [23].

However, these studies focused either on the effects of mean climate on phenology or the response of vegetation to extreme climatic conditions, rather than on the sensitivity of phenological changes in the vegetation community to extreme climatic conditions. Therefore, studying the phenology of grassland vegetation and its response to catastrophic climate change could lead to a better understanding of the mechanisms behind vegetation changes due to global warming, which is both ecologically valuable and practical. In this study, data on SOS and EOS from 2001 to 2020 and fifteen extreme climate indices were used to analyze the spatial differentiation of the sensitivity of different vegetation phenologies to climate indices. The objective of this study was to assess the sensitivity of vegetation phenological response to climate extremes by analyzing temporal, spatial, and inter-annual changes in grassland vegetation phenology. This research will serve as the scientific basis for local responses to extreme weather events, disaster prevention and mitigation, and environmental protection.

2. Materials and Methods

2.1. Study Area

Sichuan Province (97°–109° E, 26°–35° N) is the fifth largest province in China, with an area of 481,400 km^2 [24] and an altitude of 186–6304 m. The average temperature in January is 5–8 °C, the average temperature in July is 25–29 °C, and the average annual temperature is 15–19 °C. The average annual rainfall ranges from 500–1200 mm, with rainfall during the rainy season (May to October) accounting for 70% to 90% of the total annual rainfall. Sichuan is located in western China and is an important link between the southwest, northwest, and central parts of the country. It is surrounded by the Tibetan plateau to the west, the mountains of the Min, Bayankala, and Daba ranges to the north, and the Yunnan and Guizhou plateaus to the south. Since Sichuan's geography slopes from west to east and from the surrounding mountains to the core basin, climatic changes are highly dependent on altitude [25]. The northwestern Sichuan Plateau and the southwestern Sichuan Mountains are located on the eastern edge of the Tibetan Plateau. The regions have a typical monsoon climate, with the southwest monsoon from the Bay of Bengal and the Indian Ocean and the southeast monsoon from the Pacific Ocean dominating the weather. The easternmost glaciated areas of China are also located here, strongly influenced by the monsoon from the Tibetan Plateau and the western circulation [26,27]. The Eastern Sichuan Basin lies at a relatively low elevation and is surrounded by the northwestern Sichuan Plateau and the southwestern Sichuan Mountains to the west, Mount Longmen and Mount Daba to the north, and Mount Dalou and Mount Daliang to the south. The Sichuan Basin is under the influence of the East Asian monsoon, the Indian monsoon, and the circulation systems of the Tibetan Plateau, resulting in higher temperatures than other regions at the same latitude [28]. Due to its relatively large agricultural area, abundant natural resources, and high population density, Sichuan has become one of the most rapidly developing economic zones in western China. On the other hand, climate change has not only caused serious ecological problems such as floods, droughts, and other natural disasters, but also social and economic losses. Therefore, understanding and analyzing the spatial and temporal trends of extreme climate change is crucial for planning regional growth in agriculture, industry, and tourism. It will also serve as a scientific basis for tracking, diagnosing, analyzing, and predicting climate change and its consequences (Figure 1).

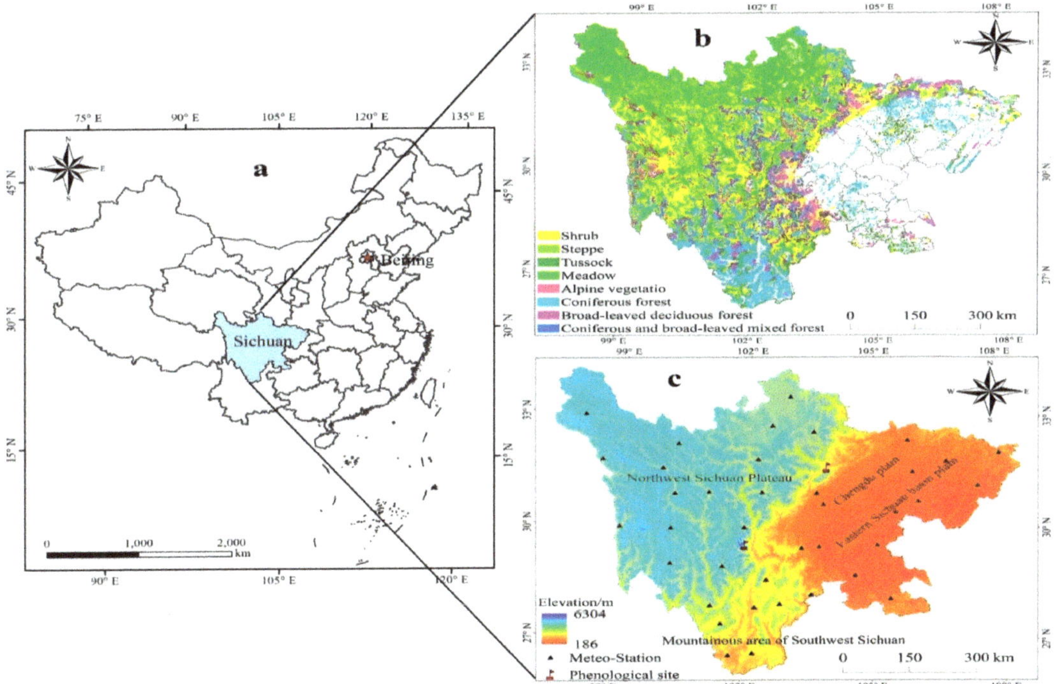

Figure 1. (**a**) The geographical position of Sichuan Province; (**b**) a map of the types of vegetation in Sichuan Province; (**c**) a topographical map and the distribution of meteorological stations in Sichuan Province.

2.2. Data Source

We selected the dataset MODIS NDVI (normalized difference vegetation index) provided by the Google Earth Engine platform (https://code.earthengine.google.com, accessed on 5 March 2021) with a period of 2001–2020 to extract the vegetation phenological parameters. The NDVI data were obtained from the National Aeronautics and Space Administration's MOD13A1 product (NASA) (Washington, DC, USA) with a spatial resolution of 250 m and a temporal resolution of 16 days. The standard level 3 product was subjected to geometric correction and atmospheric correction. In this work, the Savitzky–Golay (S–G) global filtering method was used to fit the maximum synthetic NDVI data for 16 d, using $NDVI_{mean}$ = 0.05 as the threshold for excluding non-vegetation areas.

Land surface temperature (LST) was derived using MOD11A2 of the (https://code.earthengine.google.com, accessed on 10 March 2021), with a spatial resolution of 1 km and temporal resolution of 8 days. The TVDI product was calculated using data from LST adjusted for terrain from 2001 to 2020.

Digital elevation models (DEM) were obtained from the Resource and Environmental Science Data Platform of the Chinese Academy of Sciences (http://www.resdc.cn, accessed on 15 March 2021) at 1 km resolution. Data were classified and extracted using ArcGIS to match the spatial resolution of the phenological data.

Evapotranspiration (ET) was generated using the MOD16A2 product of the Google Earth Engine platform (https://code.earthengine.google.com, accessed on 24 March 2021) with a spatial resolution of 500 m and a temporal resolution of 8 days. Inputs included vegetation cover, albedo, air temperature, barometric pressure, relative humidity, and other information, as well as daily average meteorological analysis data from MODIS and 8-d remote sensing data on vegetation attribute dynamics. The Penman–Monteith algorithm was used to obtain surface evapotranspiration data. Previous studies have shown that although the accuracy of MOD16A2 has been overestimated or underestimated in some areas, it can meet the requirements of large-scale applications [29,30].

Phenological validation data were obtained from the National Ecological Science Data Center (http://rs.cern.ac.cn/order/myDataOrders, accessed on 20 March 2021), including phenological observation data from Gongga Mountain and Maoxian Station (Figure 1), which monitor vegetation at the stages of germination, flowering, seed formation, seed dispersal, and leaf yellowing. Since the phenological period of ground observation was observed at the individual plant level, for better comparability of verification data, reference was made to Hou et al. [31].

Climate data were derived from the daily dataset of the China Meteorological Data Network (http://data.cma.cn, accessed on 20 March 2021) throughout 2001–2020. Previous studies have shown that changes in climate conditions in autumn and winter of the previous year cause hysteresis and an accumulating effect on vegetation development the following year [32–34]. The accumulating effect lasts 5 to 10 months, while the hysteresis effect lasts 2–3 months [33]. Since the vegetation EOS in Sichuan Province occurs from late September to late November, this study determined the extreme climate from September 1 to August 31 of the current year [32]. There are 158 surface meteorological stations [24] in Sichuan province. 38 stations with 20-year records were selected as candidate stations based on the record length. The collected data were first visually inspected for outliers and other problems that could lead to inaccuracies or biases in measuring changes in the seasonal cycle or variance of the data using RclimDex [35], and 38 meteorological stations in the study area were selected (Figure 1c). The indices listed in Table 1 were then selected using the common climate indices proposed by the Expert Team on Climate Change Detection and Indices (ETCCDMI). The aim was to capture the extreme state of temperature events and the marginal state of precipitation events [36,37]. The interannual time series of each index was created using RclimDex and MATLAB [38] software, and the annual scale data were interpolated into a dataset using ANUSPLINE [39].

Table 1. The definitions and classifications of climate indices.

Indicators	Description
DEM	Digital elevation model
RS	Solar radiation
ET	Evapotranspiration
TVDI	Temperature vegetation dryness index
PER	Yearly mean accumulated precipitation
RX1	Yearly maximum consecutive one-day precipitation
RX5	Yearly maximum consecutive five-day precipitation
DTR	Diurnal temperature range
TEMP_MEAN	Yearly mean value of temperature
TEMP_MAX	Yearly mean value of daily maximum temperature
TEMP_MIN	Yearly mean value of daily minimum temperature
TNN	Yearly minimum value of daily minimum temperature
TNX	Yearly maximum value of daily minimum temperature
TXN	Yearly minimum value of daily maximum temperature
TXX	Yearly maximum value of daily maximum temperature

2.3. Methods

2.3.1. Phenological Extraction Method

Since the NDVI image synthesized by Maximum Value Composite (MVC) is strongly affected by clouds and the atmosphere, it is necessary to reduce the noise and smooth the image. Filtering the NDVI time-series data by the S–G method can remove abnormal points and local mutation points as well as trends, and is not limited by the spatial and temporal scale of the data and sensor. This method is determined based on the following equation:

$$Y_j = \frac{\sum_{i=-m}^{i=m} C_i Y_{i+j}}{N} \tag{1}$$

where Y_j is the fitted sequence data, i represents each scene, Y_{i+j} is the original sequence data, C_i is the coefficient for the i-th NDVI value of the filter, and N is the number of convolved integers and is equal to the size of the smoothing window (2 m + 1), where m is half the width of the smoothing window [40].

There are numerous approaches to deriving grassland phenology from the time series of the grassland index obtained by remote sensing. The most commonly used methods are the cumulative frequency method, maximum slope method, dynamic threshold method, and the maximum ratio method [41–44]. Considering the scale of the study area, computational efficiency, feasibility of the methods, and characteristics of NDVI changes, the dynamic threshold method proposed by White et al. (1997) was used to calculate and extract the phenological information of NDVI from 2001–2020 on the grid point [45]. The dynamic thresholds for grassland SOS and EOS were set at 20% and 50%, respectively, based on the phenological information recorded in the field and repeated trials. Finally, since the TIMESAT 3.2 software can only extract $n-1$ years of phenological parameters from n years of data, the phenological data from 2001 to 2019 were used in this work.

$$NDVI_{ratios} = \frac{NDVI - NDVI_{min}}{NDVI_{max} - NDVI_{min}} \tag{2}$$

where $NDVI_{ratios}$ are the dynamic thresholds, and $NDVI_{max}$ and $NDVI_{min}$ are the maximum and minimum values of the annual NDVI, respectively. The date on which the NDVI first exceeds the threshold was defined as the start of the season (SOS), and the date on which the NDVI first falls below the threshold was defined as the end of the season (EOS).

In addition, this study evaluated the observed phenological data from the phenological sites at the ground stations (Figure 1). The phenological periods observed on the ground were observed at the individual plant level. To increase the stability of the validation data, records whose recording date was more than 30 days away from the rest of the data were excluded according to the study by Hou et al. [46]. The germination period and the blight period of the data from the ground observation stations were defined as the grassland SOS and EOS, respectively [47,48].

2.3.2. Trend Analysis

From 2001 to 2020, the Mann–Kendall (MK) trend test was used to assess the significance of phenological patterns. The M–K test is a nonparametric statistical test that evaluates the significance of a changing trend within a time series in the absence of recurrent differences or other sequences [49–51]. It is widely used to detect trends in hydrological, climatic, and ecological aspects. It is known that the M–K test, developed for independent data, rejects the null hypothesis of no trend more often than the significance level applied to autocorrelated series [52–54]. To reduce the impact of serial autocorrelation on trend detection of data series, autocorrelated series were whitened before performing the M–K test [52]. The null hypothesis (H_0) of the M–K test is that there is no trend for an independently distributed time series ($x_1, x_2, \ldots, x_n$). The significance of the pattern of change was determined using the test statistics S (if $n \leq 8$) or Z (if $n > 8$), which are calculated as follows:

$$S = \sum_{i=1}^{n-1} \sum_{j=i+1}^{n} \mathrm{sgn}(X_j - X_i) \tag{3}$$

$$\mathrm{sgn}(X_j - X_i) = \left\{ \begin{array}{c} 1 \ (X_j - X_i) > 0 \\ 0 \ (X_j - X_i) = 0 \\ -1 \ (X_j - X_i) < 0 \end{array} \right\} \tag{4}$$

when $n \geq 8$, the statistical variables Var(S) must conform to the normal distribution, the mean is 0, and the calculation method for the variance of S is as follows:

$$\mathrm{Var}(S) = \frac{n(n-1)(2n+5) - \sum_{i=1}^{m} t_i(t_i - 1)(2t_i + 5)}{18} \tag{5}$$

$$Z_s = \begin{cases} \frac{S-1}{\sqrt{\mathrm{Var}(S)}} & \text{if } S > 0 \\ 0 & \text{if } S = 0 \\ \frac{S+1}{\sqrt{\mathrm{Var}(S)}} & \text{if } S < 0 \end{cases} \tag{6}$$

Z_s, following the standard normal distribution, is an upward trend if it is greater than 0, and is a downward trend if it is less than 0.
β is used to measure the range of trend change horizontally:

$$\beta = \mathrm{Median}\left(\frac{X_i - X_j}{i - j}\right) \forall j > i \tag{7}$$

where Median is the median function and X_j and X_i are the values of the grassland index of the jth and the ith year ($j > i$) ($2001 \leq i < j \leq 2020$), respectively. Grasslands index values that were similar were combined into one level. Grassland index values that had different values were grouped. The size of a plane was denoted by t_i, while the number of planes was denoted by m. A negative value indicated a downward trend, while a positive value indicated an upward trend. When the absolute rate $|Z_s| > Z_{(1-a)/2}$, the H_0 hypothesis was rejected and the grassland index values were found to be statistically significant at a 95% confidence level. (When the absolute value of Z_s is greater than or equal to 1.615, 1.943, or 2.518, it passed the significance tests at 90%, 95%, and 98%, respectively).

2.3.3. Temperature Vegetation Dryness Index (TVDI)

When the study area has a different land cover, such as water, sparse vegetation, or lush vegetation, the scatter plot of surface temperature and vegetation index is often triangular [55] or trapezoidal [56]. Sandholt developed a temperature vegetation drought indicator based on the NDVI-Ts space (TVDI) [57]. The TVDI was calculated as follows:

$$\mathrm{TVDI} = \frac{T_s - T_{smin}}{T_{smax} - T_{smin}} \tag{8}$$

where T_s represents the surface temperature, T_{smax} was the highest temperature on the surface corresponding to the NDVI, and T_{smin} represents the lowest surface temperature corresponding to the NDVI. The calculations of T_{smax} and T_{smin} are as follows:

$$T_{smax} = a_1 + b_1 * \mathrm{NDVI} \tag{9}$$

$$T_{smin} = a_2 + b_2 * \mathrm{NDVI} \tag{10}$$

where a_1, b_1, a_2, and b_2 are the coefficients of the equation for dry and wet edge adjustment.

2.3.4. Analysis of the Persistence of Phenological Changes Using Hurst Exponent and Rescaled Range (R/S) Analysis

R/S analysis is the oldest and best-known method for estimating the Hurst index and effectively describing self-similarity and long-term dependence, widely used in the fields of climate, hydrology, seismology, and geology. In this paper, the changing trend of phenology was calculated pixel by pixel based on the Rescaled Range (R/S) analysis method, which reflects the continuity of the changing trend. The Hurst index expands from 0 to 1 according to Hurst [58] and Mandelbrot and Wallis [59,60]. If the value was equal to 0.5, the time series was a stochastic series without continuity, indicating that the future change trend of the time series was not related to that of the study period. If the value was greater than 0.5, the time series was a consistent time series, indicating the same trend of the time series in the future, with a larger value representing stronger consistency. The primary calculation methods were as follows:

$$\text{Difference sequence}: \Delta A_i = \Delta A - \Delta A_{i-1} \tag{11}$$

$$\text{Mean value series}: \overline{\Delta A} = \frac{1}{m} \sum_{i=1}^{m} \Delta A_i\,, \ (m = 1, 2, \ldots\ldots, n) \tag{12}$$

$$\text{Cumulative deviation}: X(t) = \sum_{i=1}^{m} \left(\Delta A_i - \overline{\Delta A(m)} \right), \ (1 \le t \le m) \tag{13}$$

$$\text{Range sequence}: R(m) = \max X(t)_{1 \le m \le n} - \min X(t)_{1 \le m \le n} \tag{14}$$

$$\text{Standard deviation}: S(m) = \left[\frac{1}{m} \sum_{i=1}^{m} \left(\Delta A_i - \overline{\Delta A(m)} \right)^2 \right]^{\frac{1}{2}} \tag{15}$$

2.3.5. Sensitivity Analysis

To evaluate the effects of climate indices on vegetation phenology and the significance of each variable, the random forest regression model was applied. We chose random forest model because it can be used for regression tasks, as its nonlinear nature gives it an advantage over linear algorithms. The model evaluated the impact of each independent variable on the dependent variables using the percentage increase in mean square error (%Inc MSE) [18]. First, the ntree (the set of decision trees) of the decision tree model was constructed, and the mean square error of random substitution out-of-bag (OOB) was determined, which was recorded MSE_{11}, MSE_{12}, ..., MSE_{mntree} as in the matrix below:

$$\begin{bmatrix} MSE_{11} & MSE_{12} \cdots & MSE_{1ntree} \\ MSE_{21} & MSE_{22} \cdots & MSE_{2ntree} \\ \vdots & \vdots \quad \vdots & \vdots \\ MSE_{m1} & MSE_{m2} \cdots & MSE_{mntree} \end{bmatrix} \tag{16}$$

The significant score was calculated as follows:

$$\text{scoreX}_j = S_E^{-1}\frac{\sum_{r=1}^{\text{ntree}} \text{MSE}_r - \text{MSE}_{pr}}{\text{ntree}}, (1 \leq p \leq m) \tag{17}$$

where n is the number of original data samples and m is the number of variables.

Ridge regression is often used in the real regression process as an improved approach to least squares estimation because it can minimize collinearity between independent variables and eliminate the unbiased aspect of the least squares method. Ridge regression was used to examine the sensitivity of phenology to climate indices in this study. The basic principle is as follows:

$$\hat{\beta}_{RR} = (X' \times X + K \times I) \times X' \tag{18}$$

where X is the observation matrix of the independent variable, K is the ridge parameter, I is the identity matrix, and $\hat{\beta}_{RR}$ is the sensitivity coefficient of the independent variable to the dependent variable [61].

3. Results

3.1. Assessment of the MOD13A1 Data

The results of the assessment of the accuracy of the MOD13A1 data, based on comparisons with ground observation data from 2001 to 2020, are shown in Figure 2. According to the ground observations, the SOS of grassland vegetation ranges from 70 to 115 days, while the SOS of the MOD13A1 phenological product ranges from 70 to 105 days. The r was 0.619, the significance level was $p < 0.05$, the bias and RMSE were 0.597 and 8.13 days, respectively, and the points with less than 10 days error accounted for 71.4 % of all verification points. The EOS of grassland vegetation observed on the ground was between 280 and 320 days, but the EOS that was calculated from the MOD13A1 phenological product ranged from 275 to 325 days. The correlation coefficient (r) was 0.607, the bias was 0.613, and the RMSE was 8.87 days. The validation results showed that the remotely sensed vegetation SOS and EOS had earlier dates than the ground-based observations, and the error between the remotely sensed and ground-based phenology was less than 10 days; however, the results were still well correlated. These results showed that the NDVI of phenology data from MOD13A1 can be used to calculate the SOS and EOS of grassland vegetation in Sichuan Province, and the error is within an acceptable range considering the temporal resolution of the remotely sensed images.

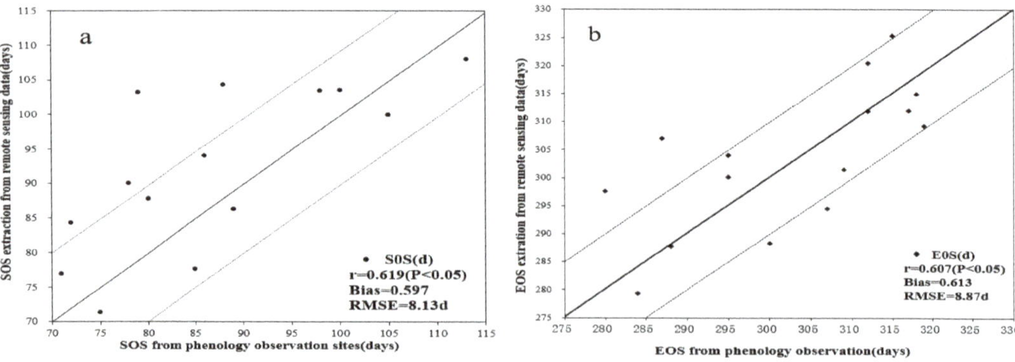

Figure 2. Correlation of grassland vegetation (**a**) SOS and (**b**) EOS between the NDVI from MOD13A1 data and ground observations.

3.2. Spatial Distribution Patterns of Grassland Vegetation Phenology

Analysis of the multi-year average spatial distribution patterns of phenology (Figure 3) indicates considerable spatial and vertical heterogeneity of grassland SOS and EOS in Sichuan Province between 2001 and 2020. The vegetation SOS of Sichuan Province was mainly concentrated in the period of 80–140 d (pixels accounted for 61.83%), and the SOS was gradually delayed with the topography from west to east of the whole grassland area (from mid-March to mid-May). The SOS of steppe, coniferous, and broad-leaved mixed forest and deciduous forest started earlier and was concentrated before the period of 100 d, with area percentages of 61.26%, 50.65%, and 53.86%, respectively. Tussock and meadow SOS were mainly concentrated in the period of 90–130 d and 100–140 d, respectively. The SOS of alpine vegetation was more advanced than that of the other vegetation types; 83.2% of the area was concentrated after the period of 120 d, while 16.7% was concentrated before the period of 120 d.

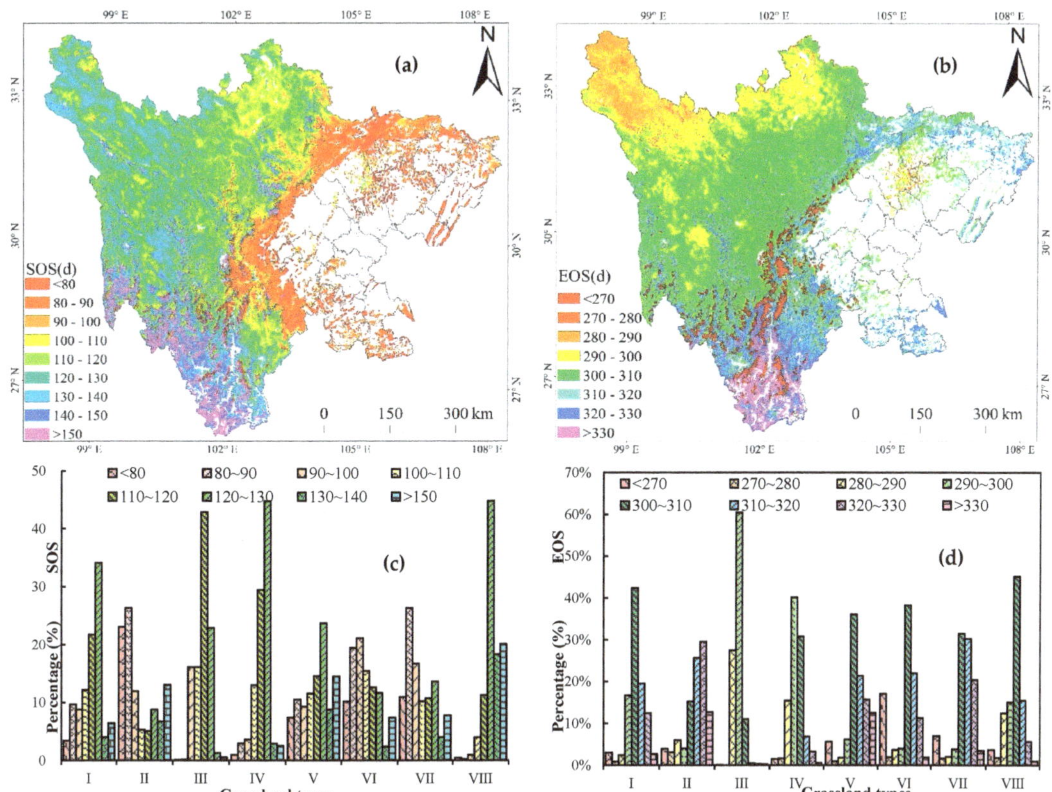

Figure 3. Spatial distribution of annual mean phenological values ((**a**) Start of the season and (**b**) End of the season) and vegetation percentages ((**c**) Start of the season and (**d**) End of the season) of the eight grassland types (I: shrub; II: steppe; III;: tussock; IV: meadow; V: coniferous forest; VI: coniferous and broad-leaved mixed forest; VII: deciduous forest; VIII: alpine vegetation) in the Sichuan Province from 2001–2020.

In the topographic region of Sichuan Province, the earliest SOS (<80 d) were mainly in the Chengdu Plain and the plain of the eastern Sichuan Basin. The vegetation types in this area are mainly deciduous forest, coniferous forest, and steppe due to the subtropical humid climate. In the northwestern Sichuan Plateau, SOS concentrated in the period of 100–130 d and the vegetation types in this area are mainly tussock, meadow, shrub, and steppe. The latest SOS was focused in the mountainous area in southwestern Sichuan; the vegetation types are mainly coniferous forest and SOS was mostly found after the 130-d period. The statistical results of SOS for different vegetation types showed that the concentration interval of SOS varied significantly. Alpine vegetation has a more advanced SOS than other grassland types.

From 2001 to 2020, the vegetation EOS of Sichuan Province was advanced and mainly concentrated in the period of 270–330 d (71.32% of pixels), and there were differences among the phenological parameters of different vegetation types. The onset of EOS for tussock vegetation was the earliest, and 88.0% concentrated before 300 d. EOS of steppe and deciduous forest vegetation were mainly concentrated after 300 d, and EOS of shrub, meadow, coniferous forest, coniferous and broad-leaved mixed forest, and alpine vegetation types were concentrated in the range of 300–320 d. In the Sichuan topographic region, the southwestern Sichuan EOS was the latest, concentrating after 310 d, followed by the Chengdu Plain and eastern Sichuan Basin, which were concentrated between 310–330 d. The earliest EOS was concentrated in the plateau region of northeastern Sichuan in the period before 310 d.

3.3. *Analysis of Interannual Phenological Variation*

To fully understand the dynamic patterns of grassland phenology in Sichuan Province, we examined the phenology of grassland vegetation at the pixel level using MATLAB software and a combination of Theil–Sen trend analysis and Mann–Kendall tests (Figure 4). From 2001 to 2020, vegetation SOS of Sichuan Province was mainly concentrated in the 80–140 d period (61.83% of pixels). SOS advanced gradually, with the most obvious trend concentrated in the northwestern Sichuan Plateau. The average advance rate throughout the study area was basically above 1.0 d/a. 38.17% of SOS exhibited a delayed trend distributed among the mountainous area of southwest Sichuan, the Chengdu Plain, and the eastern Sichuan Basin Plain. The region with the highest delay rates was the Chengdu Plain, followed by the eastern Sichuan Basin Plain and the mountainous area of southwest Sichuan. The overall average delay rate was concentrated above 1.0 d. From the perspective of different vegetation types, steppe, deciduous forest, and coniferous forest lagged at the rates of 0.656 d, 0.241 d, and 0.168 d, respectively, but SOS showed a significant trend of advance from 2006 to 2007. Shrub, tussock, meadow, coniferous and broad-leaved mixed forest, and alpine vegetation all showed an insignificant advance trend. Among them, the advance rate of tussock was the highest, with an average advance rate of 0.182 d.

In Sichuan province, 71.32% of the vegetation EOS showed a lagging trend (67.13% of the total pixels by the significance test, $p = 0.05$). The regions with the highest EOS delay rate were concentrated around the eastern Sichuan Basin Plain, and the average delay rate in 20 years was over 1.5 d. The areas where EOS showed an early trend were concentrated in the mountainous area of southwestern Sichuan and the areas bordering Tibet on the northwestern Sichuan Plateau, and the lag rate was 0–0.5 d/a. The curves of interannual variation of different vegetation types all show a significant trend of lag. The vegetation type with the most significant lag trend was steppe, with an average lag rate of 0.894 d, followed by coniferous forest, with an average lag rate of 0.542 d. The vegetation type with the least lag rate was shrub, with an average lag rate of 0.372 d, followed by meadow, with an average lag rate of 0.426 d. The EOS change curve of alpine vegetation is clearly jagged.

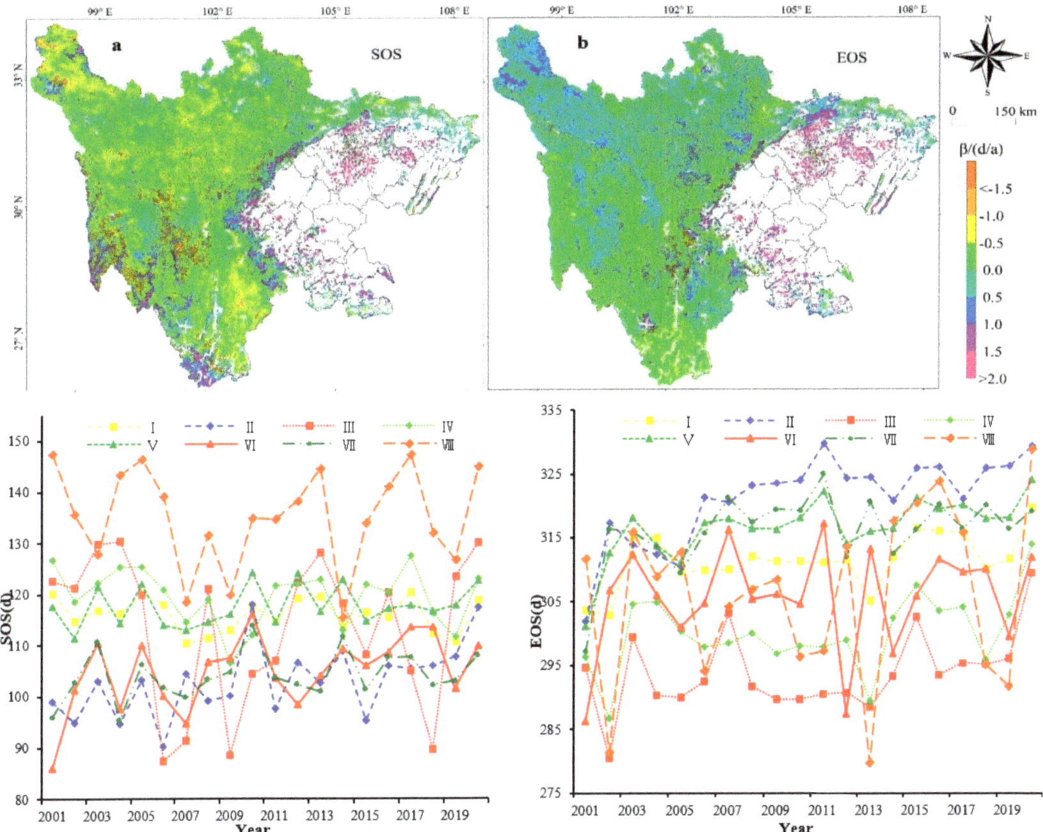

Figure 4. The trend of change in the spatial distribution of (**a**) grassland SOS, (**b**) grassland EOS, and the trend of interannual variation in vegetation phenology of SOS and EOS. (I: shrub; II: steppe; III: tussock; IV: meadow; V: coniferous forest; VI: coniferous and broad-leaved mixed forest; VII: deciduous forest; VIII: alpine vegetation).

3.4. Analysis of the Future Viability of the Phenological Period

Analysis of the sustainability of vegetation SOS and EOS in Sichuan Province from 2001 to 2020 was performed using the Hurst index (Figure 5). The future change trend of vegetation SOS in Sichuan Province exhibits a Hurst phenomenon with an average Hurst value of 0.44, i.e., the change trend of vegetation SOS in the future will have a weak opposite trend to the change trend in the period 2001–2020, i.e., vegetation SOS may be delayed. In 88.81% of the pixels, the Hurst index was between 0 and 0.5. Except for the pixels with Hurst index more than 0.5 in the eastern Sichuan Basin Plain, the Hurst index in the other regions (northwest Sichuan Plateau, Chengdu plain and mountainous area of southwestern Sichuan) was basically less than 0.5. The number of pixels with Hurst = 0.5 accounted for 3.15% of the total number of pixels, indicating that the change of vegetation SOS on a few pixels has nothing to do with the vegetation SOS in the period 2001–2020.

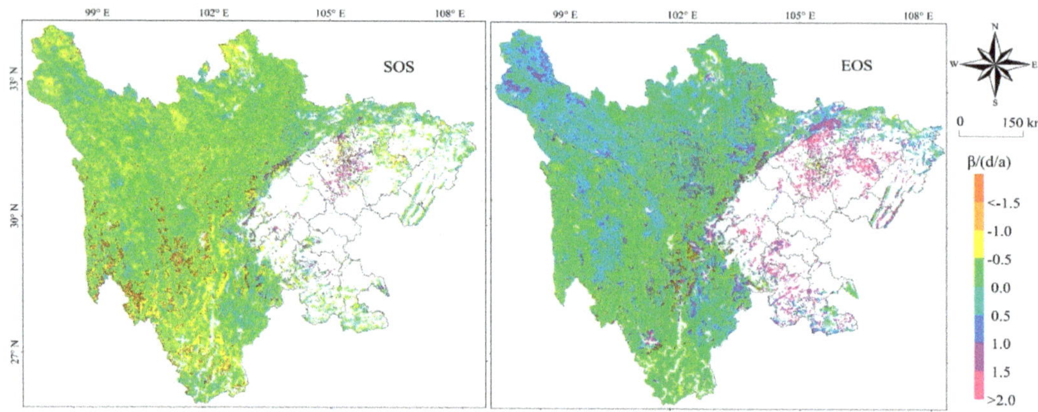

Figure 5. Analysis of the phenological sustainability of grasslands in Sichuan Province using the Hurst exponent (β(d/a).

From 2001 to 2020, pixels with a Hurst value of more than 0.5 of the trend of vegetation change EOS in Sichuan Province accounted for 23.16% of the total pixels and they were concentrated in the Sichuan basin and mountainous regions in southwest Sichuan. The sustainability of vegetation EOS indicates that in the future, Sichuan province will continue the trend of the average level of change over the past 20 years and a shift will occur. Pixels below 0.5 account for 75.54% of the total pixels, which means that 75.54% of the pixel changes of EOS are opposite to the past and will have an upward trend in the future.

3.5. The Significant Influence of Extreme Climate Indices on Phenological Variables

The random forest regression model can convert the value of a variable into an integer, with larger values indicating higher significance. The model was used in this study to see how the fifteen extreme climate indices (Table 1) affect the parameters of vegetation phenology in Sichuan Province (Figure 6). The significant effects of several extreme climate indices on the changes of different vegetation phenology characteristics were different. SOS was mainly influenced by the yearly maximum consecutive five-day precipitation (RX5), diurnal temperature range (DTR), and temperature vegetation drought index (TVDI), with over 40% Inc MSE. EOS was mainly influenced by the yearly minimum value of daily minimum temperature (TNN), yearly mean value of temperature (TEMP_MEAN), and temperature vegetation dryness index (TVDI), with over 30% Inc MSE. The model was used here to determine the effects of extreme climate indices (Table 1) on various parameters of vegetation phenology in Sichuan Province. The focus was on the response of SOS, EOS, and overall regional vegetation phenology to extreme climate indices (Figure 6).

Therefore, the five extreme climate indices (RX5, DTR, TDVI, TNN, and TEMP_MEAN) were selected for the subsequent ridge regression sensitivity analysis based on the importance of the fifteen climate indices to vegetation phenology (Figure 7).

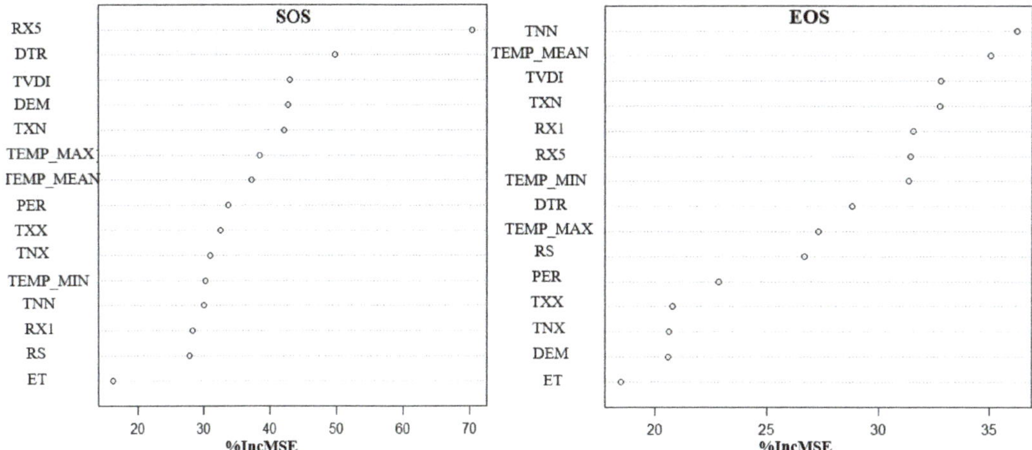

Figure 6. The importance of the climate indices for the phenological parameters start of the growing season (SOS) and end of the growing season (EOS), estimated as the percentage increase in the mean square error (% Inc MSE) of the different vegetation types.

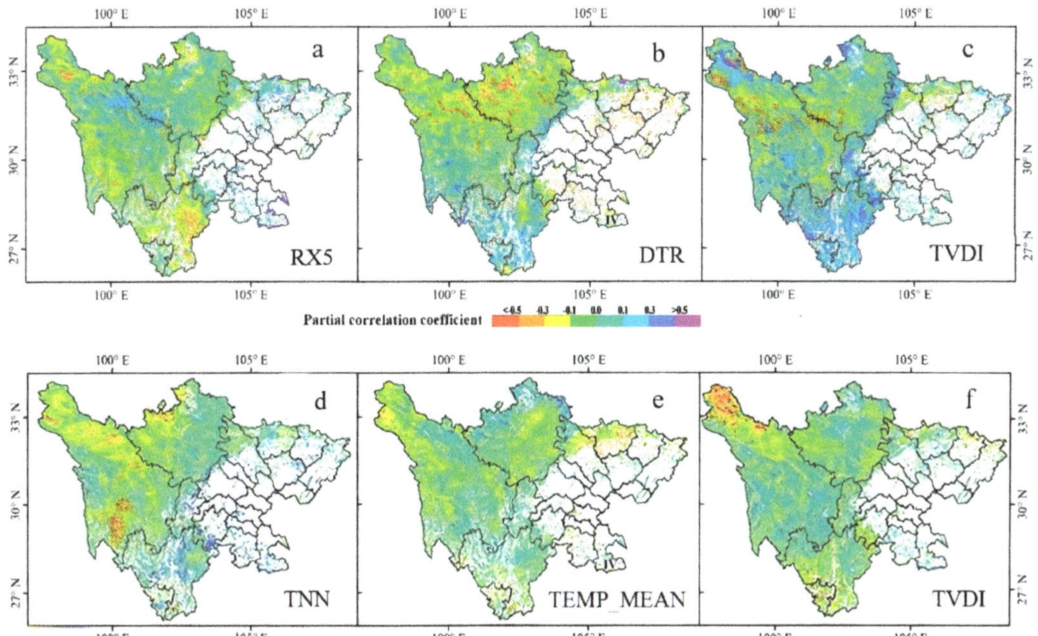

Figure 7. The sensitivity of vegetation phenology in Sichuan Province to extreme climate indices. The proportion of pixels with $p < 0.01$ is less than 10%. The sensitivity response of start of the growing season (SOS) to RX5 (**a**), the sensitivity response of SOS to DTR (**b**), the sensitivity response of SOS to TVDI (**c**), the sensitivity response of end of the growing season (EOS) to TNN (**d**), the sensitivity response of EOS to TEMP_MEAN (**e**), the sensitivity response of EOS to TVDI (**f**).

3.6. Analysis of Phenology Sensitivity to Extreme Climate Indices

The spatial distribution of phenology sensitivity to extreme climate indices (yearly maximum consecutive five-day precipitation (RX5), diurnal temperature range (DTR), yearly minimum value of daily minimum temperature (TNN), yearly mean value of temperature (TEMP_MEAN), and temperature vegetation dryness index (TVDI) in Sichuan Province from 2001 to 2020 were evaluated (Figure 7). Vegetation phenology showed different spatial differentiation concerning the five climate indices. The phenology measures SOS and EOS showed different spatial distributions with respect to the five climate indices (RX5, DTR, TVDI, TNN, TEMP_MEAN, and TVDI), respectively. The sensitivity coefficient of SOS was positive, while that of EOS was negative. The indices of yearly maximum consecutive five-day precipitation (RX5) and temperature vegetation dryness (TVDI) showed a positive sensitivity coefficient for SOS, and diurnal temperature range (DTR) showed negative sensitivity coefficient for SOS (Figure 7a). The yearly minimum value of daily minimum temperature (TNN) index showed a positive sensitivity coefficient for EOS in the mountainous area of southwestern Sichuan and a negative sensitivity coefficient for EOS in other regions (northwest Sichuan Plateau, Chengdu Plain and eastern Sichuan Basin Plain). The yearly mean values of temperature (TEMP_MEAN) and temperature vegetation dryness index (TVDI) showed a negative sensitivity coefficient for EOS in the mountainous area of southwestern Sichuan. The spatial distribution of the coefficients was relatively consistent, but not significant. The temperature vegetation dryness index (TVDI) was negative in northwest Sichuan Plateau and Chengdu Plain. The sensitivity coefficients of phenology for temperature indices were significantly higher than those for precipitation indices. The temperature vegetation dryness index (TVDI) showed positive and negative sensitivity to SOS and EOS, respectively.

4. Discussion

4.1. The Relationship between Grassland SOS and EOS, and Ground Observation and Remote Sensing Inversion Methods

According to Shen et al. [11], temporal trends in SOS differ significantly between remote sensing data and ground observations. Shen et al. [11] studied the phenological responses of plants to climate change on the Tibetan Plateau Satellite and found that remote sensing can only capture the leaf phenology of those species that are strongly represented in a community at the time of observation. If the sequence of leaf unfolding changes for different species within a community due to changing environmental conditions, the species with the highest cover at a given time may change. Satellite-based SOS is based on the greenness of a pixel, whereas ground-based observations are based on the morphological changes of individual plants. In evaluating satellite-based remote sensing of foliage phenology, the first problem is the data quality of the NDVI. Satellite images are usually acquired around midday and are contaminated by the target indicator, as well as by inaccurate atmospheric correction procedures for the sensors. In addition, the spectral configuration can be refracted and scattered by atmospheric components such as water vapor [62] and aerosols [63], leading to errors. Validation of satellite-based phenology is usually based on ground observations of phenology in cropland and deciduous forests at the species level [45]. In comparison, it is difficult to validate satellite-based phenology in natural grasslands using such species-level observations because a large number of species coexist within a pixel and have different phenological stages. Shen et al. (2015) determined the timing of spring green-up for the years 1982 to 1999 from data obtained from AVHRR NDVI using five different methods and for the years 2000 to 2011 using the same five methods from NDVIs observed by AVHRR, MODIS, and SPOT, and at MODIS EVI. The AVHRR sensor was renewed in late 2000 and since then it is suspected that the quality of the data from AVHRR NDVI had decreased from 2001 onwards [64]; the spring vegetation data of Zhang et al. [64] consist of three periods: from 1982 to 1997, obtained from the data of AVHRR NDVI, for 1998–1999, from the AVHRR and SPOT NDVI data, and from 2000–2011, from the NDVI datasets observed by MODIS and SPOT. For

1982–2011, consistent variability was found between SOS and remote sensing-based results. Che et al. [65] investigated the interannual variability of the averaged growing season end dates extracted from AVHRR LAI data on the Tibetan Plateau during 1982–2011 and Yu et al. [43] investigated the averaged growing season end dates of meadows extracted from AVHRR NDVI on the Tibetan Plateau. The extracted EOS showed a high degree of correspondence. In this study, the NDVI from MOD13A1 was used and the validation results showed that remotely sensed vegetation data SOS and EOS had earlier dates than the ground-based observations, with an error of less than 10 days between the remotely sensed and ground-based measurements of phenology; however, the results were still well correlated.

4.2. Trends in Grassland SOS and EOS

The change in the phenology of grasslands is a cyclical and continuous dynamic process. Many factors influence phenological change and there are certain relationships among the different factors. Cong et al. [66] and Slayback et al. [67] studied the phenological dynamics of different grasslands types in the northern Hemisphere from 1982 to 2009 and found that the evolution of grassland vegetation is more significant than that of forest vegetation over many years. The study found that the rate of phenological change in steppe was the largest among all vegetation types, followed by tussock vegetation, which is consistent with the conclusions of the study by Cong et al. [66]. Studies have shown that the rate of phenological change is significantly higher in grassland vegetation than in forests. The main reason for this is that grasslands are sensitive to climate change. Therefore, the increase in average temperature and the occurrence of climate extremes in recent years directly affect the productivity and carbon balance of grasslands. As a result of global warming, the SOS of grassland vegetation has changed drastically, leading to a shift in the global carbon cycle and climate adaptation [66,68]. Our research results also show that SOS of grassland vegetation was significantly delayed and EOS advanced. The differences in reported phenological changes among studies might be related to the different climatic conditions and vegetation types in the study areas [69].

4.3. The Relationship between Grassland Phenology and Extreme Climate Indices

Extreme temperature events influence vegetation growth and development by affecting the activities of photosynthetic and respiratory enzymes [18]. Extreme rainfall events inhibit growth by reducing the amount of available water in the soil where vegetation develops [69]. According to Piao et al. [41], high temperatures had a greater impact on northern vegetation than precipitation. Nagy et al. [16] found that extreme climate indices affect phenology. They pointed out that the flowering dates of temperate vegetation were almost one month earlier and some plant species could not complete their flowering cycle, but high precipitation events had little effect on the phenology of the observed plants. The sensitivity of phenological responses of different vegetation types to extreme climatic events is complicated. We found that the sensitivity of SOS was mainly influenced by the yearly maximum consecutive five-day precipitation (RX5), diurnal temperature range (DTR), and temperature vegetation dryness index (TVDI), while EOS was mainly influenced by the yearly minimum value of daily minimum temperature (TNN), yearly mean value of temperature (TEMP_MEAN), and TVDI. The SOS showed a negative sensitivity coefficient for DTR. If the diurnal temperature difference causes grassland phenology to begin earlier in the spring, this will lead to an earlier end of the growing season in the fall. SOS showed an advancing trend that could be attributed to warming, as the greater difference between TEMP_MAX and TEMP_MIN could favor growth. The increase in yearly maximum consecutive five-day precipitation (RX5) delayed the SOS, as cumulative precipitation is not a reliable measure of crop water supply. The timing and intensity of precipitation, soil freeze/thaw, snowmelt, evaporation, and permafrost decomposition can alter water availability and thus phenology. Unusual warming can boost plant activity, but this is not beneficial for vegetation development, as it leads to an early end of the growing season and shortens the vegetation growth cycle, as plants need to be exposed to lower temperatures before the end of the dormant period. Another possibility is that when vegetation is stressed by the environment, it uses its ever-changing signals—mainly hormones and signaling molecules—to regulate phenology, such as germination, leaf senescence, and growth stagnation. To protect itself from environmental stress, vegetation ends its growing season early. Moreover, a harsh climate is expected to make the seasonal trajectories of vegetative activities more vulnerable [70,71]. Vegetation phenology is both a periodic and a continuously dynamic process influenced by several interrelated influencing factors [34,72]. According to some studies, climate change is the most important element affecting vegetation phenology [72,73]. In this study, we focused on phenological changes caused by extreme climate indices, but we discovered during the analysis that the advance of vegetation SOS can indirectly lead to the advance of vegetation EOS. If vegetation EOS is affected by extreme climate indices in one year, it may also lead to changes in vegetation SOS in the following year. For this reason, we conducted a correlation analysis between SOS and EOS and the four seasons in this study (Figure 8).

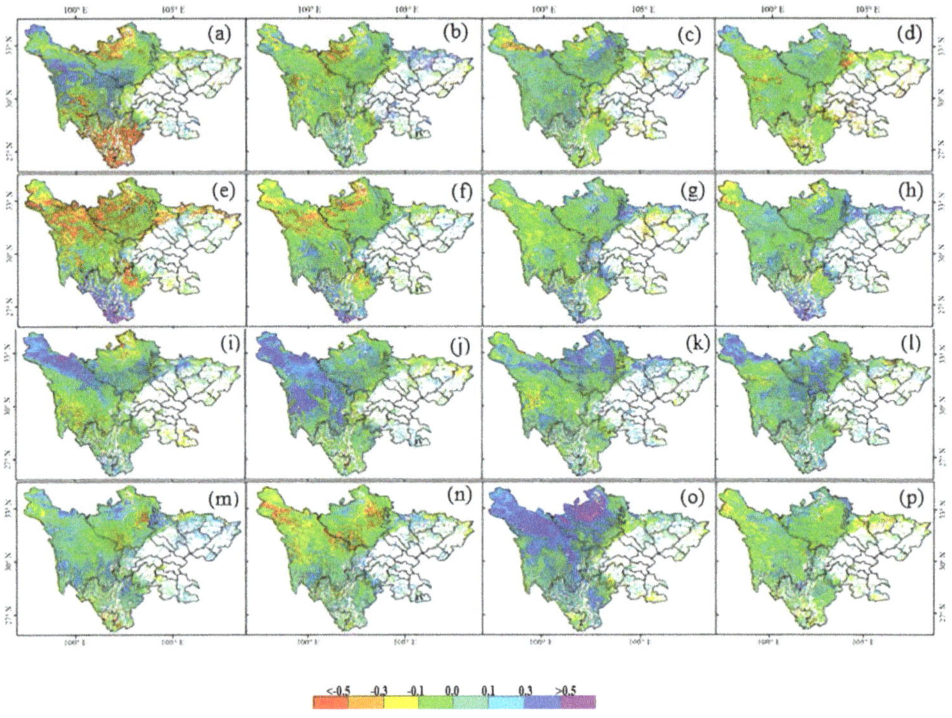

Figure 8. Partial correlation analysis between SOS and EOS and precipitation and temperature over the four seasons with significance at the $p = 0.05$ level. Partial correlation between SOS and precipitation in (**a**) spring, (**b**) summer, (**c**) autumn, (**d**) winter; partial correlation between SOS and temperature in (**e**) spring, (**f**) summer, (**g**) autumn, (**h**) winter; partial correlation between EOS and precipitation in (**i**) spring, (**j**) summer, (**k**) autumn, (**l**) winter; partial correlation between EOS and temperature in (**m**) spring, (**n**) summer, (**o**) autumn, (**p**) winter.

There was a positive correlation between vegetation SOS and precipitation in four seasons (spring, summer, autumn, and winter). SOS had a significant negative correlation with temperature in spring and summer and a positive correlation with temperature in autumn and winter. EOS was negatively correlated with temperature in summer and winter and positively correlated with precipitation in summer, autumn, and winter.

4.4. Limitations

This study examines how grassland vegetation was affected by extreme climatic factors in Sichuan Province. However, grassland types are numerous and exhibit considerable vertical zonality, and different grassland types and flowering functional groups vary in sensitivity to changes in environmental factors in Sichuan. Grasslands were more vulnerable to climate change compared to other vegetation types [74,75]. Therefore, studying areas with low climate variability may be valuable for assessing the effects of climate change on grassland vegetation phenology. In addition, the effects of precipitation and temperature accumulation on plant phenophase should be further investigated. Other factors affecting vegetation in drylands are the timing and duration of precipitation [76], the water storage capacity of the soil, and evapotranspiration [77]. In addition to meteorological considerations, grassland phenology is influenced by a variety of environmental factors, including freeze-thaw processes, extreme weather events, atmospheric CO^2 concentrations, and their combined impacts [78–80]. In this study, the future trend of vegetation dynamics is also quantified using the Hurst exponent, although the close correlation between future and past trends of vegetation dynamics was also found in the case study. However, the R/S analysis proposed in this study using the Hurst exponent could not answer the following question: how long do you think the expected trend in vegetation dynamics will last? It was more important to emphasize that time must be considered in predicting future vegetation dynamics. Future research in Sichuan Province will focus on gaining a full understanding of the combined environmental influences on grassland phenology. In addition, further case studies and theoretical research should be conducted on the detailed processes of future vegetation types.

5. Conclusions

The study discussed in this paper analyzes fifteen extreme climate indices in Sichuan Province using daily maximum and minimum temperatures and precipitation data from 38 meteorological stations from 2001 to 2020. The study highlights the spatio-temporal patterns and changing trends of vegetation phenology in the study area using NDVI from MOD13A1 data, and addresses the effects of climate events and trends on vegetation phenology. The results showed:

1. The SOS vegetation in Sichuan Province gradually evolves with the geography of the slopes from west to east and from the surrounding mountains to the core basin. Climatic changes are strongly dependent on elevation, with most vegetation in the study area showing a trend of advancing SOS. Phenological EOS values of the different vegetation types differed and showed a trend towards delayed EOS. The development of grassland vegetation SOS was gradually delayed from mid to high altitudes, while it advanced at EOS. The tendency towards delayed SOS could be caused by insufficient cooling due to the warming climate in late autumn and winter.
2. The future viability of grassland based on Hurst exponents showed that the changing trend of SOS and the changing trend between 2001 and 2020 were generally weak and opposite; i.e., grassland SOS could be delayed, which means that future trends are more random and have no clear direction. Pixels with a Hurst exponent greater than 0.5 in the Sichuan Basin indicate that the sustainability of vegetation EOS in the future will continue the same trend as the average level of change over the past 20 years, and there will be a shifting trend. To avoid the risk of ecological degradation, we need to take a series of measures to protect the plateau ecosystem: environmental protection projects should be further promoted, natural ecological reserves should be established, local people's awareness of environmental protection and nature conservation should be raised, afforestation and artificial grassland construction should take place, and excessive development should be avoided. In addition, reasonable resource planning and allocation will help to better manage environmental security problems in the province.

3. Grassland SOS was mainly influenced by the yearly maximum consecutive five-day precipitation, diurnal temperature, and temperature-vegetation dryness index, while EOS was mainly influenced by yearly minimum daily temperature, yearly mean temperature, and temperature-vegetation dryness index. Vegetation phenology in Sichuan province showed heterogeneous regional differentiation in response to different climate indices, with positive sensitivity coefficients between SOS and the climate indices compared to negative sensitivity coefficients between EOS and the climate indices. This suggests that extreme climate change leading to an advance in vegetation SOS may indirectly leads to an advance in vegetation EOS. SOS and EOS were strongly influenced by extreme climate and drought, implying that high temperatures lead to evaporation of moisture from the soil, resulting in a lack of rainfall leading to drought. SOS correlated positively with the yearly maximum consecutive five-day precipitation and temperature-vegetation dryness index, implying that precipitation and drought had a strong influence on vegetation growth.

Author Contributions: Writing—review and editing, B.A. and G.Q.; supervision, C.L. and J.W. All authors have read and agreed to the published version of the manuscript.

Funding: This work was supported by the National Natural Science Foundation of China (Grant No. 31760693).

Institutional Review Board Statement: Not applicable.

Informed Consent Statement: Not applicable.

Data Availability Statement: Not applicable.

Conflicts of Interest: The authors declare no conflict of interest.

References

1. Lieth, H. Purposes of a phenology book. In *Phenology and Seasonality Modeling*; Springer: Berlin/Heidelberg, Germany, 1974; pp. 3–19.
2. Schwartz, M.D. *Phenology: An Integrative Environmental Science*; Springer: Berlin/Heidelberg, Germany, 2003.
3. Lee, H. Intergovernmental Panel on Climate Change. 2007. Available online: https://www.ipcc.ch/site/assets/uploads/2018/03/ar4_wg2_full_report.pdf (accessed on 2 December 2021).
4. Qin, D. Climate change science and sustainable development. *Prog. Geogr.* **2014**, *33*, 874–883.
5. IPCC. *The Physical Science Basis, Summary for Policymakers, Contribution of WGI to the Fifth Assessment Report of the Intergovernmental Panel on Climate Change, 2013*; Cambridge University Press: Cambridge, UK, 24 March 2014.
6. Pei, T.; Ji, Z.; Chen, Y.; Wu, H.; Hou, Q.; Qin, G.; Xie, B. The Sensitivity of Vegetation Phenology to Extreme Climate Indices in the Loess Plateau, China. *Sustainability* **2021**, *13*, 7623. [CrossRef]
7. Huang, H.; Zhang, B.; Huang, T.; Wang, H.-j.; Ma, B.; Wang, X.-d.; Cui, Y.-q. Quantifying and predicting spatial and temporal variations in extreme temperatures since 1990 in Gansu Province, China. *Arid Land Geogr.* **2020**, *43*, 319–328.
8. Siegmund, J.F.; Wiedermann, M.; Donges, J.F.; Donner, R.V. Impact of temperature and precipitation extremes on the flowering dates of four German wildlife shrub species. *Biogeosciences* **2016**, *13*, 5541–5555. [CrossRef]
9. Duan, H.; Xue, X.; Wang, T.; Kang, W.; Liao, J.; Liu, S. Spatial and Temporal Differences in Alpine Meadow, Alpine Steppe and All Vegetation of the Qinghai-Tibetan Plateau and Their Responses to Climate Change. *Remote Sens.* **2021**, *13*, 669. [CrossRef]
10. Yang, B.; He, M.; Shishov, V.; Tychkov, I.; Vaganov, E.; Rossi, S.; Ljungqvist, F.C.; Bräuning, A.; Grießinger, J. New perspective on spring vegetation phenology and global climate change based on Tibetan Plateau tree-ring data. *Proc. Natl. Acad. Sci. USA* **2017**, *114*, 6966–6971. [CrossRef] [PubMed]
11. Shen, M.; Piao, S.; Dorji, T.; Liu, Q.; Cong, N.; Chen, X.; An, S.; Wang, S.; Wang, T.; Zhang, G. Plant phenological responses to climate change on the Tibetan Plateau: Research status and challenges. *Natl. Sci. Rev.* **2015**, *2*, 454–467. [CrossRef]
12. Richardson, A.D.; Andy Black, T.; Ciais, P.; Delbart, N.; Friedl, M.A.; Gobron, N.; Hollinger, D.Y.; Kutsch, W.L.; Longdoz, B.; Luyssaert, S. Influence of spring and autumn phenological transitions on forest ecosystem productivity. *Philos. Trans. R. Soc. B Biol. Sci.* **2010**, *365*, 3227–3246. [CrossRef] [PubMed]
13. Barichivich, J.; Briffa, K.; Osborn, T.; Melvin, T.; Caesar, J. Thermal growing season and timing of biospheric carbon uptake across the Northern Hemisphere. *Glob. Biogeochem. Cycles* **2012**, *26*. [CrossRef]
14. Keenan, T.F.; Gray, J.; Friedl, M.A.; Toomey, M.; Bohrer, G.; Hollinger, D.Y.; Munger, J.W.; O'Keefe, J.; Schmid, H.P.; Wing, I.S. Net carbon uptake has increased through warming-induced changes in temperate forest phenology. *Nat. Clim. Chang.* **2014**, *4*, 598–604. [CrossRef]

15. Piao, S.; Tan, J.; Chen, A.; Fu, Y.H.; Ciais, P.; Liu, Q.; Janssens, I.A.; Vicca, S.; Zeng, Z.; Jeong, S.-J. Leaf onset in the northern hemisphere triggered by daytime temperature. *Nat. Commun.* **2015**, *6*, 6911. [CrossRef]
16. Nagy, L.; Kreyling, J.; Gellesch, E.; Beierkuhnlein, C.; Jentsch, A. Recurring weather extremes alter the flowering phenology of two common temperate shrubs. *Int. J. Biometeorol.* **2013**, *57*, 579–588. [CrossRef] [PubMed]
17. Bokhorst, S.; Tømmervik, H.; Callaghan, T.V.; Phoenix, G.K.; Bjerke, J.W. Vegetation recovery following extreme winter warming events in the sub-Arctic estimated using NDVI from remote sensing and handheld passive proximal sensors. *Environ. Exp. Bot.* **2012**, *81*, 18–25. [CrossRef]
18. Xie, Y.; Wang, X.; Silander, J.A. Deciduous forest responses to temperature, precipitation, and drought imply complex climate change impacts. *Proc. Natl. Acad. Sci. USA* **2015**, *112*, 13585–13590. [CrossRef]
19. Ying, H.; Zhang, H.; Zhao, J.; Shan, Y.; Zhang, Z.; Guo, X.; Rihan, W.; Deng, G. Effects of spring and summer extreme climate events on the autumn phenology of different vegetation types of Inner Mongolia, China, from 1982 to 2015. *Ecol. Indic.* **2020**, *111*, 105974. [CrossRef]
20. Ciais, P.; Reichstein, M.; Viovy, N.; Granier, A.; Ogée, J.; Allard, V.; Aubinet, M.; Buchmann, N.; Bernhofer, C.; Carrara, A. Europe-wide reduction in primary productivity caused by the heat and drought in 2003. *Nature* **2005**, *437*, 529–533. [CrossRef]
21. Cleland, E.E.; Chuine, I.; Menzel, A.; Mooney, H.A.; Schwartz, M.D. Shifting plant phenology in response to global change. *Trends Ecol. Evol.* **2007**, *22*, 357–365. [CrossRef]
22. Haoming, X.; Ainong, L.; Wei, Z.; Jin, H.; Lei, G.; Bian, J.; Tan, J. Spatio-temporal variation and driving forces in alpine grassland phenology in the Zoigê plateau from 2001–2013. In Proceedings of the 2014 International Symposium on Geoscience and Remote Sensing (IGARSS), Quebec City, QC, Canada, 13–18 July 2014.
23. Luo, X.; Yang, W.; Liu, L.; Zhang, Y. Spatial-Temporal variations of vegetation and the relationship with precipitation in the summer—A case study in the hilly area of central Sichuan province. In Proceedings of the 2018 3rd International Conference on Advances in Energy and Environment Research (ICAEER 2018), Guilin, China, 10–12 August 2018; p. 03060.
24. Huang, J.; Sun, S.; Xue, Y.; Li, J.; Zhang, J. Spatial and temporal variability of precipitation and dryness/wetness during 1961–2008 in Sichuan province, west China. *Water Resour. Manag.* **2014**, *28*, 1655–1670. [CrossRef]
25. Wang, S.; Jiao, S.; Xin, H. Spatio-temporal characteristics of temperature and precipitation in Sichuan Province, Southwestern China, 1960–2009. *Quat. Int.* **2013**, *286*, 103–115. [CrossRef]
26. Li, Z.; He, Y.; Wang, C.; Wang, X.; Xin, H.; Zhang, W.; Cao, W. Spatial and temporal trends of temperature and precipitation during 1960–2008 at the Hengduan Mountains, China. *Quat. Int.* **2011**, *236*, 127–142. [CrossRef]
27. Li, J.; Feng, Z.; Kang, J. Glacial deposits and environment in the Hengduan Mountains. In *Glaciers in the Hengduan Mountains*; Li, J., Su, Z., Eds.; Science Press: Beijing, China, 1996; pp. 157–173.
28. Xu, Y.; Gao, X.; Shen, Y.; Xu, C.; Shi, Y.; Giorgi, A. A daily temperature dataset over China and its application invalidating an RCM simulation. *Adv. Atmos. Sci.* **2009**, *26*, 763–772. [CrossRef]
29. Dong, Q.; Zhan, C.; Wang, H.; Wang, F.; Zhu, M. A review on evapotranspiration data assimilation based on hydrological models. *J. Geogr. Sci.* **2016**, *26*, 230–242. [CrossRef]
30. Kong, J.; Hu, Y.; Yang, L.; Shan, Z.; Wang, Y. Estimation of evapotranspiration for the blown-sand region in the Ordos basin based on the SEBAL model. *Int. J. Remote Sens.* **2019**, *40*, 1945–1965. [CrossRef]
31. Hou, X.; Gao, S.; Niu, Z.; Xu, Z. Extracting grassland vegetation phenology in North China based on cumulative SPOT-VEGETATION NDVI data. *Int. J. Remote Sens.* **2014**, *35*, 3316–3330. [CrossRef]
32. He, Z.; Du, J.; Chen, L.; Zhu, X.; Lin, P.; Zhao, M.; Fang, S. Impacts of recent climate extremes on spring phenology in arid-mountain ecosystems in China. *Agric. For. Meteorol.* **2018**, *260*, 31–40. [CrossRef]
33. Zhao, A.; Yu, Q.; Feng, L.; Zhang, A.; Pei, T. Evaluating the cumulative and time-lag effects of drought on grassland vegetation: A case study in the Chinese Loess Plateau. *J. Environ. Manag.* **2020**, *261*, 110214. [CrossRef] [PubMed]
34. Kong, D.; Zhang, Q.; Huang, W.; Gu, X. Vegetation phenology change in Tibetan Plateau from 1982 to 2013 and its related meteorological factors. *Acta Oceanol. Sin.* **2017**, *72*, 39–52.
35. Zhang, X.; Yang, F. *RClimDex (1.0) User Manual*; Climate Research Branch Environment Canada: Gatineau, QC, Canada, 10 September 2004; p. 22.
36. Wang, H.; Pan, Y.; Li, S.; Chen, Z.; Zhao, Z.; Mi, H. Risk Assessment and Zonation of Meteorological Disasters Based on Rasterization in Jiangsu Province. *J. Liaocheng Univ. Nat. Sci. Ed.* **2019**, *32*, 99–110.
37. Ranzi, R.; Caronna, P.; Tomirotti, M. Impact of climatic and land-use changes on river flows in the Southern Alps. In *Sustainable Water Resources Planning and Management under Climate Change*; Springer: Berlin/Heidelberg, Germany, 2017; pp. 61–83.
38. Wills, A.; Mills, A.; Ninness, B. A matlab software environment for system identification. *IFAC Proc. Vol.* **2009**, *42*, 741–746. [CrossRef]
39. Hutchinson, M.F.; Xu, T. *Anusplin Version 4.2 User Guide*; Centre for Resource and Environmental Studies, The Australian National University: Canberra, Australia, 2004; p. 54.
40. Chen, J.; Jönsson, P.; Tamura, M.; Gu, Z.; Matsushita, B.; Eklundh, L. A simple method for reconstructing a high-quality NDVI time-series data set based on the Savitzky–Golay filter. *Remote Sens. Environ.* **2004**, *91*, 332–344. [CrossRef]
41. Piao, S.; Fang, J.; Zhou, L.; Ciais, P.; Zhu, B. Variations in satellite-derived phenology in China's temperate vegetation. *Glob. Chang. Biol.* **2006**, *12*, 672–685. [CrossRef]

42. Zhou, L.; Tucker, C.J.; Kaufmann, R.K.; Slayback, D.; Shabanov, N.V.; Myneni, R.B. Variations in northern vegetation activity inferred from satellite data of vegetation index during 1981 to 1999. *J. Geophys. Res. Atmos.* **2001**, *106*, 20069–20083. [CrossRef]
43. Yu, H.; Luedeling, E.; Xu, J. Winter and spring warming result in delayed spring phenology on the Tibetan Plateau. *Proc. Natl. Acad. Sci. USA* **2010**, *107*, 22151–22156. [CrossRef]
44. Kafaki, S.B.; Mataji, A.; Hashemi, S.A. Monitoring growing season length of deciduous broadleaf forest derived from satellite data in Iran. *Am. J. Environ. Sci.* **2009**, *5*, 647–652.
45. White, M.A.; Thornton, P.E.; Running, S.W. A continental phenology model for monitoring vegetation responses to interannual climatic variability. *Glob. Biogeochem. Cycles* **1997**, *11*, 217–234. [CrossRef]
46. Hou, X.-H.; Niu, Z.; Gao, S. Phenology of forest vegetation in the northeast of China in ten years using remote sensing. *Spectrosc. Spectr. Anal.* **2014**, *34*, 515–519.
47. Klosterman, S.; Hufkens, K.; Gray, J.; Melaas, E.; Sonnentag, O.; Lavine, I.; Mitchell, L.; Norman, R.; Friedl, M.; Richardson, A. Evaluating remote sensing of deciduous forest phenology at multiple spatial scales using PhenoCam imagery. *Biogeosciences* **2014**, *11*, 4305–4320. [CrossRef]
48. Zhang, X.; Friedl, M.A.; Schaaf, C.B.; Strahler, A.H.; Hodges, J.C.; Gao, F.; Reed, B.C.; Huete, A. Monitoring vegetation phenology using MODIS. *Remote Sens. Environ.* **2003**, *84*, 471–475. [CrossRef]
49. Sicard, P.; Mangin, A.; Hebel, P.; Malléa, P. Detection and estimation trends linked to air quality and mortality on French Riviera over the 1990–2005 period. *Sci. Total Environ.* **2010**, *408*, 1943–1950. [CrossRef]
50. Zhang, A.; Zheng, C.; Wang, S.; Yao, Y. Analysis of streamflow variations in the Heihe River Basin, northwest China: Trends, abrupt changes, driving factors, and ecological influences. *J. Hydrol. Reg. Stud.* **2015**, *3*, 106–124. [CrossRef]
51. Zhang, B.; He, C.; Burnham, M.; Zhang, L. Evaluating the coupling effects of climate aridity and vegetation restoration on soil erosion over the Loess Plateau in China. *Sci. Total Environ.* **2016**, *539*, 436–449. [CrossRef] [PubMed]
52. Bayazit, M.; Önöz, B. To prewhiten or not to prewhiten in trend analysis? *Hydrol. Sci. J.* **2007**, *52*, 611–624. [CrossRef]
53. Von Storch, H. Misuses of statistical analysis in climate research. In *Analysis of Climate Variability*; Springer: Berlin/Heidelberg, Germany, 1999; pp. 11–26.
54. Yue, S.; Wang, C.Y. Applicability of prewhitening to eliminate the influence of serial correlation on the Mann-Kendall test. *Water Resour. Res.* **2002**, *38*, 4-1–4-7. [CrossRef]
55. Carlson, T.N.; Gillies, R.R.; Perry, E.M. A method to make use of thermal infrared temperature and NDVI measurements to infer surface soil water content and fractional vegetation cover. *Remote Sens. Rev.* **1994**, *9*, 161–173. [CrossRef]
56. Xu, K.; Yang, D.; Yang, H.; Li, Z.; Qin, Y.; Shen, Y. Spatio-temporal variation of drought in China during 1961–2012: A climatic perspective. *J. Hydrol.* **2015**, *526*, 253–264. [CrossRef]
57. Sandholt, I.; Rasmussen, K.; Andersen, J. A simple interpretation of the surface temperature/vegetation index space for assessment of surface moisture status. *Remote Sens. Environ.* **2002**, *79*, 213–224. [CrossRef]
58. Hurst, H.E. Long-term storage capacity of reservoirs. *Trans. Am. Soc. Civ. Eng.* **1951**, *116*, 770–799. [CrossRef]
59. Mandelbrot, B.B.; Wallis, J.R. Robustness of the rescaled range R/S in the measurement of noncyclic long-run statistical dependence. *Water Resour. Res.* **1969**, *5*, 967–988. [CrossRef]
60. Mandelbrot, B.B.; Wallis, J.R. Some long-run properties of geophysical records. *Water Resour. Res.* **1969**, *5*, 321–340. [CrossRef]
61. Walker, E.; Birch, J.B. Influence measures in ridge regression. *Technometrics* **1988**, *30*, 221–227. [CrossRef]
62. Tanre, D.; Holben, B.N.; Kaufman, Y.J. Atmospheric correction algorithm for NOAA-AVHRR products: Theory and application. *Inst. Electr. Electron. Eng. Trans. Geosci. Remote Sens.* **1992**, *30*, 231–248. [CrossRef]
63. Nagol, J.R.; Vermote, E.F.; Prince, S.D. Effects of an atmospheric variation on AVHRR NDVI data. *Remote Sens. Environ.* **2009**, *113*, 392–397. [CrossRef]
64. Zhang, G.; Zhang, Y.; Dong, J.; Xiao, X. Green-up dates in the Tibetan Plateau have continuously advanced from 1982 to 2011. *Proc. Natl. Acad. Sci. USA* **2013**, *110*, 4309–4314. [CrossRef]
65. Che, M.; Chen, B.; Innes, J.L.; Wang, G.; Dou, X.; Zhou, T.; Zhang, H.; Yan, J.; Xu, G.; Zhao, H. Spatial and temporal variations in the end date of the vegetation growing season throughout the Qinghai–Tibetan Plateau from 1982 to 2011. *Agric. For. Meteorol.* **2014**, *189*, 81–90. [CrossRef]
66. Cong, N.; Shen, M.; Yang, W.; Yang, Z.; Zhang, G.; Piao, S. Varying responses of vegetation activity to climate changes on the Tibetan Plateau grassland. *Int. J. Biometeorol.* **2017**, *61*, 1433–1444. [CrossRef]
67. Slayback, D.A.; Pinzon, J.E.; Los, S.O.; Tucker, C.J. Northern hemisphere photosynthetic trends 1982–99. *Glob. Chang. Biol.* **2003**, *9*, 1–15. [CrossRef]
68. Liu, Q.; Fu, Y.H.; Zeng, Z.; Huang, M.; Li, X.; Piao, S. Temperature, precipitation, and insolation effects on autumn vegetation phenology in temperate China. *Glob. Chang. Biol.* **2016**, *22*, 644–655. [CrossRef] [PubMed]
69. Lupascu, M.; Welker, J.; Seibt, U.; Xu, X.; Velicogna, I.; Lindsey, D.; Czimczik, C. The amount and timing of precipitation control the magnitude, seasonality, and sources (14 C) of ecosystem respiration in a polar semi-desert, northwestern Greenland. *Biogeosciences* **2014**, *11*, 4289–4304. [CrossRef]
70. Butt, N.; Seabrook, L.; Maron, M.; Law, B.S.; Dawson, T.P.; Syktus, J.; McAlpine, C.A. Cascading effects of climate extremes on vertebrate fauna through changes to low-latitude tree flowering and fruiting phenology. *Glob. Chang. Biol.* **2015**, *21*, 3267–3277. [CrossRef] [PubMed]

71. Crabbe, R.A.; Dash, J.; Rodriguez-Galiano, V.F.; Janous, D.; Pavelka, M.; Marek, M.V. Extreme warm temperatures alter forest phenology and productivity in Europe. *Sci. Total Environ.* **2016**, *563*, 486–495. [CrossRef]
72. Xie, B.; Qin, Z.; Wang, Y.; Chang, Q. Monitoring vegetation phenology and their response to climate change on Chinese Loess Plateau based on remote sensing. *Trans. Chin. Soc. Agric. Eng.* **2015**, *31*, 153–160.
73. Peñuelas, J.; Rutishauser, T.; Filella, I. Phenology feedbacks on climate change. *Science* **2009**, *324*, 887–888. [CrossRef] [PubMed]
74. Liu, H.; Tian, F.; Hu, H.C.; Hu, H.P.; Sivapalan, M. Soil moisture controls on patterns of grass green-up in Inner Mongolia: An index-based approach. *Hydrol. Earth Syst. Sci.* **2013**, *17*, 805–815. [CrossRef]
75. Wang, H. The variability of vegetation growing season in northern China based on NOAA NDVI and MSAVI from 1982 to 1999. *Acta Ecol. Sin.* **2007**, *27*, 504–515.
76. Cong, N.; Wang, T.; Nan, H.; Ma, Y.; Wang, X.; Myneni, R.B.; Piao, S. Changes in satellite-derived spring vegetation green-update and its linkage to the climate in China from 1982 to 2010: A multimethod analysis. *Glob. Chang. Biol.* **2013**, *19*, 881–891. [CrossRef] [PubMed]
77. Jin, H.; He, R.; Cheng, G.; Wu, Q.; Wang, S.; Lü, L.; Chang, X. Changes in frozen ground in the Source Area of the Yellow River on the Qinghai–Tibet Plateau, China, and their eco-environmental impacts. *Environ. Res. Lett.* **2009**, *4*, 045206. [CrossRef]
78. Chen, H.; Zhu, Q.; Wu, N.; Wang, Y.; Peng, C.-H. Delayed spring phenology on the Tibetan Plateau may also be attributable to other factors than winter and spring warming. *Proc. Natl. Acad. Sci. USA* **2011**, *108*, E93. [CrossRef]
79. Reyes-Fox, M.; Steltzer, H.; Trlica, M.; McMaster, G.S.; Andales, A.A.; LeCain, D.R.; Morgan, J.A. Elevated CO_2 further lengthens growing season under warming conditions. *Nature* **2014**, *510*, 259–262. [CrossRef]
80. Yang, Y.; Guan, H.; Shen, M.; Liang, W.; Jiang, L. Changes in autumn vegetation dormancy onset date and the climate controls across temperate ecosystems in China from 1982 to 2010. *Glob. Chang. Biol.* **2015**, *21*, 652–665. [CrossRef] [PubMed]

Article

Trends of Rainfall Onset, Cessation, and Length of Growing Season in Northern Ghana: Comparing the Rain Gauge, Satellite, and Farmer's Perceptions

Winifred Ayinpogbilla Atiah [1,*,†], Francis K. Muthoni [2,†], Bekele Kotu [3], Fred Kizito [3,4] and Leonard K. Amekudzi [1]

1 Meteorology and Climate Science Unit, Department of Physics, Kwame Nkrumah University of Science and Technology (KNUST), UPO-PMB, Kumasi AK-039-5028, Ghana; leonard.amekudzi@gmail.com
2 International Institute of Tropical Agriculture (IITA), Duluti, Arusha P.O. Box 10, Tanzania; f.muthoni@cgiar.org
3 International Institute of Tropical Agriculture, Tamale NT-0000, Ghana; B.Kotu@cgiar.org (B.K.); f.kizito@cgiar.org (F.K.)
4 International Institute of Tropical Agriculture, Accra GA-184, Ghana
* Correspondence: winifred.a.atiah@aims-senegal.org
† These authors contributed equally to this work.

Citation: Atiah, W.A.; Muthoni, F.K.; Kotu, B.; Kizito, F.; Amekudzi, L.K. Trends of Rainfall Onset, Cessation, and Length of Growing Season in Northern Ghana: Comparing the Rain Gauge, Satellite, and Farmer's Perceptions. *Atmosphere* **2021**, *12*, 1674. https://doi.org/10.3390/atmos12121674

Academic Editors: Baojie He, Ayyoob Sharifi, Chi Feng and Jun Yang

Received: 17 October 2021
Accepted: 2 December 2021
Published: 13 December 2021

Abstract: Rainfall onset and cessation date greatly influence cropping calendar decisions in rain-fed agricultural systems. This paper examined trends of onsets, cessation, and the length of growing season over Northern Ghana using CHIRPS-v2, gauge, and farmers' perceptions data between 1981 and 2019. Results from CHIRPS-v2 revealed that the three seasonal rainfall indices have substantial latitudinal variability. Significant late and early onsets were observed at the West and East of 1.5° W longitude, respectively. Significant late cessations and longer growing periods occurred across Northern Ghana. The ability of farmers' perceptions and CHIRPS-v2 to capture rainfall onsets are time and location-dependent. A total of 71% of farmers rely on traditional knowledge to forecast rainfall onsets. Adaptation measures applied were not always consistent with the rainfall seasonality. More investment in modern climate information services is required to complement the existing local knowledge of forecasting rainfall seasonality.

Keywords: CHIRPS-v2; climate change adaptation; farmer perceptions; rainfall cessation; rainfall onset

1. Introduction

Global ecosystems are experiencing changes in rainfall amount, intensity, and temporal distributions [1] that poses a great challenge to agricultural production [2–4]. Trends of rainfall amount are widely documented [5,6] but shifts in rainfall seasonality (onset and cessation) have received less attention although it determines timing of cropping calendar activities. The temporal variability of rainfall is a critical determinant of timing of cropping calendar in rain-fed farming systems and is defined by three indices i.e., the onset, cessation dates, and the length of the season. The onset and cessation dates are the day of the year when rainfall starts and ends. The difference between the onset and cessation dates is the length of the growing season. There are several methods for determining the onset and cessation dates of rainfall that are applicable at different agro-ecological zones and intended uses [7–10]. Ref. [10] developed a new method that is generally applicable for estimating the onset and cessation dates across Africa continent though the study applied low resolution gridded data (approximately 110 Km). Many studies determine rainfall onsets and cessation from gauge data, e.g., [11,12] in northern Ghana. Recent advances focus on the application of gridded satellite rainfall estimates for spatial determination of onsets and cessations [10,13]. Ref. [10] mapped the timing of onset and cessation of rainfall over Africa using gridded data at 110 Km resolution and reported inconsistent deviations over West Africa when ERA-Interim and ARC-v2 data was applied. Similarly,

Ref. [14] evaluated the skill of onset forecasts for West Africa. Again, Ref. [13] demonstrated that daily climatology from CHIRPS-v2 data could forecast rainfall onset and cessation of rainfall with a bias of less than 7 days in large parts of Eastern Africa.

Agricultural production in the West Africa region is predominantly rain-fed. The region is also highly vulnerable to climate change and variability due to rampant poverty, rapidly increasing population, and low levels of technological development [15]. The increasing variability of rainfall amount and seasonality destabilizes the fragile ecosystem and threatens food production and livelihoods. Likewise, the variability of rainfall amount in northern Ghana has attracted more attention compared to seasonal distribution [5,6,16]. Moreover, there is no consensus in the literature on the three rainfall seasonality indices i.e., the onset, cessation, and length of growing period (LGP) in northern Ghana, although these are crucial determinants for cropping season activities. The LGP in northern Ghana is highly variable due to inconsistencies of rainfall onset and cessation dates Refs. [11,17]. For instance, Ref. [12] reported early-onset dates of the rain season ranging between −0.3 to −0.5 days/year (earlier onset of 7.5–12.5 days) between 1986–2010 in Tamale and Wa in northern Ghana. However, Ref. [18] reported a significant delay of up to 0.88 days/year in the Volta basin located in northern Ghana, which translates to late onset by 35 days over 40 years. Similarly, Ref. [11] reported a significant variability of onset and cessation of rainfall over a band of 2–8 years over different agro-ecological zones in Ghana. Recent studies using gauge observations have demonstrated that late onset or early cessation of rains and a high frequency of dry spells within the growing season in northern Ghana cause a significant decline in crop yields [19,20]. The variability of onset and mid-season breaks in the rain makes the agricultural calendar unpredictable and complicates decisions on sowing time, crop choice, and variety [17,21]. Erratic and delayed rainfall onset pose a severe challenge to food production and food security [11]. Lack of synergy between rainfall onsets and agronomic decisions such as sowing date is one of the main factors causing low maize production in Northern Ghana [20]. False starts of rains [22] have become more regular in northern Ghana and induce farmers to plough and plant without sufficient follow-up wet days to sustain the growth of crops [23]. Accurate prediction of the onset, cessation, and length of rainy days dates is essential for synchronized timing of cropping calendar activities. Precise information on the onset and cessation of seasonal rain can reduce the risks and costs of re-sowing seeds due to the season's false onset [24]. The late onset of rains provides the first outlook of a rainy season and is a reliable early warning of food insecurity several months before harvesting [25]. A ten-day delay of onset of the rainy season makes drought conditions more likely. Early identification and warning of drought conditions can inform preparedness of interventions to save lives and livelihoods.

Inadequate weather observation networks hamper timely and accurate rainfall forecasts that can guide farmers on cropping calendar decisions [26]. The observation gauge network in Ghana is characterized by low density, skewed distribution, short-term records, and significant data gaps [6,26,27]. The information generated from a few gauge stations with long-term data is applied to generate agro-advisories over a large area beyond the (>50 Km^2) recommended by World Meteorological Organization (WMO). Most gauge networks are in the main urban centers, resulting in inadequate coverage in rural areas where agricultural activities take place [26]. Satellite-derived rainfall can complement the sparse gauge network to produce spatially explicit layers representing the onset, cessation, and length of the rain season [10,13,28]. In this manner, the big data generated from remote sensing platforms is applied to identify hotspots where agricultural production experiences a high risk of climate change and variability. Identifying locations that are more vulnerable to shifts in seasonal calendars associated with climate change and variability is essential to guide the evidence-based targeting of appropriate climate-smart technologies [17].

Farmer perceptions on changing rainfall patterns determine annual cropping decisions. If farmers' perceptions agree with the trends recorded by the observation network, it means more awareness of prevalent trends and a higher likelihood of applying appropriate adaptive measures [28,29]. Proper knowledge of local trends of rainfall seasonality is required to guide the implementation of appropriate adaptation measures that reduce the negative effects of climate change. Ref. [30] showed that adaptation measures implemented without considering local climate reality led to maladaptive outcomes that exacerbated the vulnerability to climate change and variability in northern Ghana. Integrating knowledge from meteorological observations with local perceptions is essential for developing locally relevant and sustainable adaptation strategies. Several studies reported agreement between farmer perceptions of rainfall trends with observation data [29,31]. However, other studies reported deviation between farmers' perceptions of climate change compared to observation networks [32,33]. Therefore, evaluating the seasonal trends of rainfall from farmers' perception, observation network, and satellite time series can reduce uncertainties on climatic trends.

This study uses daily rainfall data from satellites to map the spatial variations of the onset, cessation, and length of the rain season in Ghana for 39 years (1981–2019). We validated the three indices generated from the satellite with rain gauge data. We examined the long-term trends of the three indices for almost four decades. Moreover, survey data are used to explore the farmer's perceptions of changes on the three seasonal indices and their coping strategies. The specific objectives of this study are to; (1) generate maps of the onset, cessation, and LGP from daily rainfall from satellite and gauge networks over 39 years in northern Ghana, (2) map the variability and trends of the onset, cessation, and LGP over 39 years, (3) validate the three indices with gauge networks, (4) examine the methods farmers apply to forecast the onset of rainy seasons and (5) examine the crop management practices applied by farmers as adaptation to the observed trends of the three seasonal indices in northern Ghana.

2. Study Area and Data

2.1. Study Area

The study area covers three administrative regions in northern Ghana, namely the Upper East (UER), Upper West (UWR), and Northern (NR) regions (Figure 1). Ghana lies in the tropics and is characterized by a tropical monsoonal climate system with two dominant seasons (wet and dry) [11]. Rainfall is controlled by the West African Monsoon (WAM) and convective activities due to the movements of the Inter-tropical Discontinuity (ITD) [11]. The ITD oscillates from South to North and retreats to the South annually. The three regions in Northern Ghana experience a unimodal rainfall pattern between May and September ranging between 500 mm and 1200 mm [5,34]. The LGP ranges from 140 to 240 days and increases in a North to South and East to West gradient [11].

2.2. Data Source

The Climate Hazards Center InfraRed Precipitation with Station data (CHIRPS-v2) has four decades of quasi-global rainfall data set [35]. Data records start from 1981 to near-present with an area coverage between 50° S–50° N. CHIRPS-v2 incorporates 0.05o resolution satellite imagery with in-situ station data to create a gridded rainfall time series for trend analysis and seasonal drought monitoring [35]. CHIRPS-v2 uses the Tropical Rainfall Measuring Mission Multi-Satellite Precipitation Analysis version 7 (TMPA-v7) to calibrate global Cold Cloud Duration (CCD) rainfall estimates. Validation studies over the region have shown that the CHIRPS-v2 dataset correlated well with gauge observations, especially on monthly to seasonal scales [6,36,37]. Additionally, CHIRPS V2 have exhibited a good skill at representing these rainfall indices as well as rainfall extremes compared to other globally available data sets such as, GPCC, TRMM 3B42 and CMORPH. This is because CHIRPS blends thermal infrared and passive microwave tend to perform better than IR-only or PM-only products [36,37]. For this analysis, the daily rainfall of CHIRPS-v2 at a spatial resolution of $0.05° \times 0.05°$ was used. Rain gauge data for six available stations located in UER (Garu and Zuarungu), UWR (Wa and Babile), NR (Bole and Tamale) from 1981 through 2016 were obtained from Ghana Meteorological Agency (GMet). A survey was conducted in December 2020 to elicit farmers' perception of the trends of the three rainfall indices, their impacts on cropping calendar activities, and the adaptation measures implemented to adapt to the changes (Table 1). A total of 400 farmers were interviewed in the UER (94), UWR (146), and NR (160). The average age of respondents was 50 years. The farmers were selected using a stratified random sample from a list of members involved in the ongoing Africa research in sustainable intensification for the next generation (Africa RISING; https://africa-rising.net/, accessed on 20th August 2021) program. Farmer responses were conducted using a structured interview and recorded with tablets using the KoboCollect toolbox. Farmers reported the onset dates in weekly intervals (the week of the specific month) because they could not recall the precise dates.

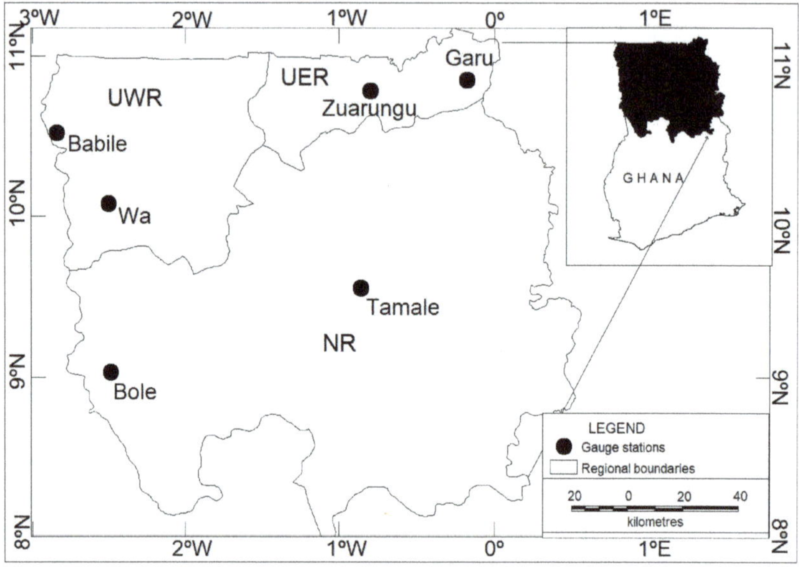

Figure 1. The Location of the three administrative regions and the synoptic gauge stations in Northern Ghana.

Table 1. The Parameters recorded during the survey on farmer's perceptions on rainfall seasonality in northern Ghana.

ID	Parameter	Choices
1	Sex	1 = Male
		2 = Female
2	Age	Age of respondent
3	Educational Level	0 = Informal
		1 = Primary
		2 = Secondary
		3 = Certificate
		4 = Diploma
		5 = Bachelor
		6 = Masters or above
		7 = Others (specify)

Table 1. *Cont.*

ID	Parameter	Choices
4	Main economic activity	1 = Crop Farming
		2 = Livestock Farming
		3 = Fishing
		4 = Formal Employment
		5 = Petty business
		6 = Others (specify)
5	Main Crop	Maize
		Groundnuts
		Cowpea
		Soybean
6	Onset/Cessation trend	Same
		Early
		Late
7	Method of forecasting rainfall onset	Birds/insect movements
		Meteorological agency
		Pattern of clouds
		Change in temperature
		Traditionally first week of May
		Vegetation phenology
		Wind direction
		Other (specify)
8	Importance of onset on yield of the main crop	Somehow important
		Important
		Very important
9	Replant frequency due to false onset of rainfall	Never
		Once
		Twice
		Thrice
		Four times
		Five times
10	Adaptation measures	None
		Drought-tolerant cultivars
		Early maturing cultivars
		Increase seed rate
		Intercropping
		Replanting
		Other (specify)

3. Methodology

The methods presented below aim at computing the three seasonal rainfall indices i.e., onset and cessation of rains and the LGP from gauge, satellite and farmers perceptions. The rainfall onset and cessation dates were determined using the percentage mean cumulative rainfall amount (PMCR; [11]). Daily rainfall fields for all grids over northern Ghana from 1981 to 2019 were extracted from the CHIRPS-v2 gridded rainfall data. The percentage mean annual rainfall for 5-day intervals was calculated for all grids. This was followed by accumulating the percentages of the 5-day periods. When the cumulative percentages are plotted against time through the year, the first point of maximum positive curvature of the graph corresponds to rainfall onset, and the last point of maximum negative curvature corresponds to the rainfall cessation. The points of maximum curvatures corresponding to the onset and cessation of rainfall are respectively 7–8% and over 90% of the annual rainfall (Figure A1). The length of the growing season is then the difference of the onset and cessation dates of rainfall (Equation (1)).

$$LGP = RC - RO + 1 \tag{1}$$

where LGP is the length of the growing period, RC is the rainfall cessation, RO is the rainfall onset. The time series rasters of the three seasonal rainfall indices (onset, cessation, LGP) were applied to test null hypothesis (no trend) from the entire time series data. The trends of the three seasonal rainfall indices were determined using Theil Sen's slope estimator ([38]; Equation (2)). There are several methods for testing the significance of climatic trends such as the Mann–Kendall test [39,40], Spearman's rho test [41,42] and graphical method [43]. Existence of serial autocorrelation and ties in time series of climate data influence the magnitude of variance of the test statistic [44]. We plotted the autocorrelation function (ACF) to check if the time series of the three rainfall indices had serial dependency. Since the dataset did not show significant serial-dependence (Figure A2), the significance of the trends was then tested using Mann–Kendall test [39] at a significant value of 0.05 (Equations (3)–(5)). The ability of the CHIRPS-v2 satellite data to capture the onset and cessation dates in the UER, UWR, and NR was assessed using data from six-gauge stations (Figure 1). A point to grid approach was applied, whereby rainfall for the six-gauge stations was extracted and matched with geolocated satellite data. Where a particular station's data were missing, nearby station's data were used to gap-fill. This is because stations that are close to each other, not greater than 4 Km apart, have been revealed to have similar rainfall patterns. After that, the onsets, cessations, and LGP for both gauge and satellite at these stations were computed and compared. To validate the indices' captured by satellite data in the UER, UWR, and NR, the index's average for the two stations that fell in each region was computed for gauge and satellite and compared.

$$Q = \frac{Y_i' - Y_i}{I' - I} \tag{2}$$

where Q is a Theil Sen's slope estimator. Y_i' are Y_i the values at times i' and i, where i' is greater than i, and n' is all data pairs for which i' is greater than i.

$$\begin{aligned} Z_{MK} = \frac{S-1}{\sqrt{VAR(S)}} \text{ if } S > 0 \\ = 0 \text{ if } S = 0 \\ = \frac{S+1}{\sqrt{VAR(S)}} \text{ if } S < 0 \end{aligned} \tag{3}$$

$$S = \sum_{k-1}^{n-1} \sum_{j-k+1}^{n} sgn(x_j - x_k) \tag{4}$$

and

$$VAR(S) = \frac{1}{18}\left[n(n-1)(2n+5) - \sum_{p-1}^{g} t_p(t_p-1)(2t_p+5)\right] \quad (5)$$

Z_{MK} is the MK test statistic, S is the number of positive differences minus the number of negative differences, and $VAR(S)$ is the variance of S.

The satellite's performance against the gauge station data were assessed using the Pearson correlation coefficient (r), the Root-Mean-Square Error (RMSE), and the bias. The Pearson correlation coefficient (r) measures the linear relationship between the satellite and the gauge estimates and ranges between −1 to +1 (Equation (6)). RMSE is the mean deviation of the estimates from the observations (Equation (7)). Bias estimates the extent to which the satellite under or overestimates the gauge observations (Equation (8)). The Rainfall seasonal indices and their trends were generated using shell script and Python packages.

$$r = \frac{\sum_{i=0}^{n}(G_i - \bar{G})(S_i - \bar{S})}{\sqrt{\sum_{i=0}^{n}(G_i - \bar{G})^2}\sqrt{\sum_{i=0}^{n}(S_i - \bar{S})^2}} \quad (6)$$

$$RMSE = \frac{\sqrt{\frac{1}{n}\sum_{i=0}^{n}(G_i - S_i)}}{\bar{G}_i} \quad (7)$$

$$Bias = \frac{\sum_{i=0}^{n} S_i}{\sum_{i=0}^{n} G_i} \quad (8)$$

where S is satellite (CHIRPs) data, G is the gauge data.

4. Results and Discussion

4.1. Satellite-Derived Seasonal Rainfall Indices (Onsets, Cessations, and Length of Growing Period (LGP)

Figures 1–4 show maps of the three seasonal rainfall indices representing the onset, cessation, and the LGP, respectively, derived from the CHIRPS-v2 data. Only maps for 2000 to 2019 seasons are shown for illustration purposes but a summary for all the years is shown in Figure A3. The rainy season's onset showed a progression along the south-west to northeast direction, with rains starting earlier in the former (Figure 2). Exceptionally early onsets were recorded in northern Ghana in 2013 that can potentially disorient farmers given it rained at the time that farmers usually prepare their fields. Figure 2 shows cases in 2009, 2011, 2013, and 2018 where early rainfall onsets (mid-March) were observed in the NR. In 2007, and 2014 seasons we observed an early onset of rain (21st March to 15th April) covering the entire region. However, the UWR experienced more instances of the early start of rainfall than the UER. Moreover, the NR has shown the earliest onset among the three regions. The transition agro-ecological zone in the NR showed the early start of rainfall (21st March) compared to areas located closer to the UER and UWR. This could also be due to the CHIRPS-v2 poor strength along the zonal boundaries, as reported in [36], which could lead to false onsets at these locations. On average, rainfall onsets occurred between 28th April to 25th May in the UER, 10th April–15th May in the UWR, and 21st March to 25th April in the NR except for exceptional seasons. Generally, the results revealed that the region's rainfall indices have substantial latitudinal variability, with early onsets (21st March–15th April) south of 10oN latitude and late onsets (15th April–25th May) at the north of 10° N latitude.

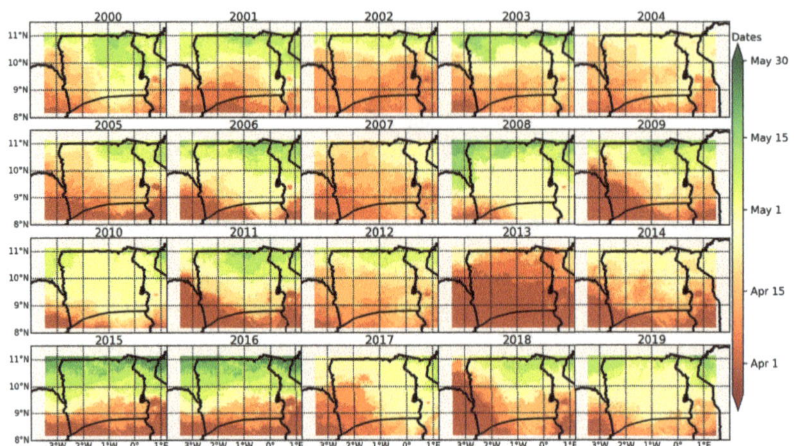

Figure 2. Variability of rainfall onsets dates from 2000 to 2019 over Northern Ghana generated from CHIRPS-v2 data.

Figure 3 represents the rainfall cessation over the northern portions of Ghana from 2000 to 2019. Rainfall cessation dates occurred between 14th September to 10th November and showed a seasonal progression from the North to South. However, for some years, almost all parts of the region showed early (2001, 2004, 2005, 2007, 2011, and 2017) and late (2009, 2012, 2019) cessation dates. These reflect the high level of uncertainty farmers encounter when making cropping decisions between seasons. Early cessations may interrupt grain filling, thus reducing the yield, while too late cessations may delay harvesting leading to increased pre-harvest losses and infestations of moulds that cause aflatoxins. On average, the UER and UWR had relatively earlier cessations (14th September–2nd October) compared to the NR (12th October–10th November). Generally, the UWR has relatively earlier rainfall cessations than UER, except in 2018. In 2012, the entire NR experienced very late cessations (1st–10th November).

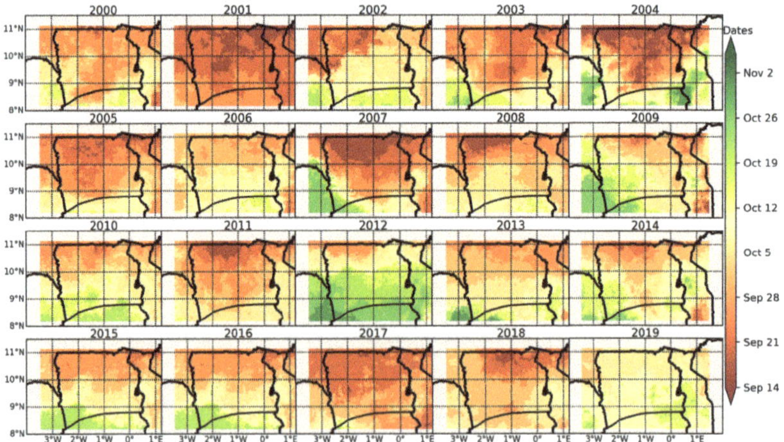

Figure 3. Maps on variability of rainfall cessations dates from 2000 to 2019 over Northern Ghana generated from CHIRPS-v2.

The length of the growing period ranged between 120–210 days throughout the study area (Figure 4). The NR experienced longer LGP (180–210 days) than the UER and UWR (120–150 days). The southwestern of the NR exhibited relatively longer LGP (over 200 days), particularly during the 2009, 2013, and 2018 seasons. In contrast to the onset dates, the LGP showed a seasonal progression along the North-East to South-west direction, with some years having exceptionally shorter (2000–2001) or longer (2012–2013) rainy seasons (Figure 4). Similar to rainfall onset, LGP has a strong latitudinal variability with longer and shorter LGP observed south and north of the 10° N latitude, respectively. The 2012 and 2013 had exceptionally long LGP (180 to 210 days) in almost the entire region. Long LPGs are a consequence of early onset and late rainfall cessation and vice versa.

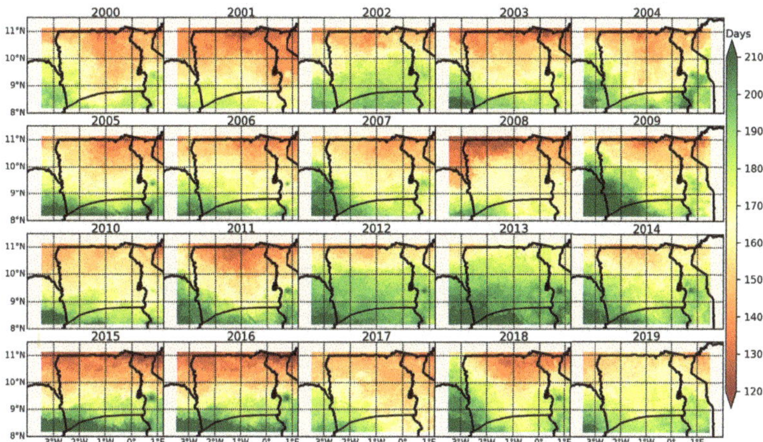

Figure 4. Maps showing the variability of LGP from 2000 to 2019 over Northern Ghana derived from CHIRPS-v2.

4.2. Trends of Rainfall Onset, Cessation, and Length of the Growing Season

The CHIRPS-v2 data revealed earlier, and late onsets date over the West and East of 1.5° W longitude, respectively (Figure 5a). The observed trends of rainfall onset dates confirm and contradict recent studies in the same region. For example, [12] reported earlier onset dates of the rain season in Tamale and Wa gauge stations between 1986–2010, ranging from −0.3 to −0.5 days/year 7.5 to 12.5 days earlier in the 25 years period. However, Ref. [18] reported a significant delay of onset dates up-to 0.8 days/year in the Volta basin in Ghana (35 days delay in 40 years). Ref. [17] reported a 16-day delay of rainfall onset in the UER, although they monitored a shorter period (1997–2014) with very coarse resolution gridded data (110 Km). Rainfall showed late-cessation dates over most parts of Northern Ghana (Figure 5b). Similarly, longer LGP were observed East of 1.5° W longitude of northern Ghana due to late cessation of rains (Figure 5a,c). The contrasting trends of onset and cessation dates reported in different studies could result from different datasets, the temporal span and the methods applied in the analysis. Nevertheless, several studies have demonstrated the accuracy of the PMCR method that we applied [10,11]. However, one weakness of the method is that it mostly does not take into consideration false onsets. Moreover, a weakness identified with the CHIRPS-v2 data was that it tends to be biased towards capturing false onsets near boundaries of the study area.

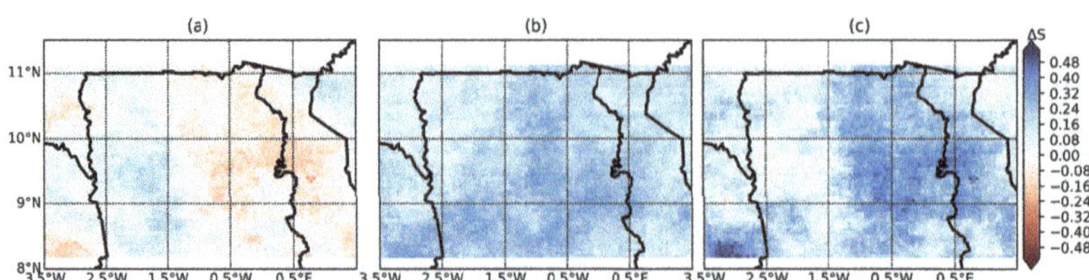

Figure 5. Spatial-temporal trends (days/year) of rainfall onsets (**a**), cessation dates (**b**), and length of growing season (**c**) over Northern Ghana for 39 years period (1981–2019). The blue and red tones represent grid cells with early (shorter) and late (longer) trends, respectively.

4.3. Validation of Seasonal Rainfall Indices Derived from Satellite and Farmer's Perceptions

Gauge vs. Satellite Data

This section assessed the CHIRPS-v2 satellite data's ability to capture the three seasonal rainfall indices (onset, cessation and LGP) in the UER, UWR, and NR of Northern Ghana. Figure 6 represents the variability values of the rainfall indices from satellite and gauge data at the six stations from 1981 to 2016 in, UER (Garu and Zuarungu), and UWR (Wa and Babile), and NR (Bole and Tamale). The rainfall onsets as captured by gauge occurred between 80–140 days, while the onsets captured by satellite occurred between 80–135 days in the six stations, except for exceptionally early onsets observed at some locations. It is observed that satellite data generally captured the earliest onsets in UWR (Wa) compared to gauge. Specifically, Bole showed the earliest onset recorded on 58 days for gauge and 78 days for satellite. Exceptionally early onsets were also observed at some locations as shown by outliers (in black circles). The gauge data showed that rainfall cessation dates generally occurred between 255–305 days compared to between 255–295 days from the CHIRPS-v2 except for exceptional early (late) cessations at Tamale (Wa) (Figure 6). Early cessation of rainfall was observed at the Garu and Zuarungu stations located in the Upper East region. The LGP at the location of all stations from the gauge and CHIRPS-v2 data ranged between 105–205 days and 130–205 days, respectively. The shortest LGP observed from gauge was 105 days at the Zuarungu station. On average, CHIRPS-v2 captured longer LGP compared to gauge at Babile, Garu and Zuarungu stations. Generally, CHIRPS-v2 consistently returned early (late) onset (cessation) dates across the six stations. The onset dates derived from CHIRPS-v2 had less temporal variability compared to the gauge. The cessations date had less variability over time compared to the onset dates. CHIRPS-v2 returned longer LGP as a consequent of early-onset and late-onset dates than the gauge stations.

The statistical performance of the satellite in capturing the rainfall indices at the stations (Bole, Babile, Garu, Tamale, Zuarungu, and Wa) is presented in Figure 7. CHIRPS-v2 showed an average performance for the rainfall onsets compared to the gauge at the two stations located at the UER and the Wa and Tamale stations with r values between 0.4 and 0.60. However, a very poor correspondence ($r < 0.22$) was observed for the Babile and Bole stations located at the UWR and NR, respectively. The rainfall onsets' bias values range from −15 to +6 days at the six stations. Generally, the onset days are under-estimated (earlier onset dates) in all stations except in the Tamale station, where onsets were slightly over-estimated (delayed onset dates) by six days. The Root means square error values of onsets at the stations were in the range of 0.09–0.25. The Bole station in the Northern region recorded the highest value (RMSE = 0.25) while the least was observed in Wa located in the Upper West region. In general, for rainfall cessations, CHIRPS-v2 showed a good agreement with gauge (0.4–0.74) except in the Wa station where a weak correlation coefficient ($r = 0.12$) was recorded. The bias values are in the range of 4–9 days, indicating a late estimate of rainfall cessations dates. The minimal RMSE values (0.03–0.05) observed in all stations reveal that CHIRPS-v2 data generally captured the rainfall cessations. The correlation coefficients of LGP between gauge and CHIRPS-v2 at the stations were in the range of 0.4–0.53 except in the Bole station. The low correlation in Bole ($r = 0.04$) could be because this station had more than 20% of missing data that was gap-filled. CHIRPS-v2 produced longer LGP in all but Tamale station, where the estimate was shorter by 4 days than the gauge. The RMSE for the LGP from CHIRPS-v2 at all the stations ranges from 0.1–0.18. CHIRPS-v2 in the Wa and Tamale stations recorded the lowest RMSE value of 0.1.

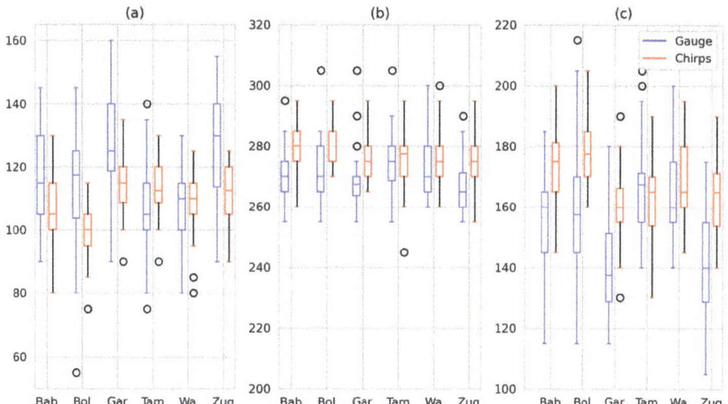

Figure 6. Variability (day of year) of rainfall onset (**a**), cessation (**b**) and the LGP (**c**) derived from CHIRPS-v2 (red) and gauge data (blue) at the location of the six stations in Northern Ghana from 1981 through 2016. Black circles indicate outliers. Station locations are shown in Figure 1 and their labels are Babile (Bab), Bole (Bol), Garu (Gar), Tamale (Tam), Wa (Wa) and Zug (Zuarungu).

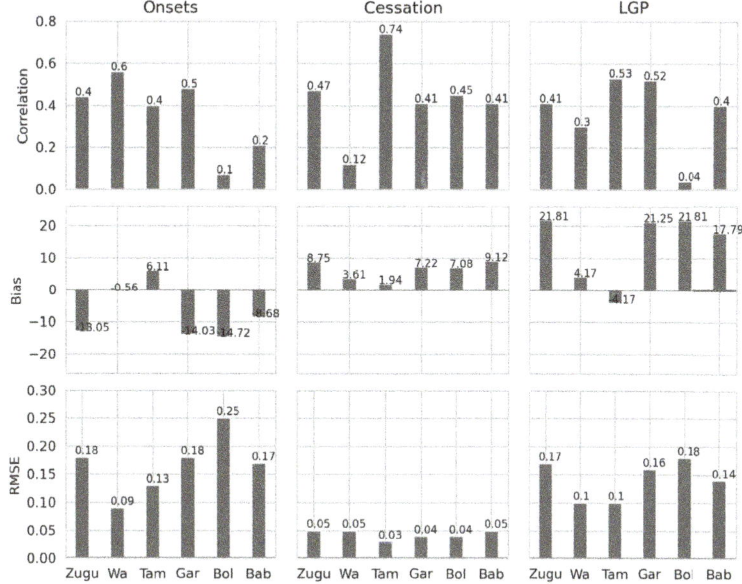

Figure 7. Agreement between CHIRPS-v2 and gauge data in representing the rainfall onset, cessation, and LGP at the six stations from 1981 to 2016. Performance was measured using Pearson correlation (r), Bias, and root-mean-square errors (RMSE). Station names are like Figure 6.

In contrast, relatively high RMSE values (0.14–0.18) were observed in the remaining stations. Although CHIRPS-v2 showed good skill in capturing the rainfall indices over the study region, high biases were detected in some cases. According to [26], the skill of CHIRPS rainfall estimates has high spatial variability due to the influence of climate, topography, and seasonal rainfall patterns. CHIRPS-v2 is known to overestimate and under-estimate low and high-intensity rains in this region, respectively [6]. The inherent systematic biases in CHIRPS-v2 emanates from low density and decreasing Global Telecommunication System (GTS) data over time, leading to insufficient representation of rainfall [26,27,36]. The systematic bias has significant implications for agricultural production, considering that crops' success or failure is more dependent on accurate estimating of onsets and cessations of rains. The accuracy of estimating the three rainfall indices can be improved by applying robust method for correcting systematic biases in CHIRPS-v2 data such as the bias correction and spatial disaggregation (BCSD; [45]) or the Bayesian bias correction method [46].

Figure 8 represents the temporal trends of rainfall onset, cessation, and length of the growing period during 1981–2016 derived from gauge and CHIRPS-v2 dataset in the UWR (Wa), UER (Zuarungu), and NR (Tamale). At Zuarungu station, there was a fair agreement ($r > 0.4$) between gauge and satellite for all rainfall indices, and the RMSE was relatively low (RMSE < 0.19; Figure 8a,d,g). Both the gauge and satellite captured the decreasing trend of rainfall onsets and increasing cessations and LGP in Zuarungu, although with differing slopes. At Wa station, the rainfall onset dates derived from CHIRPS-v2 data showed a good correlation with gauge ($r = 0.57$), but the direction and magnitude of slopes differed substantially (Figure 8b). At Tamale station, the satellite agreed better with the gauge for the rainfall cessation ($r = 0.74$, Figure 8f), followed by LGP ($r = 0.53$; Figure 8i), and lastly, the onsets ($r = 0.4$, Figure 8c). Moreover, both satellite and gauge showed earlier onset dates but late or longer cessation and LGP in Tamale. The observed trends of rainfall onset contrast and confirms some findings in [12], who reported earlier onset dates of 7.5 days in Tamale and Wa. Our results reveal early (late) onsets dates of 8 days in Tamale (Wa) as captured by CHIRPS-v2 from 1981 to 2016. Therefore, our results agree with [12] at Tamale but contrast at Wa station.

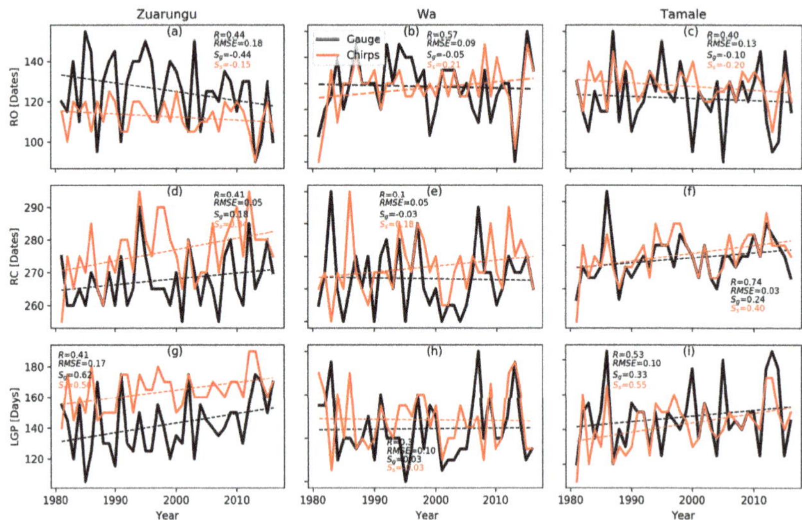

Figure 8. Trends of rainfall onset (RO), cessation (RC) date and LGP captured by gauge and satellite data in the UER, UWR, and NR from 1981 through 2016. Sg and Ss are the slope of linear regression for the gauge, and satellite, respectively.

4.4. Farmers Perceptions on Timing of Rainfall Indices and Agronomic Adaptive Measures

Over 95% of the UWR and UER region farmers reported late-onset dates of the rainy season (Figure 9a). However, responses were more divergent in the NR, where 76% and 16% of farmers reported early and late-onset dates, respectively. A total of 248 (62%) and 134 (33%) of farmers reported late and early onset of the rains, respectively (Figure 9b). In contrast, over 80% of farmers in UWR and UER reported early cessation of rainfall (Figure 9c). Farmers' experiences on the cessation of rains in NR were split between early (49%), late (41%), and no change (11%). The regional aggregates showed that a total of 288 (72%) and 86 (22%) farmers reported early and late cessation of rain, respectively (Figure 9d). The late onset and early cessation of rains in the UWR and UER imply the shortening of the growing season, but the NR status is divergent.

Table 2 shows a comparison of the trends of onset and cessation dates of rainfall from the gauge and satellite at representative stations in Figure 5 and the aggregated farmers perceptions per region (Figure 9a–c). The farmer perceptions relatively matched the long time series of the gauge and satellites because the household survey elicited farmer perceptions of the trends of onset and cessation dates by comparing the last growing season (2020) with the situation over three decades ago. However, the farmer perceptions were aggregated per region and not necessary at the location of the station, a fact that may reduce the precision. Table 2 shows an agreement between the observation network and farmer perceptions in Tamale station. However, most farmers in UER perceived late onset and early cessation dates of rainfall which completely differed from observation network. Likewise, the satellite and farmer perceptions on onset dates differed from gauge in Wa station although the cessation dates were converging. Therefore, the agreement between gauge, satellite, and farmer perceptions had wide spatial-temporal variability. Our results agree with [33] that farmers' perceptions can vary with location, which decreases their spatial reliability compared to observation networks. Similarly, Refs. [33,47] observed that farmers' perceptions of climate change in northern Ghana deviated from the meteorological records. One possible explanation of the deviations of farmer perceptions and the gauge observations could be their failure to differentiate between climate variability and change [29]. Climate variability has weakened farmers' altitude on traditional forecasting methods and has become more open to scientific measurements. Farmers' perceptions are shaped mainly by short-term variability of climate parameters and the frequency of extreme events than slow long-term changes in the average conditions [29]. Farmers are more perceptive of changes in temperature than rainfall, and their perceptions depend on location, age, and indigenous knowledge [29,33]. In areas that experience land degradation and climate change, farmers' perceptions can confound changes in rainfall seasonality with changes in soil fertility [31].

Table 2. Summary of the trends of onset and cessation dates of rainfall from the gauge and satellite at representative stations from Figure 5 and the aggregated farmer's perceptions per region obtained from Figure 9a–c.

Region	Gauge Station	Correlation Gauge vs. Satellite (r)	Satellite	Gauge	Farmer's Perceptions
		Onset			
UER	Zuarungu	0.44	Early	Early	Late
UWR	Wa	0.57	Early	Late	Late
NR	Tamale	0.40	Early	Early	Early
		Cessation			
UER	Zuarungu	0.40	Late	Late	Early
UWR	Wa	0.10	Early	Early	Early
NR	Tamale	0.74	Late	Late	Divergent

Over 68% of farmers across all regions replanted the main crop seeds at least once in the last five seasons (Figure 3g,h). This reflects the high frequency of false onsets of the rains in the previous five cropping seasons. Replanting increases the cost of seeds, therefore reducing profitability. Studies had indicated that farmers planted up-to seven times after repeated crop failures during drought seasons in Burkina Faso [48]. Repeated replanting is done late in the growing season, making crops flower after a shortened vegetative period that eventually reduces the crop yield. Evidently, 79% of farmers identified the timing of the onset of rains as a crucial determinant of the crop yield (Figure 9e,f).

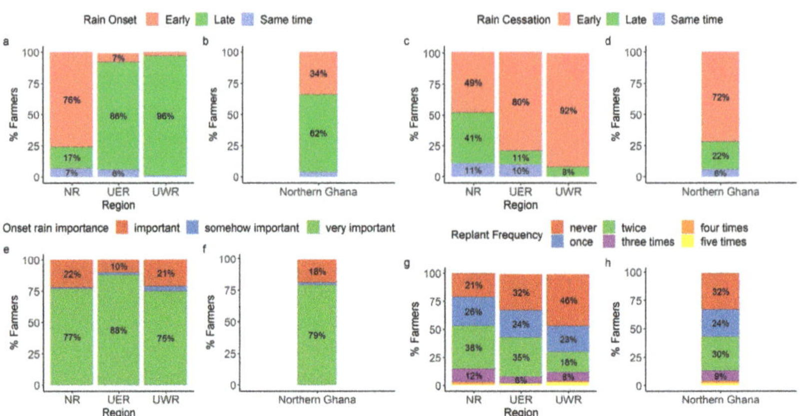

Figure 9. Farmers' perceptions of the rainfall onset trend per region (**a**) and aggregated in entire Northern Ghana (**b**). Cessation for each region (**c**) and aggregated for the entire Northern Ghana (**d**). The importance of the onset date of the rain on the yield of the main crop per region (**e**) and whole Northern Ghana (**f**). The frequency of replanting of seeds of the main crop in the last 5 growing seasons due to the false start of the rain season per region (**g**) and aggregate for northern Ghana (**h**).

Results showed that only 29% of farmers rely on data from meteorological agencies to forecast the start of the season (Figure 10a). The use of meteorological agency data was lowest in UWR. Most farmers (>70%) rely on traditional methods to forecast the onset of the rainy season, such as a change in temperature, the pattern of clouds, vegetation phenology, movement pattern of insects/birds, and wind direction (Figure 10a). The traditional knowledge of forecasting rainfall onsets is easy to use and affordable to local farmers, but they are becoming less reliable due to increasing climate variability [48]. The traditional methods are available to different socio-economic and demographic groups over space and time, e.g., herders are more likely to observe movement and nesting of birds in the bush. At the same time, rural women are more likely to note the change of water levels or behavior of insects at water sources where they fetch water [48]. Similarly, Ref. [29] observed that farmers in Ethiopian highlands rarely used scientific climate information despite being in the frontline of implementing the adaptation measures. Our research highlights the need to invest in modern climate services such as the automated gauge network within the farming communities in northern Ghana to complement the existing local knowledge of forecasting the onset of rainy seasons. This will support the provision of tailored weather information services to farmers. Farmers in the study area are willing to pay for climate information disseminated through mobile services [49]. Moreover, by integrating the scientific and traditional methods, our study improves the understanding of the current climate knowledge systems that can help to enhance the modern observation networks in a culturally and locally relevant manner. Our study helps to enhance the observation network by identifying the degree and locations where the satellite estimates mimic the gauge data. The accuracy assessment is important considering that the satellite data are increasingly relied upon for agro-advisory due to the prevalent sparse gauge network in this region. The traditional forecasting methods observe the changes in temperature, wind and clouds that are integral variables of interest to modern meteorology. Therefore, as suggested by [48], scientific meteorologists could build on local understanding between temperature and seasonal rainfall to explore technical aspects of scientific forecasts based, for example, on sea surface temperature. The results further points to the uncertainties arising from relying solely on traditional methods and existing daily satellite rainfall estimates.

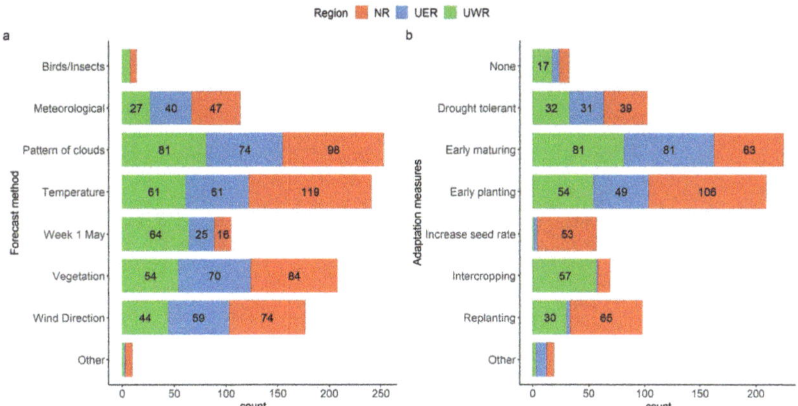

Figure 10. Farmers' responses on the methods applied to forecast the season's onset (**a**) and the adaptation measures to cope with observed trends on the onset and cessation of rainfall (**b**) in northern Ghana.

Figure 10b shows the crop management practices applied by farmers to adapt to shifts of onset and cessation dates of rainfall in the three regions. The early maturing cultivars, early planting, drought-tolerant cultivars and replanting emerged the most common but with different intensities across the three regions (Figure 10b). Planting early maturing cultivars is a strategy to cope with shortened LGP. Early planting enables crops to take advantage of every drop of soil moisture at the onset of season. The UER and eastern NR experienced longer LGP resulting from early-onset and late-cessation dates, therefore, planting early and medium-late maturing varieties are viable adaptation measures in that part. However, in the UER fewer farmers applied early planting (49) compared to early maturing varieties (81) reflecting adaptations that are mis-aligned to local reality. However, the situation was the opposite in the NR. Therefore, adaptation measures were not always consistent with the rainfall seasonality. Similarly, Ref. [30] noted the blanket recommendations policies on the adoption of early maturing crop varieties where seasons are becoming longer in northern Ghana led to maladaptation outcomes that increased vulnerability to climate change and variability. Farmers in UER respond to delayed onset and shortening of LGP by spreading out the sowing of crops across the first three months of the season through a wait-and-see or delay strategy [17,24]. They sow the drought-tolerant crops (sorghum and millet) in April to take advantage of early rains. Farmer's shift sowing of drought-sensitive crops (maize, rice, and groundnut) from May to June or July to reduce the risk of exposure to early season drought. However, this can increase the risk of exposure to the late-season dry spell since these crops mature after 3–6 months. In addition to, the timing of rainfall seasons, selecting appropriate adaptation measures needs to consider the significant spatial-temporal trends of rainfall amount and temperature reported in northern Ghana [5,6].

5. Conclusions

Spatio-temporal variability of rainfall seasonality in northern Ghana, poses a challenge to food security and other socio-economic activities. This study presents a comprehensive analysis of the variability and trends of three rainfall indices (rainfall onsets, cessations, and length of the growing period) using the spatially high-resolution CHIRPS-v2 daily rainfall series for the period of 39 years (1981–2019) over Northern Ghana. The study further assesses the satellite and farmers' perceptions of the start of the rainfall season over three Northern regions (Upper East, Upper West, and Northern) using gauge data obtained from the Ghana Meteorological Agency (GMet) during 2020. Our findings show that the region's rainfall indices have substantial latitudinal variability, with late onsets at the North of 10° N latitude and early onsets south of 10° N latitude annually. Conversely, early (short) cessions (LGP) are seen to occur at the South of 10oN latitude, while late (long) cessions (LGP) are observed at the North of 10oN latitude. On average, CHIRPS-v2 captured rainfall onsets between 21st March and 25th May, rainfall cessations, 17th September to 10th November, and the LGP is usually between 120–210 days annually in Northern Ghana. Significant late cessations and longer LGP were observed over most parts of Northern Ghana. CHIRPS-v2 data revealed slightly, but significant late, and early onsets date at the West and East of 1.5° W longitude, respectively. Our findings indicated a trend towards late-cessation dates in most parts of the region. CHIRPS-v2 was biased towards capturing early onsets and late onsets, resulting in a relatively longer LPG. CHIRPS-v2 agreed better with observation for the rainfall cessation, followed by LGP, and lastly, the onsets. The satellite-generated rainfall onset dates agreed better with the gauge at Wa and Bole stations. In contrast, farmers' perceptions were more accurate than satellite stations in the Tamale station. Therefore, farmers' perceptions and CHIRPS-v2 to accurately estimate rainfall onsets are time and location-dependent. Approximately 29% of farmers rely on meteorological agencies' data to forecast the rainfall season's start, while the remainder depends on traditional knowledge. Adaptation measures were not always consistent with the rainfall seasonality. CHIRPS-v2 has inherent systematic biases that could come from low density and decreasing gauge observations over time in Ghana, leading to insufficient representation of rainfall indices. CHIRPS-v2 data were biased towards capturing false onsets near boundaries that have significant implications for agricultural production, considering that crops' success or failure is more dependent on accurate estimating of onsets. Thus, the study recommends the correction of systematic biases in CHIRPS-v2 to improve agro-advisories on the timing of seasonal calendar activities. Moreover, our study highlights the need to invest in modern climate information services such as the automated gauge network to complement the existing local knowledge of forecasting the rainfall seasonality in in northern Ghana.

Author Contributions: Conceptualization, W.A.A. and F.K.M.; Methodology, W.A.A. and F.K.M.; Software, W.A.A.; Validation, W.A.A. and F.K.M.; Visualization, W.A.A.; Writing—original draft preparation, W.A.A. and F.K.M.; Data curation, W.A.A. and F.K.M; Supervision, W.A.A., F.K.M., B.K., F.K. and L.K.A.; Writing—Reviewing and Editing, W.A.A., F.K.M., B.K., F.K. and L.K.A. All authors have read and agreed to the published version of the manuscript.

Funding: This research was funded by the United States Agency for International Development with grant number: AID-BFS-G-11-00002.

Institutional Review Board Statement: The study was conducted according to guidelines of the declaration of Helsinki and approved by Institutional Review Board of International Institute of Tropical Agriculture (protocol code IRB/AF/001/2021 approved on 10 December 2020.

Informed Consent Statement: Informed consent was obtained from all subjects involved in the study.

Data Availability Statement: Remote sensing data is open source available at https://www.chc.ucsb.edu/data/chirps, accessed on 12 June 2021. The survey and gauge stations can be provided by authors upon request.

Acknowledgments: We thank Benedict Boyubie and the enumerators for organizing and conducting a farmer's survey.

Conflicts of Interest: The authors declare no conflict of interest. The funders had no role in the design of the study; in the collection, analyses, or interpretation of data; in the writing of the manuscript, or in the decision to publish the results.

Appendix A

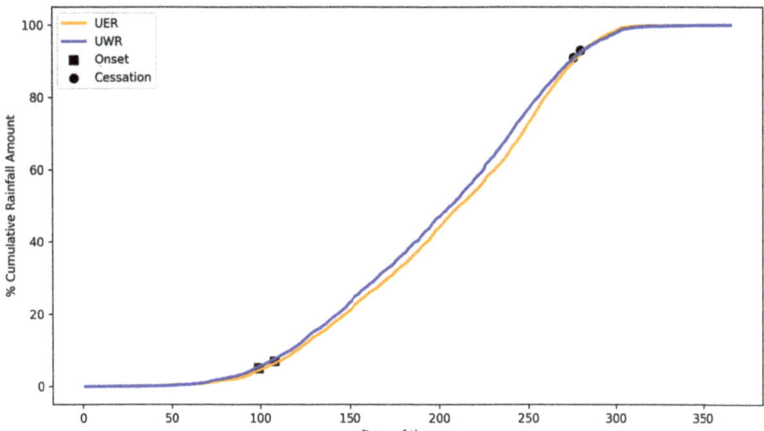

Figure A1. The percentage cumulative rainfall amount averaged over 1981–2019 in the Upper East (UER) and Upper West (UWR) regions.

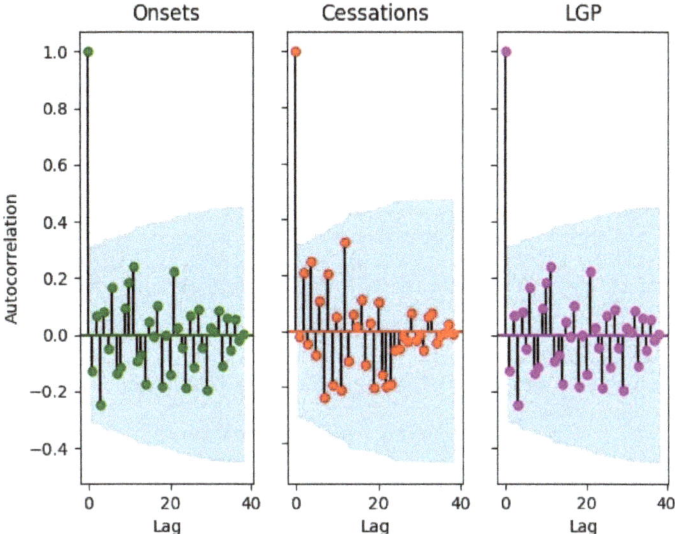

Figure A2. Autocorrelation function of rainfall onset (green), cessation (red) and length of growing season (magenta) over Northern Ghana. The blue shaded region is the confidence interval with a value of $\alpha = 0.05$. Anything within this range represents a value that has no significant correlation which implies no serial autocorrelation.

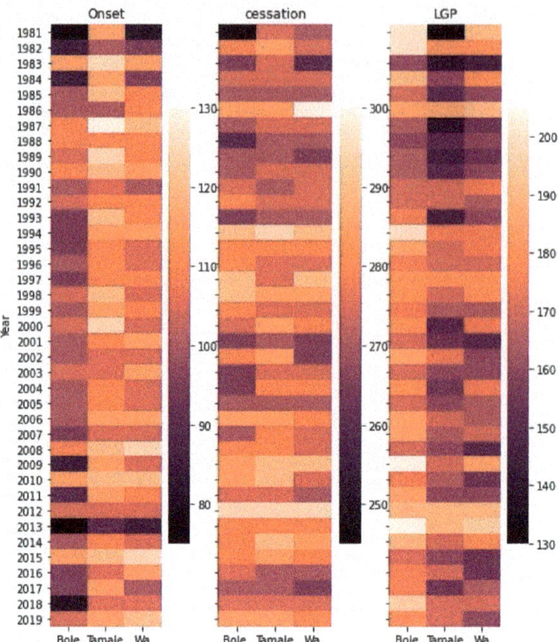

Figure A3. Summary of rainfall onset, cessation and length of growing season over three-gauge stations in northern Ghana. The onset and cessation are days of the year (DOY) while LPG is the number of days.

References

1. Harrison, L.; Funk, C.; Peterson, P. Identifying changing precipitation extremes in Sub-Saharan Africa with gauge and satellite products. *Environ. Res. Lett.* **2019**, *14*, 085007. [CrossRef]
2. Atiah, W.A.; Amekudzi, L.K.; Akum, R.A.; Quansah, E.; Antwi-Agyei, P.; Danuor, S.K. Climate Variability and Impacts on Maize (Zea Mays) Yield in Ghana, West Africa. Available online: https://rmets.onlinelibrary.wiley.com/doi/full/10.1002/qj.4199 (accessed on 1 December 2021).
3. Adhikari, U.; Nejadhashemi, A.P.; Woznicki, S.A. Climate change and eastern Africa: a review of impact on major crops. *Food Energy Secur.* **2015**, *4*, 110–132. [CrossRef]
4. Ray, D.K.; Gerber, J.S.; MacDonald, G.K.; West, P.C. Climate variation explains a third of global crop yield variability. *Nat. Commun.* **2015**, *6*, 1–9. [CrossRef]
5. Atiah, W.A.; Tsidu, G.M.; Amekudzi, L.; Yorke, C. Trends and interannual variability of extreme rainfall indices over Ghana, West Africa. *Theor. Appl. Climatol.* **2020**, *140*, 1393–1407. [CrossRef]
6. Muthoni, F. Spatial-temporal trends of rainfall, maximum and minimum temperatures over west Africa. *IEEE J. Sel. Top. Appl. Earth Obs. Remote Sens.* **2020**, *13*, 2960–2973. [CrossRef]
7. Liebmann, B.; Marengo, J. Interannual variability of the rainy season and rainfall in the Brazilian Amazon Basin. *J. Clim.* **2001**, *14*, 4308–4318. [CrossRef]
8. Liebmann, B.; Bladé, I.; Kiladis, G.N.; Carvalho, L.M.; Senay, G.B.; Allured, D.; Leroux, S.; Funk, C. Seasonality of African precipitation from 1996 to 2009. *J. Clim.* **2012**, *25*, 4304–4322. [CrossRef]
9. Akinseye, F.M.; Agele, S.O.; Traore, P.; Adam, M.; Whitbread, A.M. Evaluation of the onset and length of growing season to define planting date—'A case study for Mali (West Africa)'. *Theor. Appl. Climatol.* **2016**, *124*, 973–983. [CrossRef]
10. Dunning, C.M.; Black, E.C.; Allan, R.P. The onset and cessation of seasonal rainfall over Africa. *J. Geophys. Res. Atmos.* **2016**, *121*, 11–405. [CrossRef]
11. Amekudzi, L.K.; Yamba, E.I.; Preko, K.; Asare, E.O.; Aryee, J.; Baidu, M.; Codjoe, S.N. Variabilities in rainfall onset, cessation and length of rainy season for the various agro-ecological zones of Ghana. *Climate* **2015**, 3, 416–434. [CrossRef]
12. Gbangou, T.; Ludwig, F.; van Slobbe, E.; Hoang, L.; Kranjac-Berisavljevic, G. Seasonal variability and predictability of agro-meteorological indices: Tailoring onset of rainy season estimation to meet farmers' needs in Ghana. *Clim. Serv.* **2019**, *14*, 19–30. [CrossRef]

13. MacLeod, D. Seasonal predictability of onset and cessation of the east African rains. *Weather Clim. Extrem.* **2018**, *21*, 27–35. [CrossRef]
14. Vellinga, M.; Arribas, A.; Graham, R. Seasonal forecasts for regional onset of the West African monsoon. *Clim. Dyn.* **2013**, *40*, 3047–3070. [CrossRef]
15. Antwi-Agyei, P.; Fraser, E.D.; Dougill, A.J.; Stringer, L.C.; Simelton, E. Mapping the vulnerability of crop production to drought in Ghana using rainfall, yield and socioeconomic data. *Appl. Geogr.* **2012**, *32*, 324–334. [CrossRef]
16. Baidu, M.; Amekudzi, L.K.; Aryee, J.N.; Annor, T. Assessment of long-term spatio-temporal rainfall variability over Ghana using wavelet analysis. *Climate* **2017**, *5*, 30. [CrossRef]
17. Boansi, D.; Tambo, J.A.; Müller, M. Intra-seasonal risk of agriculturally-relevant weather extremes in West African Sudan Savanna. *Theor. Appl. Climatol.* **2019**, *135*, 355–373. [CrossRef]
18. Laux, P.; Kunstmann, H.; Bárdossy, A. Predicting the regional onset of the rainy season in West Africa. *Int. J. Climatol. J. R. Meteorol. Soc.* **2008**, *28*, 329–342. [CrossRef]
19. Chemura, A.; Schauberger, B.; Gornott, C. Impacts of climate change on agro-climatic suitability of major food crops in Ghana. *PLoS ONE* **2020**, *15*, e0229881. [CrossRef]
20. Owusu Danquah, E.; Beletse, Y.; Stirzaker, R.; Smith, C.; Yeboah, S.; Oteng-Darko, P.; Frimpong, F.; Ennin, S.A. Monitoring and Modelling Analysis of Maize (Zea mays L.) Yield Gap in Smallholder Farming in Ghana. *Agriculture* **2020**, *10*, 420. [CrossRef]
21. Mkonda, M.Y.; He, X. Climate variability and crop yields synergies in Tanzania's semiarid agroecological zone. *Ecosyst. Health Sustain.* **2018**, *4*, 59–72. [CrossRef]
22. Ocen, E.; de Bie, C.; Onyutha, C. Investigating False start of the Main Growing Season: A Case of Uganda in East Africa. Available online: https://www.sciencedirect.com/science/article/pii/S2405844021025317 (accessed on 1 December 2021)
23. Van de Giesen, N.; Liebe, J.; Jung, G. Adapting to climate change in the Volta Basin, West Africa. *Curr. Sci.* **2010**, *98*, 1033–1037.
24. Sarku, R.; Dewulf, A.; van Slobbe, E.; Termeer, K.; Kranjac-Berisavljevic, G. Adaptive decision-making under conditions of uncertainty: The case of farming in the Volta delta, Ghana. *J. Integr. Environ. Sci.* **2020**, *17*, 1–33. [CrossRef]
25. Shukla, S.; Husak, G.; Turner, W.; Davenport, F.; Funk, C.; Harrison, L.; Krell, N. A slow rainy season onset is a reliable harbinger of drought in most food insecure regions in Sub-Saharan Africa. *PLoS ONE* **2021**, *16*, e0242883. [CrossRef]
26. Dinku, T. Challenges with availability and quality of climate data in Africa. In *Extreme Hydrology and Climate Variability*; Elsevier: Amsterdam, The Netherlands, 2019; pp. 71–80.
27. Contractor, S.; Donat, M.G.; Alexander, L.V.; Ziese, M.; Meyer-Christoffer, A.; Schneider, U.; Rustemeier, E.; Becker, A.; Durre, I.; Vose, R.S. Rainfall Estimates on a Gridded Network (REGEN)—A global land-based gridded dataset of daily precipitation from 1950 to 2016. *Hydrol. Earth Syst. Sci.* **2020**, *24*, 919–943. [CrossRef]
28. Salerno, J.; Diem, J.E.; Konecky, B.L.; Hartter, J. Recent intensification of the seasonal rainfall cycle in equatorial Africa revealed by farmer perceptions, satellite-based estimates, and ground-based station measurements. *Clim. Chang.* **2019**, *153*, 123–139. [CrossRef]
29. Darabant, A.; Habermann, B.; Sisay, K.; Thurnher, C.; Worku, Y.; Damtew, S.; Lindtner, M.; Burrell, L.; Abiyu, A. Farmers' perceptions and matching climate records jointly explain adaptation responses in four communities around Lake Tana, Ethiopia. *Clim. Chang.* **2020**, *163*, 481–497. [CrossRef]
30. Antwi-Agyei, P.; Dougill, A.J.; Stringer, L.C.; Codjoe, S.N.A. Adaptation opportunities and maladaptive outcomes in climate vulnerability hotspots of northern Ghana. *Clim. Risk Manag.* **2018**, *19*, 83–93. [CrossRef]
31. Diem, J.E.; Hartter, J.; Salerno, J.; McIntyre, E.; Grandy, A.S. Comparison of measured multi-decadal rainfall variability with farmers' perceptions of and responses to seasonal changes in western Uganda. *Reg. Environ. Chang.* **2017**, *17*, 1127–1140. [CrossRef]
32. Esayas, B.; Simane, B.; Teferi, E.; Ongoma, V.; Tefera, N. Climate variability and farmers' perception in southern Ethiopia. *Adv. Meteorol.* **2019**, *2019*, 7341465. [CrossRef]
33. Guodaar, L.; Bardsley, D.K.; Suh, J. Integrating local perceptions with scientific evidence to understand climate change variability in northern Ghana: A mixed-methods approach. *Appl. Geogr.* **2021**, *130*, 102440. [CrossRef]
34. Atiah, W.A.; Amekudzi, L.K.; Quansah, E.; Preko, K. The spatio-temporal variability of rainfall over the agro-ecological zones of Ghana. *Atmos. Clim. Sci.* **2019**, *4*, 527–544. [CrossRef]
35. Funk, C.; Peterson, P.; Landsfeld, M.; Pedreros, D.; Verdin, J.; Shukla, S.; Husak, G.; Rowland, J.; Harrison, L.; Hoell, A.; et al. The climate hazards infrared precipitation with stations—A new environmental record for monitoring extremes. *Sci. Data* **2015**, *2*, 1–21. [CrossRef]
36. Atiah, W.A.; Amekudzi, L.K.; Aryee, J.N.A.; Preko, K.; Danuor, S.K. Validation of satellite and merged rainfall data over Ghana, West Africa. *Atmosphere* **2020**, *11*, 859. [CrossRef]
37. Atiah, W.A.; Tsidu, G.M.; Amekudzi, L.K. Investigating the merits of gauge and satellite rainfall data at local scales in Ghana, West Africa. *Weather. Clim. Extrem.* **2020**, *30*, 100292. [CrossRef]
38. Theil, H. A rank-invariant method of linear and polynomial regression analysis. *Indag. Math.* **1950**, *12*, 173.
39. Mann, H.B. Nonparametric tests against trend. *Econom. J. Econom. Soc.* **1945**, *13*, 245–259. [CrossRef]
40. Kendall, K. Thin-film peeling-the elastic term. *J. Phys. D Appl. Phys.* **1975**, *8*, 1449. [CrossRef]
41. Spearman, C. The proof and measurement of association between two things. *Am. J. Psychol.* **1987**, *100*, 441–471. [CrossRef]

42. Lehmann, E.L.; D'Abrera, H.J. *Nonparametrics: Statistical Methods Based on Ranks*; Holden-Day: Hoboken, NJ, USA, 1975. Available online: https://rss.onlinelibrary.wiley.com/doi/abs/10.2307/2344536 (accessed on 10 May 2021).
43. Onyutha, C. Graphical-statistical method to explore variability of hydrological time series. *Hydrol. Res.* **2021**, *52*, 266–283. [CrossRef]
44. Onyutha, C. Statistical uncertainty in hydrometeorological trend analyses. *Adv. Meteorol.* **2016**, *2016*, 266–283. [CrossRef]
45. Wood, A.W.; Leung, L.R.; Sridhar, V.; Lettenmaier, D. Hydrologic implications of dynamical and statistical approaches to downscaling climate model outputs. *Clim. Chang.* **2004**, *62*, 189–216. [CrossRef]
46. Kimani, M.W.; Hoedjes, J.C.; Su, Z. Bayesian bias correction of satellite rainfall estimates for climate studies. *Remote Sens.* **2018**, *10*, 1074. [CrossRef]
47. Dakurah, G. How do farmers' perceptions of climate variability and change match or and mismatch climatic data? Evidence from North-west Ghana. *GeoJournal* **2021**, *86*, 2387–2406. [CrossRef]
48. Roncoli, C.; Ingram, K.; Kirshen, P. Reading the rains: Local knowledge and rainfall forecasting in Burkina Faso. *Soc. Nat. Resour.* **2002**, *15*, 409–427. [CrossRef]
49. DONKOH, S.A. Farmers' willingness-to-pay for weather information through mobile phones in northern Ghana. *Ghana J. Sci. Technol. Dev.* **2019**, *6*, 19–36. [CrossRef]

atmosphere

MDPI

Article

Heavy Lake-Effect Snowfall Changes and Mechanisms for the Laurentian Great Lakes Region

Oleksandr Huziy [1,2,*], Bernardo Teufel [1,2], Laxmi Sushama [1,2] and Ram Yerubandi [3]

1 Department of Civil Engineering, McGill University, Montreal, QC H3A 0C3, Canada; bernardo.teufel@mail.mcgill.ca (B.T.); laxmi.sushama@mcgill.ca (L.S.)
2 Trottier Institute for Sustainability in Engineering and Design, McGill University, Montreal, QC H3A 0C3, Canada
3 Water Science and Technology Division, Environment and Climate Change Canada, Burlington, ON L7R 4A6, Canada; ram.yerubandi@ec.gc.ca
* Correspondence: guziy.sasha@gmail.com

Abstract: Heavy lake-effect snowfall (HLES) events are snowfall events enhanced by interactions between lakes and overlying cold air. Significant snowfall rates and accumulations caused during such events disrupt socioeconomic activities and sometimes lead to lethal consequences. The aim of this study is to assess projected changes to HLES by the end of the century (2079–2100) using a regional climate model for the first time with 3D representation for the Laurentian Great Lakes. When compared to observations over the 1989–2010 period, the model is able to realistically reproduce key mechanisms and characteristics of HLES events, thus increasing confidence in future projections. Projected changes to the frequency and amount of HLES suggest decreasing patterns, during the onset, active and decline phases of HLES. Despite reduced lake ice cover that will allow enhanced lake–atmosphere interactions favouring HLES, the warmer temperatures and associated increase in liquid to solid precipitation ratio along with reduced cold air outbreaks contribute to reduced HLES in the future climate. Analysis of the correlation patterns for current and future climates further supports the weaker impact of lake ice fraction on HLES in future climates. Albeit the decreases in HLES frequency and intensity and projected increases in extreme snowfall events (resulting from all mechanisms) raise concerns for impacts on the transportation, infrastructure and hydropower sectors in the region.

Keywords: lake-effect snowfall; regional climate modelling; climate change; Great Lakes; mechanisms

Citation: Huziy, O.; Teufel, B.; Sushama, L.; Yerubandi, R. Heavy Lake-Effect Snowfall Changes and Mechanisms for the Laurentian Great Lakes Region. *Atmosphere* **2021**, *12*, 1577. https://doi.org/10.3390/atmos12121577

Academic Editors: Baojie He, Ayyoob Sharifi, Chi Feng and Jun Yang

Received: 15 October 2021
Accepted: 26 November 2021
Published: 27 November 2021

1. Introduction

Heavy lake-effect snowfall (HLES) events are winter snowstorms resulting from the rapid modification of cold air masses passing over the relatively warm waters of large lakes. First, the overlying atmosphere is destabilised through strong vertical fluxes of heat and moisture from water to air. Once the moisture-laden air hits land, the increased surface roughness causes convergence and upward vertical motion, resulting in HLES. These extreme events are frequent during the cold season downwind of the Laurentian Great Lakes, which provide large areas for lake–atmosphere interactions and are often on the way of the north-westerly flow of cold air coming from the Canadian North. The frequency of cold air outbreaks (CAO) [1,2] is one of the important factors controlling HLES [3,4]. The high snow rates and accumulations during HLES [3,5,6] often disrupt the transportation, infrastructure and hydropower sectors, and other socioeconomic activities [3]. For example, in November 2014, one of the heaviest lake-effect snowfall events on record resulted in the death of more than a dozen people, as thousands of motorists were trapped in vehicles, falling trees triggered power outages and hundreds of roofs and structures collapsed under the weight of the snow.

Given their impacts, it is important to understand how the frequencies, intensities and timings of these events will change in future climate. Previous studies have quantified projected changes to HLES for the Great Lakes region using one-dimensional (1D) lake models [3]. Notaro et al. [3] reported decreases in HLES during the 21st century, although slight increases are noted by the middle of the century around Lake Superior under Representative Concentration Pathway 8.5 (RCP8.5) for one of the two considered driving global climate models (GCMs). Under doubled CO_2 concentrations, Janoski et al. [7] found decreases in mean snowfall across the Great Lakes region, but no significant changes to extreme snowfall, using a GCM at ~50 km resolution.

One-dimensional lake models do not adequately simulate mixing processes [8,9], leading to biases in lake surface temperature and ice concentration. Furthermore, 1D models cannot capture the circulation patterns and currents, which also determine the lake ice onset and offset. Offline simulations, in which lake and atmospheric runs are performed separately, cannot capture the two-way feedbacks between the lake and the atmosphere. Given that HLES is generally confined to a narrow region on the lee shores of the lakes, the spatial resolution of GCMs is too coarse to resolve this phenomenon, and downscaling using regional climate models (RCMs) is required. Therefore, this study considers a coupled system consisting of the Nucleus for European Modelling of the Ocean (NEMO; Madec et al. [10]) coupled to the limited-area version of the Global Environment Multiscale (GEM) model to study HLES and associated mechanisms in current and future climates over the Laurentian Great Lakes region (Figure 1). Projected changes to HLES amounts and frequencies produced by the coupled system are analysed and linked to the projected warming, lake ice cover and precipitation changes in this study. The NEMO model coupled with GEM was used for the same region in previous studies in weather forecast mode [11], which showed that it is able to reproduce surface currents, lake ice fractions and lake surface temperatures well.

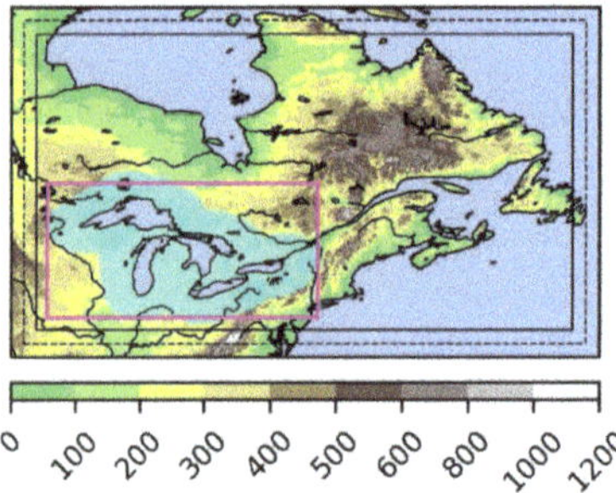

Figure 1. Experimental domain with topography (m) shown in color. The region between the inner blue solid and dashed lines is the blending zone (10 grid points wide) and the region outside the dashed rectangle is the pilot zone (10 grid points wide). The region around the Great Lakes, highlighted in cyan color, is the 200 km near-shore zone and the magenta rectangle shows the region used for analysis.

The paper is organised as follows: Section 2 describes the model, simulations, and methods. Section 3 deals with model evaluation. Projected changes to HLES are presented in Section 4. Finally, the summary and conclusions are presented in Section 5.

2. Models and Methods

The regional climate model used in this study is the limited-area version of the Global Environment Multiscale (GEM) model used for numerical weather prediction by Environment and Climate Change Canada [12]. The following physical parameterisations are used in GEM: deep convection by Kain and Fritsch [13], shallow convection by Kuo [14], large-scale condensation by Sundqvist et al. [15], correlated-K solar and terrestrial radiations by Li and Barker [16], subgrid-scale orographic gravity-wave drag by McFarlane [17], low-level orographic blocking parameterisation by Zadra et al. [18,19] and planetary boundary layer parameterisation by Benoit et al. and Delage et al. [18,20–22]. The land surface scheme is the Canadian LAnd Surface Scheme (CLASS), version 3.5 [23,24]. This version of CLASS uses a flexible soil layering scheme, i.e., the number of soil layers and their thickness can be adjusted as required. CLASS includes prognostic equations for energy and water conservation for the soil layers and a thermally and hydrologically distinct snowpack (treated as a variable-depth layer) where applicable. The sub-grid lakes are represented by the 1D Hostetler lake model [25,26], while resolved lakes such as the Laurentian Great Lakes are represented by the 3D ocean model NEMO with the Louvain-la-Neuve sea Ice Model (LIM3) for lake ice processes [27]. The timestep used for GEM in this study is 300 s, and information is exchanged between GEM and NEMO every 30 min, which is the timestep used for NEMO. The horizontal resolution used for both GEM and NEMO is 0.1 degrees. In NEMO, a 23-level z-coordinate with partial step bathymetry is used, in which layer thickness increases with depth. In LIM3, five ice categories are used to represent the sub-grid-scale ice thickness distribution. Each category is further vertically divided into one layer of snow and several ice layers of equal thicknesses.

To assess projected changes to heavy lake-effect snowfall, two 22-year-long simulations for the current 1989–2010 (GEM_NEMOc) and future 2079–2100 (GEM_NEMOf) periods are performed, driven by the second generation of the Canadian Earth System Model (CanESM2) for RCP8.5, a high-warming scenario characterised by increasing emissions throughout the 21st century. Water temperatures for the Great Lakes are initialised from a continuous (1950–2100) offline NEMO simulation driven by atmospheric fields produced by a previous CanESM2-driven transient climate change GEM simulation for the same emissions scenario at 0.44° resolution. Land surface conditions are initialised from the same 0.44° resolution GEM simulation. The temperature profiles for sub-grid lakes are initialised to 4.2 °C, except for the surface and the top layer, where the temperatures are set to 5 °C and 10 °C, respectively to represent the effects of surface cooling and residual heat from the preceding warm season, given the sparsity of observation data for smaller lakes in the study region (the simulations in the current and future climates were initialised in winter). Prior to assessing projected changes to HLES, the ability of GEM_NEMOc in simulating such events and their drivers is assessed through comparison with observations.

To objectively diagnose HLES from model outputs, the following fields are used: snowfall, lake ice fraction and wind fields. The algorithm for the HLES diagnostic is inspired by the work of Notaro et al. [3], which probes the 200-km region around the Great Lakes shorelines for intensive snowfall events (higher than 10 cm/day), that are associated with sufficient fetch over unfrozen lakes. Furthermore, to be considered an HLES, the snowfall within the 200 km zone surrounding the lake must be above the average snowfall for the 500 km zone around the lakes by 4 cm/day.

To determine projected changes to the temporal distribution of HLES, area-averaged values over the target region are analysed for current and future climates for the November–April HLES period. The link between projected changes to HLES and 2m air temperature, total precipitation and lake ice cover is established by studying the spatial change patterns during selected phases of the November–April period, as well as correlation maps for current and future climates. The linear correlation coefficients of HLES with 2m air temperature and total precipitation are estimated using the collocated time series of seasonal mean values. To assess relations between HLES and lake ice cover, spatial mean lake ice fraction time series were correlated with HLES time series at each grid cell within the HLES region around the Great Lakes.

Given that extreme snowfall events (including, but not limited to HLES) have the potential to cause significant disruptions, projected changes to the intensity-frequency distribution of daily HLES and total snowfall are also assessed. To this end, the daily HLES and total snowfall over all gridcells of the analysis region (within 200 km of the Great

Lakes shorelines) are used. The probability of daily snowfall exceeding diverse thresholds is calculated for both HLES and total snowfall, in both current and future climates. For the current climate, a bootstrapping procedure (consisting of 100 samples) based on selecting random days (with replacement) is employed to construct a confidence interval around the intensity-frequency curve, which allows for identifying statistically significant changes.

Since CAO are an important triggering mechanism for HLES, their climatology and projected changes are assessed following Wheeler et al. [1], where a grid cell is considered to be affected by CAO when the temperature is lower than 1.5 times the standard deviation below the 31-day running mean. Furthermore, a spatial continuity criterion is applied to ensure that the CAO affected grid points are contiguous and are associated with the same cold air mass. The applied criterion will also filter out regions smaller than $2^\circ \times 2^\circ$ (20 by 20 grid points).

3. Model's Ability in Simulating HLES and Its Drivers

To assess the ability of GEM-NEMO in simulating the factors controlling HLES, the lake ice fraction simulated by GEM_NEMOc is compared to observations from Canadian Ice Service (CIS) and National Ice Center (NIC). Additionally, the CAO frequency in GEM_NEMOc is compared to that obtained from the Daymet [28] dataset, which is a 1 km horizontal resolution dataset derived from daily observations at weather stations. Figure 2 shows that GEM_NEMOc has a negligible bias in lake ice fraction for the onset (November–December; ND) and active (January–February; JF) phases, while it underestimates lake ice fraction during the decline (March–April; MA) phase. CAO occur in all three phases but are more frequent during the onset (ND) and active (JF) phases, peaking during the latter, in both observations and GEM_NEMOc. GEM_NEMOc slightly underestimates CAO frequency during the onset phase and slightly overestimates CAO frequency during the active and decline phases. In general, the GEM-NEMO modelling system performs reasonably well at simulating the factors and mechanisms leading to HLES.

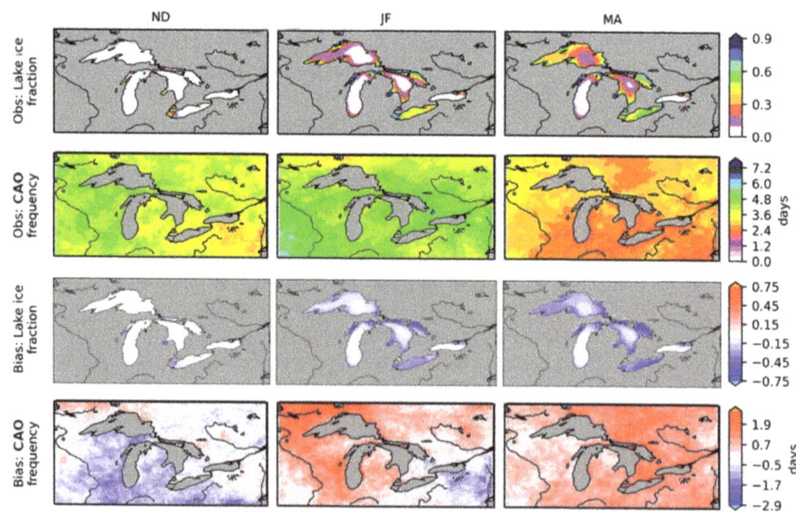

Figure 2. Observed fields (upper rows) and corresponding biases (lower rows) in the GEM_NEMOc simulated mean climatologic lake ice fraction and CAO frequency for the November–December (ND), January–February (JF), March–April (MA) periods of the 1989–2010 period. CAO frequency is not shown over the Great Lakes due to lack of reliable observation data.

To assess the ability of GEM-NEMO in simulating HLES realistically in current climate, GEM_NEMOc simulation spanning the 1989–2010 period, driven by CanESM2 outputs at the lateral boundaries, is compared to observations. The observed HLES is obtained by applying the algorithm described in the Methods section; the observed data used for the detection of HLES are as follows: total precipitation and 2m air temperature fields from Daymet dataset [28], wind field from ERA-Interim reanalysis and lake ice fraction observations from Canadian Ice Service (CIS) and National Ice Center (NIC).

The GEM_NEMOc simulation captures well the spatial variability of HLES (Figure 3) but underestimates the frequency and magnitude of HLES, particularly downwind of Lake Superior and Huron. Investigation of the 2m air temperature, total precipitation and lake ice fraction suggests that this underestimation is due to warm biases in the northern part of the study domain. The monthly distribution of the HLES amounts during the cold season is captured well by the model (figure not shown). The spatial correlations of modelled and observed HLES for ND, JF, and MA phases are, respectively, 0.74, 0.75 and 0.75. The correlation values for HLES frequencies are similar, i.e., 0.76 for the three phases.

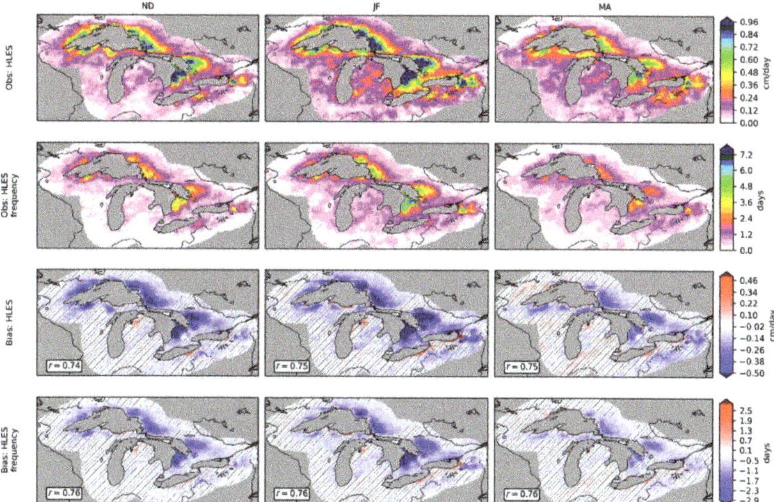

Figure 3. Observed fields (upper rows) and corresponding biases (lower rows) in the GEM_NEMOc simulated mean climatologic HLES (cm/day) and HLES frequency (days) for the November–December (ND), January–February (JF), March–April (MA) periods of the 1989–2010 period. Biases not significant at 10% significance level are hatched. Biases are not shown for the Great Lakes due to lack of reliable observation data.

4. HLES Characteristics in Current and Future Climates

The coupled system used in this study captures well HLES as demonstrated by the comparison between observed and GEM simulated HLES when driven by CanESM2 outputs at the lateral boundaries (see Figure 3) for the 1989–2010 period. The model reproduces the temporal and spatial patterns realistically, with maximum HLES occurring in January. Spatial correlations of simulated and observed HLES and that of other related variables for the HLES onset (ND), active (JF) and decline (MA) phases are larger than 0.6. This gives confidence to consider the coupled system to assess future changes. Comparison of GEM_NEMOf with GEM_NEMOc suggests decreases to the total cold season HLES in the region by 71%. The spatial and temporal differences in HLES are analysed further in detail in the following sections to gain a better understanding of the processes leading to the changes.

4.1. Spatio-Temporal Changes to HLES

In simulations of the current climate, HLES occurs during the November to April period, with maximum HLES occurring during the month of January (Figure 4), in agreement with observations. The HLES amounts for the months of October and May are less than 1% of the total received and are therefore not considered. In future climates, HLES amounts are significantly reduced during the entire HLES period, with most of the HLES occurring during the December to March period. For a detailed analysis of processes, the HLES onset, active and decline phases are considered. Spatial patterns of projected changes to HLES for these three phases (Figure 5) suggest decreases with the largest absolute decreases for JF. These decreases are due to decreases in the HLES frequency as reflected in the spatial patterns of changes for HLES rates and frequencies.

Analysis of the projected changes to lake ice fraction reveals important reductions for the active and decline phases. Changes are minimal for the onset phase as lakes are generally ice-free even in the current climate. The decrease in HLES during the onset phase in future climate, despite the availability of open water, is primarily due to the decrease in CAO days and also due to the higher liquid to solid precipitation ratio in a future warmer climate; the warmer air temperature in future climate leads to rainfall as opposed to snowfall in the current climate. This is reflected in the spatial plots of projected changes to snowfall, which show decreases, despite an increase in precipitation. As for the active JF phase, even though lake ice fraction is reduced in future climate, the decrease in HLES seems to be more related to the reduced snowfall associated with increased rainfall to snowfall ratio in future climate as significant changes to CAO are not noted. The decline phase HLES decreases are very similar to that of the onset ND phase, with both precipitation phase and CAO decreases contributing.

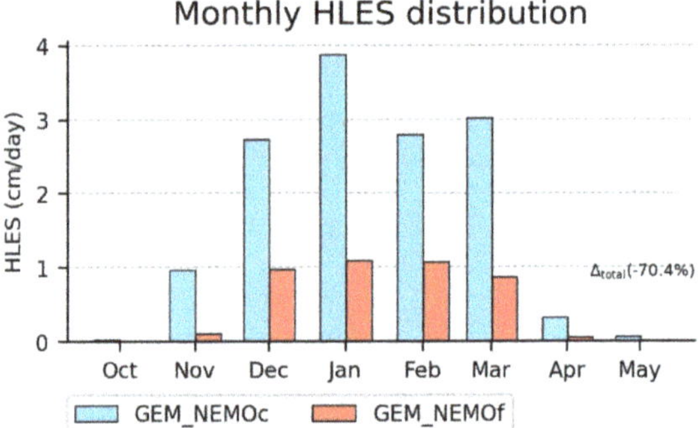

Figure 4. Modelled monthly distribution of area-averaged HLES (cm/day) for the current 1989–2010 (blue, GEM_NEMOc) and future 2079–2100 (red, GEM_NEMOf) periods.

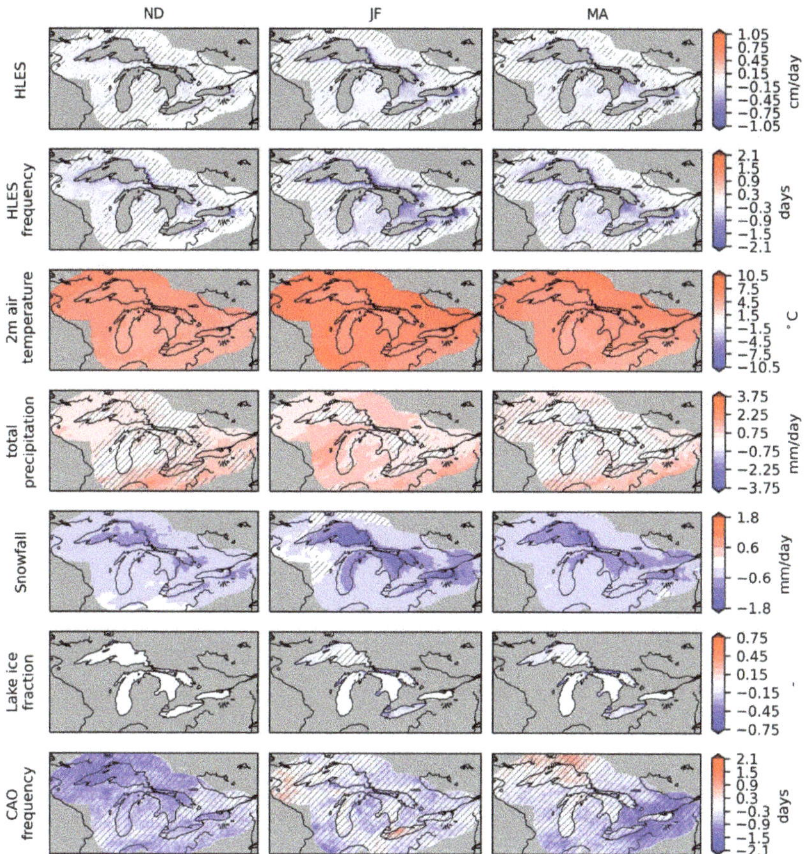

Figure 5. Projected changes to HLES (cm/day), HLES frequency (days), 2m air temperature (°C), total precipitation (mm/day), snowfall (mm/day), lake ice fraction, and cold air outbreaks (days) for the onset (ND), active (JF) and decline (MA) phases, for the 2079–2100 period with respect to the 1989–2010 period, for RCP8.5 scenario. Grid cells with no significant changes at 10% significance level are hatched.

4.2. Relative Importance of HLES Mechanisms in Current and Future Climates

To further investigate the links of HLES amount with air temperature, total precipitation and lake ice cover, the corresponding correlation maps are analysed for current and future climates for the three HLES phases. The correlations of HLES with surface air temperature in the current climate are predominantly negative for all three phases (Figure 6). There are different ways in which temperature can influence HLES. The importance of these influences on HLES is the direct effect of temperature on the precipitation phase and the indirect effect of temperature through its impact on lake ice fraction. The former is suggestive of negative HLES-temperature correlations, while the latter may lead to positive correlations. These correlations stay negative in future climate, particularly during the active JF phase, suggesting that the impact of 2m temperature on HLES is mostly through its direct effect on the precipitation phase. For the shoulder ND and MA phases, a mix of positive and negative correlations are noted for future climate suggesting that 2m temperatures influence both lake ice fraction and precipitation phase are important for HLES.

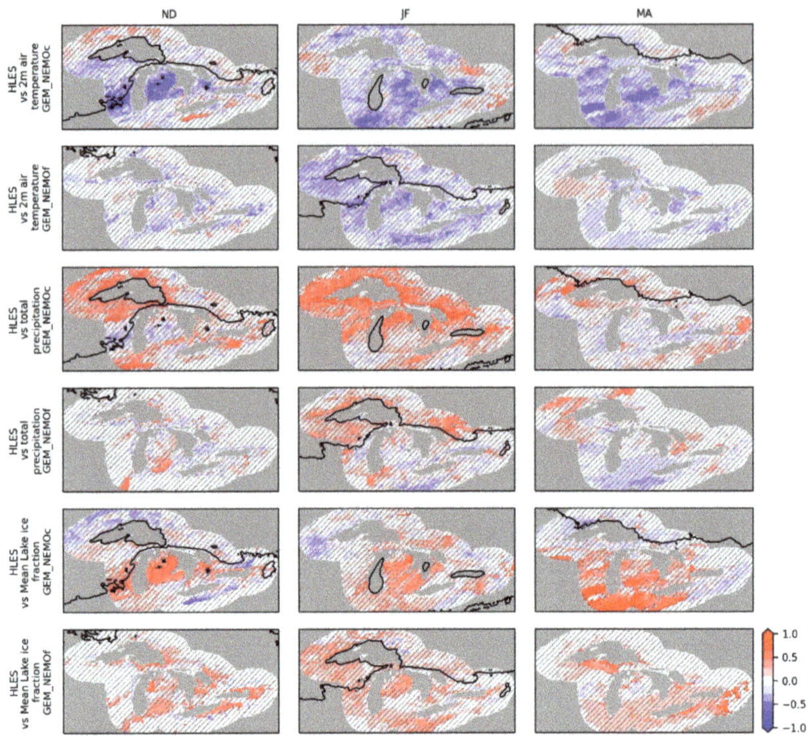

Figure 6. Correlation maps of HLES amount with respect to 2m air temperature, total precipitation and lake ice fraction for the onset (ND), active (JF) and decline (MA) phases, for the 1989–2010 period (GEM_NEMOc) and the 2079–2100 period (GEM_NEMOf). Grid cells with no significant correlations at 10% significance level are hatched.

The positive HLES–precipitation correlations in the current climate weaken or become negative in future climate due to the higher liquid to solid precipitation ratios in the future and is consistent with the more negative HLES-2m temperature correlations discussed above. The negative correlations of the spatial mean lake ice cover with HLES during the current climate, mostly noted to the north and east of the lakes, suggest the dominant influence of lake ice in those regions. The obtained positive correlations for future climate suggest the weakening influence of lake ice fraction on HLES.

4.3. Intensity of HLES and Total Snowfall in Current and Future Climates

Given that extreme snowfall events (irrespective of the driving mechanism) have the potential to cause significant disruptions, projected changes to the intensity-frequency distribution of daily HLES and total snowfall are assessed over the region within 200 km of the Great Lakes shorelines. The ~70% projected decrease in both HLES frequency and amount (see Figures 4 and 5) is also reflected in the intensity of HLES events, which is projected to decrease for extreme HLES events producing daily accumulations in excess of 40 cm (Figure 7). While total snowfall (see Figure 5), and snowy days (daily snowfall > 1 cm) are projected to decrease over the analysis domain, the most extreme snowfall events (daily snowfall > 50 cm) are projected to increase in frequency and intensity, with a number of future events projected to exceed the heaviest event in the current climate (Figure 7), suggesting that extreme snowfall-related impacts will continue and might increase in the future climate.

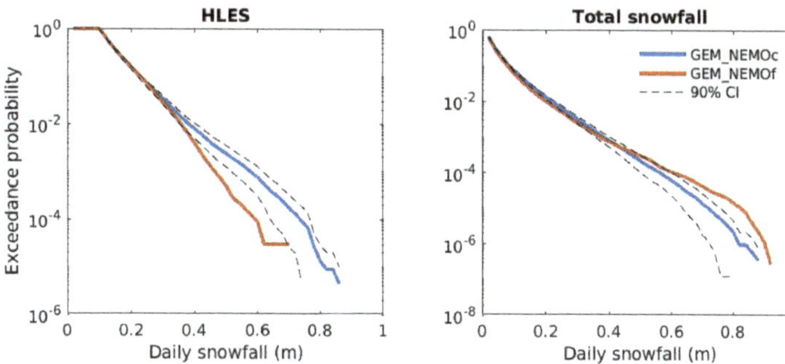

Figure 7. Intensity-frequency diagrams for HLES (**left**) and total snowfall (**right**) over the region within 200 km of the Great Lakes shorelines, for the 1989–2010 period (GEM_NEMOc, blue lines) and the 2079–2100 period (GEM_NEMOf, red lines). The 90% confidence interval for the 1989–2010 period is shown with black dashed lines.

5. Conclusions

This work presents a systematic analysis of the spatiotemporal characteristics of projected changes to heavy lake-effect snowfall for the Laurentian Great Lakes region, based on current and future simulations of a regional climate model coupled with NEMO for the Laurentian Great Lakes region. The analysis considers the onset, active and decline phases of HLES to study contributions of involved mechanisms: indirect temperature effect via lake ice, and direct temperature impact on the precipitation phase. Results suggest that HLES will decrease during the entire November to April HLES period, similar to Notaro et al. [3], despite the reduced lake ice cover fractions in the warmer future climate due to a lower fraction of precipitation falling as snow. Cold air outbreaks also generally decrease, except for the active phase, but do not contribute to increasing HLES due to the direct effect of warmer temperatures on the precipitation phase. A particularly interesting result, showing how the link between air temperature, lake ice cover and HLES can be modulated by the changing climate, suggests that the dominant role of lake ice fraction on HLES in current climate weakens in future climate, particularly for the active and decline phases of HLES.

While the number of lake-effect snow events is projected to decrease significantly because of fewer below freezing days by the end of the 21st century, it is possible that HLES might increase in the near future as lake temperatures rise while winter air temperatures remain cold enough to produce snow, continuing the trend of increasing HLES observed in recent decades. Both increases and decreases to HLES would have significant implications for communities and hydrologic systems that are influenced by lake-effect precipitation. Since snowfall-related impacts occur regardless of the driving mechanism, this study also analysed all extreme snowfall events, which are projected to increase in frequency and intensity in future climates. The implications include expenditures for snow removal, changes in soil moisture, lake and river levels and spring flooding, along with impacts on the agricultural, hydropower, recreational and transportation sectors.

It is recognised that this study does not cover all possible uncertainties related to the regional climate model formulation and driving data. Therefore, it is desirable to extend this study to include an ensemble of RCMs, driven by multiple GCMs for a variety of scenarios of future climates. This study nevertheless provides useful insights regarding future HLES characteristics and controlling mechanisms and represents the first study to assess projected changes to HLES and its mechanisms using a 3D lake model.

Author Contributions: Conceptualisation, L.S. and O.H.; methodology, L.S. and O.H.; software, O.H.; formal analysis, O.H. and B.T.; investigation, O.H. and B.T.; resources, L.S.; writing—original draft preparation, O.H. and L.S.; writing—review and editing, O.H., B.T., L.S. and R.Y.; visualisation, O.H. and B.T.; supervision, L.S.; project administration, L.S.; funding acquisition, L.S. and R.Y. All authors have read and agreed to the published version of the manuscript.

Funding: This research was funded by the Natural Sciences and Engineering Research Council (NSERC) of Canada (Grant Number: RGPCC 433915 – 2012) and Environment and Climate Change Canada (ECCC) (Grant Number: GCXE18S029).

Institutional Review Board Statement: Not applicable.

Informed Consent Statement: Not applicable.

Data Availability Statement: The data that support the findings of this study along with the analysis scripts are openly available in PANGAEA [29].

Acknowledgments: The research was enabled by the computing resources provided by Calcul Québec and Compute Canada.

Conflicts of Interest: The authors declare no conflict of interest.

References

1. Wheeler, D.D.; Harvey, V.L.; Atkinson, D.E.; Collins, R.L.; Mills, M.J. A climatology of cold air outbreaks over North America: WACCM and ERA-40 comparison and analysis. *J. Geophys. Res. Atmos.* **2011**, *116*. [CrossRef]
2. Yang, G.; Leung, L.R.; Jian, L.; Giacomo, M. Persistent cold air outbreaks over North America in a warming climate. *Environ. Res. Lett.* **2015**, *10*, 044001.
3. Notaro, M.; Bennington, V.; Vavrus, S. Dynamically Downscaled Projections of Lake-Effect Snow in the Great Lakes Basin. *J. Clim.* **2015**, *28*, 1661–1684. [CrossRef]
4. Vavrus, S.; Walsh, J.E.; Chapman, W.L.; Portis, D. The behavior of extreme cold air outbreaks under greenhouse warming. *Int. J. Clim.* **2006**, *26*, 1133–1147. [CrossRef]
5. Vavrus, S.; Notaro, M.; Zarrin, A. The Role of Ice Cover in Heavy Lake-Effect Snowstorms over the Great Lakes Basin as Simulated by RegCM4. *Mon. Weather. Rev.* **2013**, *141*, 148–165. [CrossRef]
6. Wright, D.M.; Posselt, D.J.; Steiner, A.L. Sensitivity of Lake-Effect Snowfall to Lake Ice Cover and Temperature in the Great Lakes Region. *Mon. Weather. Rev.* **2013**, *141*, 670–689. [CrossRef]
7. Janoski, T.P.; Broccoli, A.J.; Kapnick, S.B.; Johnson, N.C. Effects of Climate Change on Wind-Driven Heavy-Snowfall Events over Eastern North America. *J. Clim.* **2018**, *31*, 9037–9054. [CrossRef]
8. Bennington, V.; Notaro, M.; Holman, K.D. Improving Climate Sensitivity of Deep Lakes within a Regional Climate Model and Its Impact on Simulated Climate. *J. Clim.* **2014**, *27*, 2886–2911. [CrossRef]
9. Gula, J.; Peltier, W.R. Dynamical Downscaling over the Great Lakes Basin of North America Using the WRF Regional Climate Model: The Impact of the Great Lakes System on Regional Greenhouse Warming. *J. Clim.* **2012**, *25*, 7723–7742. [CrossRef]
10. Madec, G.; Delécluse, P.; Imbard, M.; Lévy, C. *OPA 8.1 Ocean General Circulation Model Reference Manual*; Notes du pôle de mode′lisation; laboratoire d′oce′anographie dynamique et de climatologie, Institut Pierre Simon Laplace des sciences de l′environnement global: Paris, France, 1998; Volume 11.
11. Dupont, F.; Chittibabu, P.; Fortin, V.; Rao, Y.R.; Lu, Y. Assessment of a NEMO-based hydrodynamic modelling system for the Great Lakes. *Water Qual. Res. J.* **2012**, *47*, 198–214. [CrossRef]
12. Cote, J.; Gravel, S.; Methot, A.; Patoine, A.; Roch, M.; Staniforth, A. The operational CMC-MRB Global Environmental Multiscale (GEM) model. Part I: Design considerations and formulation. *Mon. Weather. Rev.* **1998**, *126*, 1373–1395. [CrossRef]
13. Kain, J.S.; Fritsch, J.M. A One-Dimensional Entraining Detraining Plume Model and Its Application in Convective Parameterization. *J. Atmos. Sci.* **1990**, *47*, 2784–2802. [CrossRef]
14. Kuo, H.-L. On formation and intensification of tropical cyclones through latent heat release by cumulus convection. *J. Atmos. Sci.* **1965**, *22*, 40–63. [CrossRef]
15. Sundqvist, H.; Berge, E.; Kristjánsson, J.E. Condensation and cloud parameterization studies with a mesoscale numerical weather prediction model. *Mon. Weather. Rev.* **1989**, *117*, 1641–1657. [CrossRef]
16. Li, J.; Barker, H.W. A radiation algorithm with correlated-k distribution. Part I: Local thermal equilibrium. *J. Atmos. Sci.* **2005**, *62*, 286–309. [CrossRef]
17. McFarlane, N.A. The Effect of Orographically Excited Gravity Wave Drag on the General Circulation of the Lower Stratosphere and Troposphere. *J. Atmos. Sci.* **1987**, *44*, 1775–1800. [CrossRef]
18. Zadra, A.; McTaggart-Cowan, R.; Roch, M. Recent changes to the orographic blocking. In Proceedings of the Seminar presentation, RPN, Dorval, QC, Canada, 30 March 2012.
19. Zadra, A.; Roch, M.; Laroche, S.; Charron, M. The subgrid-scale orographic blocking parametrization of the GEM Model. *Atmos Ocean* **2003**, *41*, 155–170. [CrossRef]

20. Benoit, R.; Cote, J.; Mailhot, J. Inclusion of a Tke Boundary-Layer Parameterization in the Canadian Regional Finite-Element Model. *Mon. Weather. Rev.* **1989**, *117*, 1726–1750. [CrossRef]
21. Delage, Y. Parameterising sub-grid scale vertical transport in atmospheric models under statically stable conditions. *Bound. -Layer Meteorol.* **1997**, *82*, 23–48. [CrossRef]
22. Delage, Y.; Girard, C. Stability Functions Correct at the Free-Convection Limit and Consistent for Both the Surface and Ekman Layers. *Bound. -Layer Meteorol.* **1992**, *58*, 19–31. [CrossRef]
23. Verseghy, D.L. CLASS—A Canadian Land Surface Scheme for GCMs.1. Soil Model. *Int. J. Clim.* **1991**, *11*, 111–133. [CrossRef]
24. Verseghy, D.L. Local climates simulated by two generations of Canadian GCM land surface schemes. *Atmos Ocean* **1996**, *34*, 435–456. [CrossRef]
25. Hostetler, S.W.; Bates, G.T.; Giorgi, F. Interactive Coupling of a Lake Thermal-Model with a Regional Climate Model. *J. Geophys. Res.-Atmos* **1993**, *98*, 5045–5057. [CrossRef]
26. Martynov, A.; Sushama, L.; Laprise, R.; Winger, K.; Dugas, B. Interactive Lakes in the Canadian Regional Climate Model, Version 5: The Role of Lakes in the Regional Climate of North America. 2012. Available online: https://www.tandfonline.com/doi/full/10.3402/tellusa.v64i0.16226 (accessed on 23 November 2021).
27. Vancoppenolle, M.; Bouillon, S.; Fichefet, T.; Goosse, H.; Lecomte, O.; Morales Maqueda, M.A.; Madec, G. *LIM The Louvain-la-Neuve Sea Ice Model*; Note du Pôle de Modélisation de l'Institut Pierre-Simon: Laplace, France, 2012.
28. Thornton, P.E.; Running, S.W.; White, M.A. Generating surfaces of daily meteorological variables over large regions of complex terrain. *J. Hydrol.* **1997**, *190*, 214–251. [CrossRef]
29. Huziy, O. HLES projected changes study dataset. *Pangaea* **2020**. [CrossRef]

atmosphere

MDPI

Article

Offshore Wind and Wave Energy Complementarity in the Greek Seas Based on ERA5 Data

Kimon Kardakaris [1], Ifigeneia Boufidi [1] and Takvor Soukissian [2,*]

[1] School of Naval Architecture and Marine Engineering, National Technical University of Athens, Zografou Campus, 9, Iroon Polytechniou, 157 80 Zografou, Greece; kimonkardakaris@mail.ntua.gr (K.K.); ifigeneiaboufidi@mail.ntua.gr (I.B.)

[2] Institute of Oceanography, Hellenic Centre for Marine Research, 190 13 Anavyssos, Greece

* Correspondence: tsouki@hcmr.gr

Abstract: In this work, 20 years (2000–2019) of ERA5 wave and wind data are analyzed and evaluated for the Greek Seas by means of in-situ measurements derived from the POSEIDON marine monitoring system. Four different statistical measures were used at six locations, where in-situ wind and wave measurements are available from oceanographic buoys. Furthermore, the ERA5 wind and wave datasets were utilized for the estimation of the available wind and wave energy potential for the Greek Seas, as well as for the assessment of complementarity and synergy between the two resources. In this respect, an event-based approach was adopted. The spatial distribution of the available wind and wave energy potential resembles qualitatively and quantitatively the distributions derived from other reanalysis datasets. Locations with high synergy and complementarity indices were identified taking into account water depth. Finally, taking into consideration a particular offshore wind turbine power curve and the power matrix of the PELAMIS wave energy converter, the estimation of the combined energy potential on a mean annual basis is performed.

Keywords: offshore wind energy; wave energy; hybrid wind–wave energy; complementarity–synergy; Greek seas; ERA5

Citation: Kardakaris, K.; Boufidi, I.; Soukissian, T. Offshore Wind and Wave Energy Complementarity in the Greek Seas Based on ERA5 Data. *Atmosphere* **2021**, *12*, 1360. https://doi.org/10.3390/atmos12101360

Academic Editor: Baojie He

Received: 13 September 2021
Accepted: 13 October 2021
Published: 18 October 2021

1. Introduction

Over recent decades, the global demand for energy has been rapidly increasing. This, in combination with the decline of fossil fuels, makes the need for renewable energy more urgent than ever. The marine environment is a vast source of renewable energy. Among the marine renewable energy (MRE) technologies, the most mature in terms of technological development and large-scale deployment is offshore wind energy [1]. On the other hand, onshore wind farms face strong social opposition and the most favorable siting locations have been occupied [2].

The main advantage the marine environment offers, regarding offshore wind turbines, is that the offshore winds are generally stronger and less variable, thus allowing operation at maximum capacity for a larger percentage of the time. However, the installation in deeper waters is an important factor, and mainly a current disadvantage, that determines to a great extent the high construction and maintenance costs [3,4]. Among the other forms of MRE, the exploitation of wave energy is also promising in areas with relatively calm wave climates, characterized thus by intermediate levels of power availability like the Mediterranean Sea [5]. In the relevant literature, it is usually considered that wave energy has some particular advantages such as small energy loss, better predictability, and higher energy density; see, for example [6–8].

1.1. Hybrid Offshore Wind–Wave Energy

A potential solution in order to compensate for the high installation and maintenance costs of offshore wind energy is through the development of hybrid systems that combine offshore wind and an alternative marine renewable energy source (e.g., wave energy or offshore solar energy, etc.). Hybrid offshore wind–wave (WW) energy farms refer to wave energy extraction at the same marine space or on a shared platform with offshore wind turbines; see, e.g., [9,10]. See also the reviews in [11,12]. In this respect, there are several advantages as regards the development of hybrid WW energy farms that can provide important benefits such as:

(i) The better utilization of the available marine space, whereas marine spatial planning, is an important prerequisite in this direction. For example [13] studied the joint exploitation of the WW resource for the Italian seas based on a marine spatial planning framework. A relevant assessment was performed for the island of Tenerife, where the optimal positions for collocating WW energy devices were examined taking into consideration the bathymetry and the distance from ports [14]. In [15], a review on the multiple-use of marine space is presented.
(ii) The reduced power variability, especially for locations where wind and wave resources are not strongly correlated. These aspects were recently investigated in a multisite analysis in [16], using field observations of met-ocean conditions. It was found that the reduction in variability depends on the magnitude and lag of resource correlation and the wind–wave capacity mix of the particular location, rendering thus hybrid systems more beneficial in certain locations than others; see also [17,18]. Furthermore, Ref. [19] studied the WW resource for the Black Sea, Ref. [20] for specific locations at the coasts of Ireland, and [21] for the European seas.
(iii) The enlargement of weather windows for operation and maintenance [11,22]. Moreover, wave energy converters (WECs), acting as wave barriers, can create a wave shadow area where wind turbines can be installed in milder sea state conditions [22].
(iv) The decrease in the potential environmental impacts compared to the impacts of the stand-alone installations [23].

Collocation of WECs with offshore wind turbines is also advantageous in financial terms (e.g., by using the same grid infrastructure and port facilities, following a common consenting process, increasing the capacity factor of the installation, etc.). See also the relevant discussion in [11,24].

In order to identify suitable offshore areas for a hybrid WW installation, it is necessary to evaluate the impact of the combined use on the variability of the final total energy output. This is usually quantified through exploitability indices, which combine information on the availability of wind and wave energy, along with the degree of correlation between the two resources [25]. The assessment of co-located WW farms off the Danish coast was studied in [18]. An approach based on the assessment of the available resources and technical constraints was developed by implementing the Co-Location Feasibility (CLF) index that takes into account not only the available power, but also the correlation between the resources and the power variability.

Sustainability indices have also been investigated for the co-location of offshore wind farms and aquaculture plans [26], whereas wave energy, offshore wind energy and aquaculture activities were studied in the Canary Archipelago, resulting in a methodology and a useful tool to mapping the co-location opportunities [27]. In order to examine a combined WW energy farm, Ref. [17] examined real meteorological data to determine the difference in power performance between a wind turbine combined with a number of WECs than a single wind turbine. They concluded that the power variability (i) is reduced in the case of the combined resources for any number of WECs and (ii) the two resources exhibit really good complementarity that is, however, strongly dependent on the site selection.

1.2. Synergy and Complementarity

In the feasibility studies for hybrid WW energy farm development, complementarity is one of the most important aspects.

Regarding complementarity and synergy assessment between renewables in general, there are several studies in the relevant literature, examining in particular wind and solar power onshore or offshore. Such studies were conducted in Italy using high-resolution data, while apart from the assessment of spatial and temporal complementarity, its scale is approached via Monte Carlo simulation [28]. In [29], the possibility of the combined use of solar and wind energy over Europe is assessed, examining also the complementarity at different time scales using 3-year long time series. In [30], a study for the complementarity of wind and solar resources over Mexico is performed using GIS-based software with high-resolution maps. A systematic quantitative analysis of the complementarity characteristics of solar and wind resources on the Australian continent is performed in [31], exhibiting the spatio-temporal synergy of the two sources, while in [32], the spatio-temporal complementarity of solar and wind power is studied at different time scales for Germany. Relevant assessments were also carried out for offshore regions of China, where wind power is always negatively correlated with photovoltaic power on the hourly, daily and monthly time scales [33]. Recently [24] assessed in-depth offshore wind and solar complementarity aspects for the Mediterranean basin using ERA 5 reanalysis data. A review as regards complementarity/synergy between renewable energy sources is provided in [34].

However, for the assessment of the synergistic characteristics of WW power, there is very limited related work, as opposed to the ones about wind and solar, especially over the marine environment of the Greek seas [35]. Specifically, an investigation of the combined use of WW power for remote islands in Greece, was conducted in [36], using stochastic processes, without, however, a further assessment on their complementarity. The site selection for hybrid offshore WW energy systems in the Greek seas was studied in [37] based on the Analytic Hierarchy Process (AHP) method where environmental impacts have also been assessed.

The structure of this paper is the following: In Section 2, the ERA5 reanalysis wind and wave data for the Greek Seas that was used in the analysis are described and evaluated against in-situ measured data obtained from oceanographic buoys. In Section 3, the theoretical background of an event-based complementarity and synergy analysis is presented in brief. In Section 4, the numerical results of this work are presented and discussed. As a first step, the offshore wind and wave power potential is estimated and then, the complementarity and synergy indices are calculated for the examined area, while the water depth was also taken into consideration. In addition, a realistic application is presented, considering an actual wind turbine and an attenuator WEC at the most favorable locations regarding synergy/complementarity. Finally, in Section 5 some concluding remarks are provided.

2. ERA 5 Reanalysis Data and Initial Evaluation

In this section, the ERA5 reanalysis wind and wave dataset is described and evaluated for the Greek Seas, by using in-situ measured data obtained from six oceanographic buoys.

2.1. ERA5 Dataset

Produced by the European Centre for Medium-Range Weather Forecasts (ECMWF), the ERA5 reanalysis dataset combines vast amounts of historical observations into global estimates using advanced modeling and data assimilation systems [38]. Moreover, it provides hourly estimates of a large number of atmospheric, land and oceanic climate variables, covering the Earth on a ~30 km grid and resolving the atmosphere using 137 levels from the surface up to a height of 80 km. The data can be freely accessed from the Copernicus Climate Data Store (https://cds.climate.copernicus.eu/, (accessed on 3 March 2021) see also https://www.ecmwf.int/en/forecasts/datasets/reanalysis-datasets/era5, (accessed on 3 March 2021) [38,39].

Regarding ERA5 wind and wave data, in [40] sea surface wind speed data over the Caspian Sea were evaluated in comparison with measurements from offshore platforms and showed good agreement for measurements greater than 2 m/s. In [41] wind speed data over the South and Southeast Brazilian coastline from ERA5 and two more reanalysis datasets were compared against in-situ measurements and concluded that ERA5 has a better performance. Furthermore, in [42], ERA5 wave data were compared to an observed wave dataset collected offshore in the swell-dominated region of the Oman coast in the western Arabian Sea. It was concluded that a finer grid than the one provided by the ERA5 wave dataset is necessary to fully model the complexity of the region.

In this work, 20 years (1 January 2000–31 December 2019) of available wind and wave data were utilized for the Greek Seas (defined by a rectangle with the top left corner at 42° N, 19° E and bottom right corner at 33° N, 30° E). For the significant wave height and the wave energy period the data are provided on a $0.50 \times 0.50°$ spatial grid, while for the wind speed, the data are available at 100 m height (i.e., at a typical wind turbine hub height) on a $0.25° \times 0.25°$ spatial grid; see [39].

2.2. In-Situ Measurements

In-situ measurements are provided by the POSEIDON marine monitoring network in the Greek seas. POSEIDON system was established in 1999 and comprises of oceanographic buoys, deployed in deep water locations, https://poseidon.hcmr.gr/ (accessed on 7 June 2021); see also [43,44]. The buoys measure the most important spectral wave parameters (significant wave height, spectral peak period and mean wave direction) as well as wind speed and direction at a height of 3 m above the sea surface. Buoy measurements were widely considered as the primary reference data source for model-generated and satellite wind validation; see, e.g., [45–48]. It is worth mentioning that buoys measure temporal averages at a specific location while model-generated data refer to instantaneous spatial averages. An important disadvantage of the measurements obtained from buoys is their limited spatial coverage, a drawback which is partly compensated by the increased measurement accuracy [49].

For the evaluation of ERA5 data, in-situ wind and wave measurements are obtained and analyzed from six buoys deployed in the following locations: Mykonos [37.51° N, 25.46° E], Lesvos [39.15° N, 25.81° E], Santorini [36.26° N, 25.50° E], Athos [39.96° N, 24.72° E], Pylos [36.83° N, 21.62° E], and Zakynthos [37.95° N, 20.60° E]; see Figure 1 and Table 1. Wind measurements are averaged over a recording period of 600 s with a sampling frequency of 1 Hz and a recording interval of 3 h. Wave measurements have a recording period of 1024 s, sampling frequency 1 Hz and recording interval 3 h. Buoy records exhibit gaps of various lengths due to software malfunctioning, hardware damage, etc.; nevertheless, the number of available records is sufficient to carry out the statistical analysis.

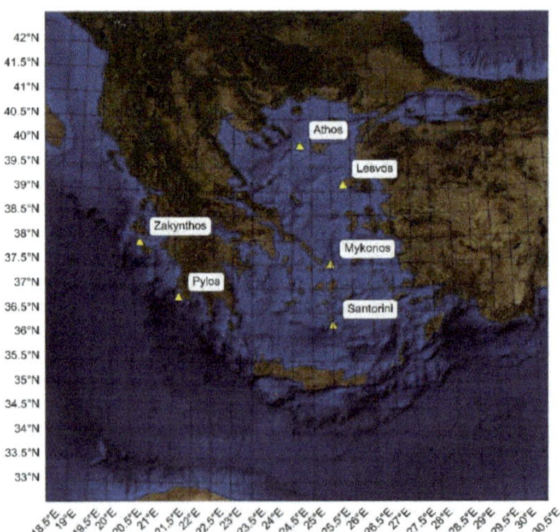

Figure 1. Location of the POSEIDON buoys.

Table 1. POSEIDON buoy locations and measurement period.

Buoy Location	Coordinates	Measurement Period
Mykonos	37.51° N, 25.46° E	1999–2012
Lesvos	39.15° N, 25.81° E	1999–2012
Santorini	36.26° N, 25.50° E	1999–2012
Athos	39.96° N, 24.72° E	2000–2015
Pylos	36.83° N, 21.62° E	2007–2015
Zakynthos	37.95° N, 20.61° E	2008–2011

2.3. Data Reparation

In order to compare wind and wave data from both sources, the corresponding time series have to be co-located in the spatial and temporal domain. Therefore, all the data were fixed to the 3-h resolution (corresponding to the recording interval of the in-situ data). Subsequently, the ERA5 reanalysis dataset was spatially co-located with the buoy data via the nearby grid point values by using a simple form of inverse squared distance weighting interpolation function based on the values of the four nearest grid points; see, e.g., [49]. Denoting x_1, x_2, x_3 and x_4 the respective variables (wind or wave parameters) at the four grid points surrounding the buoy location, and r_1, r_2, r_3 and r_4 the corresponding distances from that location, the requested data for each variable at the specific buoy location can be estimated as follows:

$$x = \frac{\sum_{i=1}^{4} \frac{x_i}{r_i^2}}{\sum_{i=1}^{4} \frac{1}{r_i^2}} \tag{1}$$

It is worth mentioning that for the buoy in Pylos, the estimation of the examined wave parameters took into consideration only three grid points, as one of the four nearby locations was onshore and thus it was impossible to include it in the calculation procedures.

For wind speed comparison and evaluation purposes, the common reference height above the sea surface was set at 3 m for both wind data sources. To enable the comparison, the ERA5 wind speed data were adjusted to 3 m height from the surface by using the following equation:

$$u_{h_2} = u_{h_1} \frac{\ln\left(\frac{h_2}{z_0}\right)}{\ln\left(\frac{h_1}{z_0}\right)} \tag{2}$$

where u_{h_2} (m/s) is the calculated wind speed at height h_2 (m), u_{h_1} (m/s) is the known wind speed at height h_1 (m), and z_0 (m) is the roughness length equal to 0.0002 m for neutral atmospheric conditions.

Regarding wave data, the in-situ wave period measurements refer to the spectral peak period T_p. Under the JONSWAP spectrum with a peak enhancement $\gamma = 3.3$, T_p and the wave energy period T_e are approximately related as follows:

$$\frac{T_e}{T_p} \approx 0.9 \tag{3}$$

see also [50].

2.4. Data Evaluation

In order to evaluate the performance of the ERA5 reanalysis dataset, four different statistical measures were adopted for every variable and measuring station, e.g., Root Mean Squared Error (*RMSE*), Mean Bias Error (*MBE*), or simply bias, Pearson correlation coefficient (r), and Mean Absolute Error (MAE). $RMSE$, MBE and MAE are, respectively, defined as follows:

$$RMSE = \sqrt{\frac{1}{n}\sum_{i=1}^{n}(X_i - \hat{X}_i)^2} \tag{4}$$

$$MBE = bias = \frac{1}{n}\sum_{i=1}^{n}(X_i - \hat{X}_i) \tag{5}$$

$$MAE = \frac{1}{n}\sum_{i=1}^{n}|X_i - \hat{X}_i| \tag{6}$$

where X_i denotes the measured parameter obtained from the buoy and $\hat{X}_i$ the corresponding parameter obtained from the ERA5 dataset. $RMSE$ is actually the standard deviation of the errors (also referred to as prediction errors), between the "true" values of a variable (i.e., the X_i's) and the corresponding values obtained from experiments (i.e., the $\hat{X}_i$'s). In this respect, $RMSE$ provides a measure of the spread of $X_i - \hat{X}_i$. MBE (bias) is the mean error between the X_i's and $\hat{X}_i$'s, and MAE is the corresponding mean absolute error (absolute bias). The above-mentioned measures provide different forms of absolute error between the two data sources.

For the estimation of the corresponding relative errors, the relative error RE and the scatter index SI are also introduced:

$$RE = \frac{100}{n}\sum_{i=1}^{n}\frac{(X_i - \hat{X}_i)}{X_i} \tag{7}$$

$$SI = \frac{\sqrt{\frac{1}{n}\sum_{i=1}^{n}(X_i - \hat{X}_i - MBE)^2}}{\overline{X}} \tag{8}$$

respectively, where $\overline{X}$ denotes the mean value of X.

The lower the values of the metrics provided in Equations (4)–(8), the better the agreement between X and $\hat{X}$.

As an indicative result, in Figures 2–4, the scatter diagrams of H_S, T_e and u_W at 100 m asl, as obtained from buoys and ERA5 dataset along with the corresponding regression lines for two locations in the northern Aegean Sea (Athos) and Ionian Sea (Pylos) are, respectively, provided. The estimation of the regression lines was performed using the ordinary least squares method, after excluding outliers. The identification of the outliers was based on the Cook's distance criterion, i.e.,

$$CD_i = \frac{\sum_{j=1}^{n}\left(\hat{y}_j - \hat{y}_{j(i)}\right)^2}{s^2}, \ i = 1, 2, \cdots, n, \tag{9}$$

where n is the number of data points, $\hat{y}_{j(i)}$ is the fitted response value (i.e., the values of ERA5 dataset) obtained after excluding the i-th observation, and s^2 is the mean squared error.

The corresponding statistical results for H_S, T_e and u_W are summarized in Tables 2–4, respectively.

The ERA5 reanalysis dataset, in general, tends to underestimate the actual values of the examined parameters. Specifically, MBE is positive in all locations (except for Santorini) and for all wind and wave parameters examined (except for u_W in Athos). Furthermore, the correlation coefficient values take values well above 0.8 for every case (except for Santorini as regards T_e and Pylos, Zakynthos as regards u_W), which indicates a rather strong correlation between the two data sources.

The largest relative errors regarding wave parameters are encountered in Pylos (26.26% for H_S and 23.6% for T_e), while the scatter index SI takes its maximum values (0.308 for H_S and 0.183 for T_e) at Santorini. For wind speed, the largest relative error, in absolute terms, corresponds to Athos (−34.69%) and the largest scatter index (0.406) to Zakynthos.

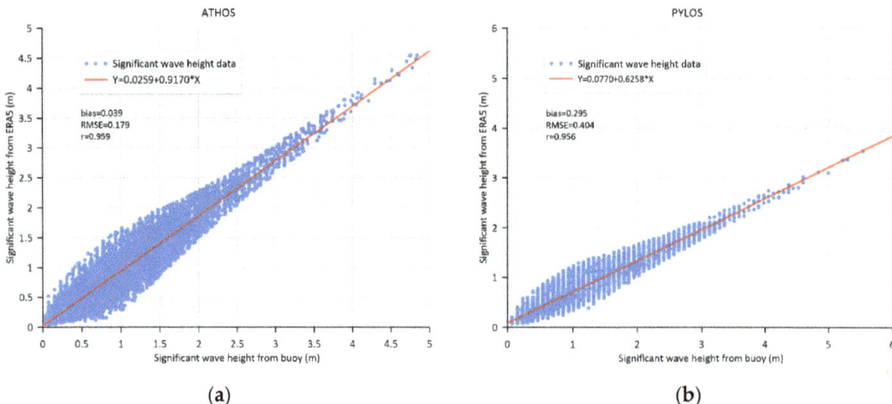

Figure 2. Scatter plots of H_S obtained from buoy and ERA5 dataset for Athos (**a**) and Pylos (**b**).

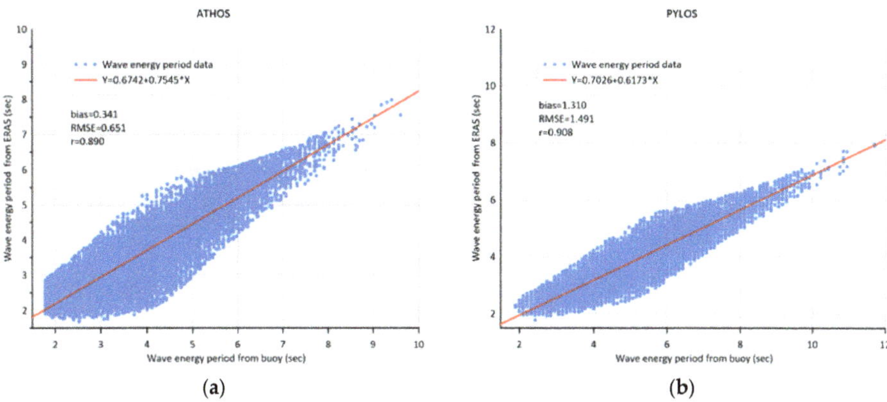

Figure 3. Scatter plots of T_e obtained from buoy and ERA5 dataset for Athos (**a**) and Pylos (**b**).

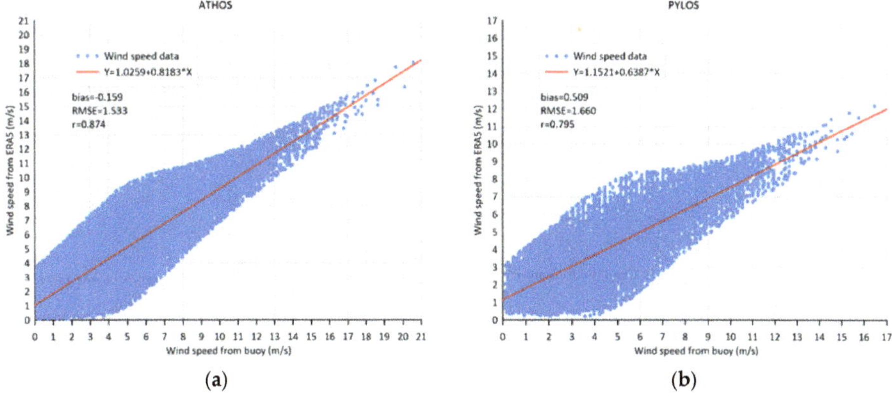

Figure 4. Scatter plots of u_W obtained from buoy and ERA5 dataset for Athos (**a**) and Pylos (**b**).

Table 2. Results of data evaluation of the significant wave height H_S obtained from ERA5 and in-situ measurements.

Parameter	Athos	Mykonos	Pylos	Santorini	Lesvos	Zakynthos
$RMSE$	0.179	0.272	0.404	0.306	0.201	0.186
MBE	0.039	0.025	0.295	−0.157	0.030	0.024
r	0.959	0.922	0.956	0.873	0.905	0.935
MAE	0.132	0.205	0.307	0.232	0.148	0.142
RE (%)	−1.304	−11.969	26.263	−23.392	2.868	2.692
SI	0.224	0.274	0.278	0.308	0.276	0.234

Table 3. Results of data evaluation of the wave energy period T_e obtained from ERA5 and in-situ measurements.

Parameter	Athos	Mykonos	Pylos	Santorini	Lesvos	Zakynthos
$RMSE$	0.651	0.808	1.491	0.822	0.638	0.898
MBE	0.341	0.200	1.310	−0.131	0.294	0.478
r	0.890	0.832	0.908	0.731	0.870	0.877
MAE	0.522	0.641	1.318	0.672	0.508	0.743
RE (%)	6.328	−0.741	23.628	−7.238	4.826	6.553
SI	0.134	0.176	0.136	0.183	0.139	0.158

Table 4. Results of data evaluation of the wind speed u_W obtained from ERA5 and in-situ measurements.

Parameter	Athos	Mykonos	Pylos	Santorini	Lesvos	Zakynthos
$RMSE$	1.533	1.627	1.660	1.641	2.025	1.959
MBE	−0.159	0.232	0.509	−0.249	0.829	0.309
r	0.874	0.879	0.795	0.821	0.807	0.716
MAE	1.220	1.326	1.350	1.309	1.649	1.566
RE (%)	−34.687	−19.405	−9.060	−25.459	−6.471	−25.932
SI	0.320	0.235	0.344	0.287	0.305	0.406

It should be noted here that a part of the discrepancies derived from the above analysis, might be attributed to the following reasons: (i) the measured energy wave period T_e, is not a direct result from wave spectral analysis, but is obtained by the approximate relation (3); (ii) the wind speed u_W at 100 m asl, is a result obtained from the measured wind speed at 3 m asl by using the approximate relation (2). On the other hand, it can be seen that the performance of ERA5 is strongly site-dependent. As a general conclusion, it can be stated that the ERA5 dataset underestimates the wind and wave conditions in the Greek Seas. However, for the scope of this work, the ERA5 dataset performs satisfactorily compared to buoy measurements, since the deviations are almost all acceptable; the detailed evaluation of ERA5 wind and wave datasets is an undergoing activity of the authors.

3. Wind and Wave Synergy and Complementarity

Let us first define the most important wind and wave power characteristics. The wind power incident on a surface A is provided by the following equation:

$$WP = \frac{1}{2}\rho A u_W^3 \tag{10}$$

where ρ is the atmospheric density that is considered constant and equal to 1.225 kg/m^3, A is the vertical to the wind speed surface for which the power is calculated, and u_W is the wind speed in m/s. For $A = 1\ \text{m}^2$, the wind power density is obtained; this quantity will be used in the rest analysis and will be simply denoted as WP.

The wave energy flux per unit length of wave front (also known as wave power density or wave energy potential) is defined as follows:

$$VP = \frac{\rho g}{64\pi} H_S^2 T_e \cong 0.49 H_S^2 T_e \text{ (in kW/m)}, \quad (11)$$

where ρ is the density of seawater that is considered constant and equal to 1025 kg/m^3, g is the gravitational acceleration equal to 9.8066 m/s^2, H_S is the significant wave height in m, and T_e is the wave energy period in s.

For the complementarity/synergy analysis, we will follow the event-based approach that was presented in [24]; see also [31].

Specifically, some lower thresholds regarding the mean annual offshore wind and wave power potential, WP_L and VP_L, respectively, should be introduced. In principle, the locations that are of interest are the ones that satisfy the following conditions: $WP_{AN} > WP_L$ and $VP_{AN} > VP_L$, where WP_{AN} and VP_{AN} denote the mean annual wind and wave power density, respectively.

Then, the wind to wave complementarity index WCV, is defined as follows:

$$WCV = \Pr\left[[WP_{AN} > WP_L] \cap [VP_{AN} \leq VP_L]\right] \quad (12)$$

In this case, it is clear that if the mean annual wave power density is below the corresponding lower threshold VP_L, but the mean annual wind power density is above the corresponding lower threshold WP_L, then energy from wind complements energy from wave. For the estimation of this index, the mean annual wind and wave densities (for all examined years) should be estimated and compared to the corresponding lower thresholds. Then WCV can be estimated as the frequency of occurrence of the compound event $[WP_{AN} > WP_L] \cap [VP_{AN} \leq VP_L]$. In a similar way, the wave to wind power complementarity index VCW, can be defined as follows:

$$VCW = Pr[[WP_{AN} \leq WP_L] \cap [VP_{AN} > VP_L]] \quad (13)$$

Large values of the above indices suggest that the corresponding resources are strongly complementary, while low values denote poor complementarity.

Moreover, the following events can be also defined:

$$E_W = [WP > WP_L],\ E_V = [VP > VP_L] \quad (14)$$

Using the above events E_W and E_V, and the exclusive OR operator, the synergy index of wind and wave power, SWV is defined as follows:

$$SWV = Pr[E_W \oplus E_V] \quad (15)$$

where $\oplus$ denotes the exclusive OR operator. This operator yields true if exactly one of E_W, E_V is true. The operator yields false if both E_W, E_V are true or both are false. For the estimation of the synergy index, the instantaneous events WP and VP should be firstly compared with respect to the corresponding lower thresholds WP_L, VP_L. Then SWV can be estimated as the frequency of occurrence of the compound event $[WP > WP_L] \oplus [VP > VP_L]$ for the examined time series. Values of SWV close to 1 suggest areas that are highly synergetic, while low synergetic areas are characterized by values of SWV close to zero.

Moreover, the joint non-availability of wind and wave power index UWV, is defined as follows:

$$UWV = Pr[[WP_{AN} \leq WP_L] \cap [VP_{AN} \leq VP_L]] \quad (16)$$

where a value of UWV close to 1 indicates that the specific area is out of consideration regarding the joint development of offshore wind and wave energy.

The adopted lower thresholds WP_L and VP_L, are 280 W/m^2 and 5 kW/m, respectively.

4. Results

4.1. Offshore Wind and Wave Power Potential

In order to identify potentially more promising grid points for the co-exploitation of WW energy, it is necessary to assess first the two sources separately.

The spatial distributions of the mean annual wind and wave power potential of the Greek seas based on the ERA 5 datasets are presented in Figure 5a,b, respectively.

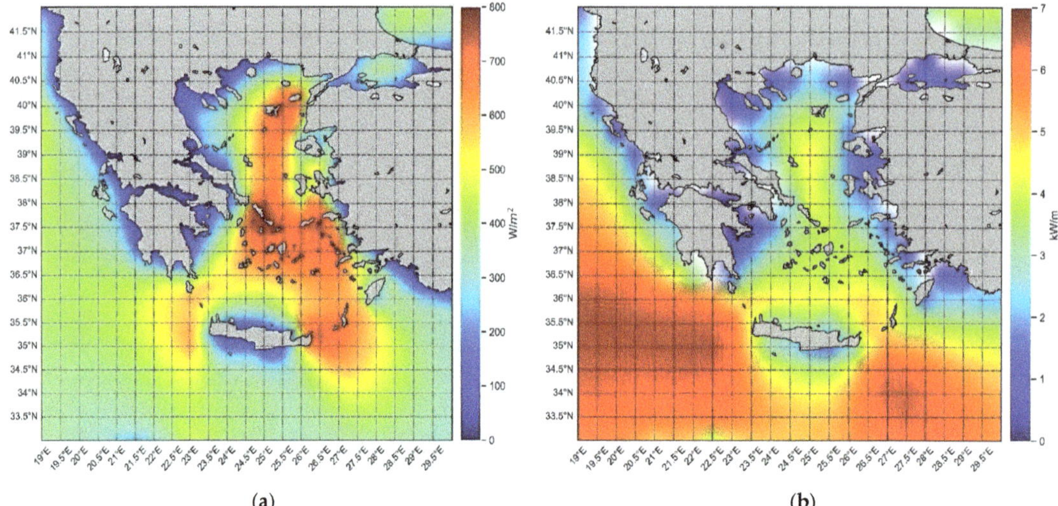

Figure 5. Offshore wind (**a**) and wave energy (**b**) potential.

From Figure 5a it is evident that the highest wind power density values (650–800 W/m^2), are encountered across the central axis of Aegean Sea, namely from Limnos and Samothraki Isl. up to the Cyclades complex, and from Samos and Ikaria Isl. up to the straits between Crete and Kasos Isl. These areas are characterized by strong winds during winter and the Etesian winds (seasonal winds of northern direction) during summer. Wind power potential values ranging between 500–600 W/m^2 are observed mainly in the straits of Crete and Kythira Isl. These results are in fair agreement with those provided in [51], where an Eta-based numerical atmospheric model of the POSEIDON system with a higher resolution (0.1 deg $\times$ 0.1 deg) was used.

Wave energy flux (see Figure 5b), ranges between 5–7 kW/m in west, southwest and southeast areas of Crete Isl. (areas between Crete and Kithira Isl., and Karpathos and Kasos Isl., respectively). These areas are characterized by large fetch lengths that lead to larger wind waves and swells. Although the Aegean Sea is an area characterized by strong winds, the presence of many islands limits the wind fetch blocking swells from being developed. Consequently, wave energy flux values are relatively low (3–5 kW/m). These results are quantitatively and qualitatively in fair agreement with the results of [52], who studied ten years of data obtained from the WAM-Cycle 4 numerical wave simulation model with a higher resolution (0.1 deg $\times$ 0.1 deg).

4.2. Synergy and Complementarity between Offshore Wind and Wave Energy

In this section, the assessment of the most favorable locations in terms of complementarity and synergy is examined. Specifically, in Figure 6, the complementarity indices WCV and VCW are provided, along with the synergy index SWV and the joint non-availability of wind and wave power index UWV. The areas characterized by strong wind to wave complementarity (i.e., high values of WCV) are encountered across the central and eastern

Aegean Sea, as well as in straits between Crete and Rhodes Isl. and between Crete Isl. and Peloponnesus. The overall maximum value of WCV is 45.143% and corresponds to the location (38.00° N, 26.50° E), at a water depth of ~325 m. The areas characterized by strong wave to wind complementarity (i.e., high values of VCW) are located very offshore in the southern Ionian Sea. The overall maximum value of VCW is 9.987% and corresponds to the location (33.00° N, 21.50° E), at a water depth of ~1008 m. Relatively high values of synergy between wind and wave are encountered across the central Aegean Sea, the western and eastern offshore areas of Crete Isl. and in offshore areas of the Ionian Sea. The overall maximum value of SWV is 31.072% and corresponds to the location (35.00° N, 27.00° E), at a water depth of ~3151 m. Finally, the joint non-availability of wind and wave power index takes very high values across almost all coastal areas of Greece.

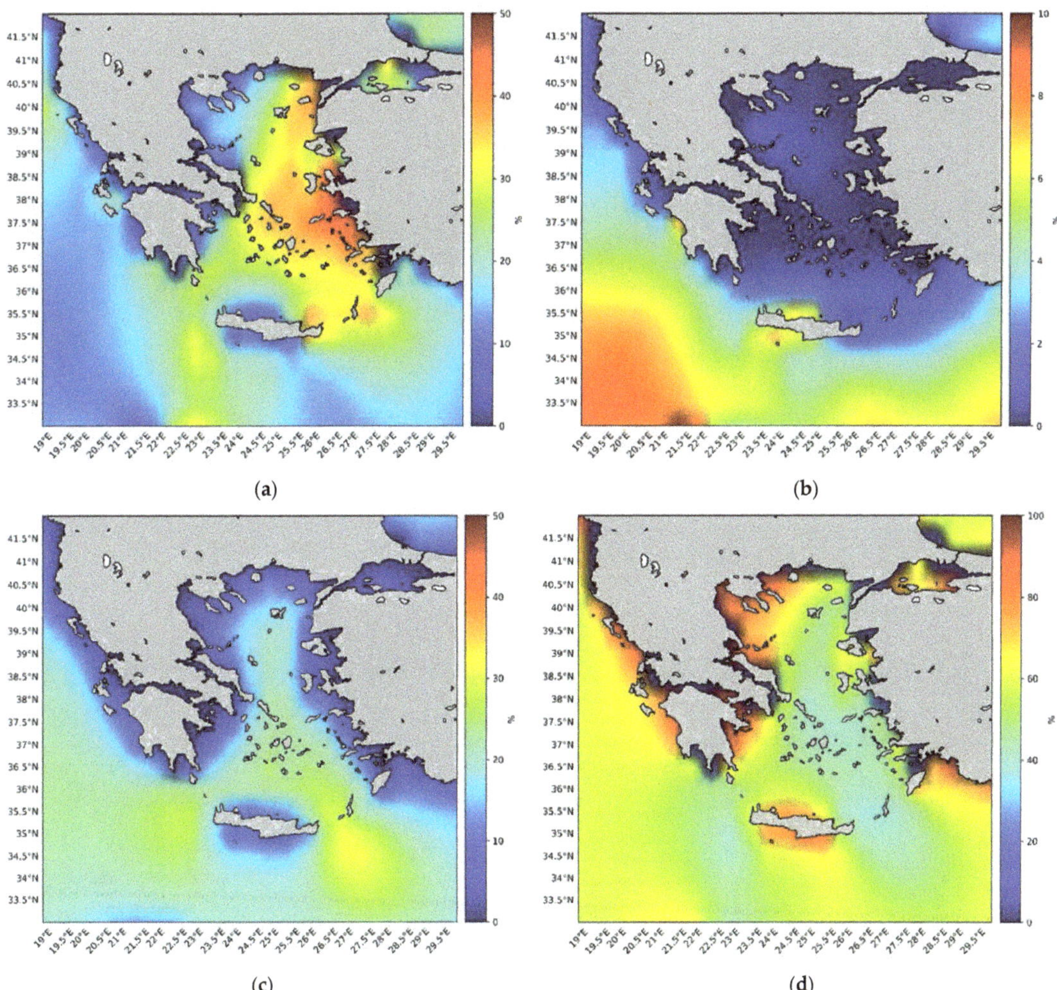

(a) (b) (c) (d)

Figure 6. Spatial distribution of WCV (**a**), VCW (**b**), SWV (**c**) and UWV (**d**).

Based on the theory presented in Section 3 and the results depicted in Figure 6, 50 locations characterized by the highest values of wind to wave complementarity (Figure 6 a) and synergy (Figure 6c) indices were identified. From these locations, the ones with water depths greater than 300 m were excluded from further analysis. This is in agreement with [53], where it is stated that "*The considered water depth is between 200 and 300 m, which is the deep-water range used in the current floating offshore wind turbine (FOWT) industry*". Let it be noted however that, regarding the offshore oil and gas sector, the relevant water depths refer to the level of thousands of meters [53].

The bathymetry data of the examined region were derived from the European Marine Observation and Data Network (EMODnet—https://portal.emodnet-bathymetry.eu/ (accessed on 21 July 2021)). Moreover, locations very close to the shore were also excluded from further analysis.

In Figure 7a, the grid points (of depths less than 300 m), exhibiting the highest values of complementarity index WCV between offshore wind and wave energy are shown. There are 23 locations with depths less than 300 m exhibiting the highest values of wind to wave complementarity. Note that all points are located in the Aegean Sea, while the location with the overall highest value of complementarity index (45.143%) is located southeast of Chios Isl. (38° N, 26.5° E). In Figure 7b eight grid points (of depths less than 300 m) that exhibit the highest values of synergy index SWV are depicted. All points are located in the southern Aegean Sea, except from point (35° N, 27.5° E) that exhibits the highest value of synergy index (29.261%) and is located southeast of Karpathos Isl.

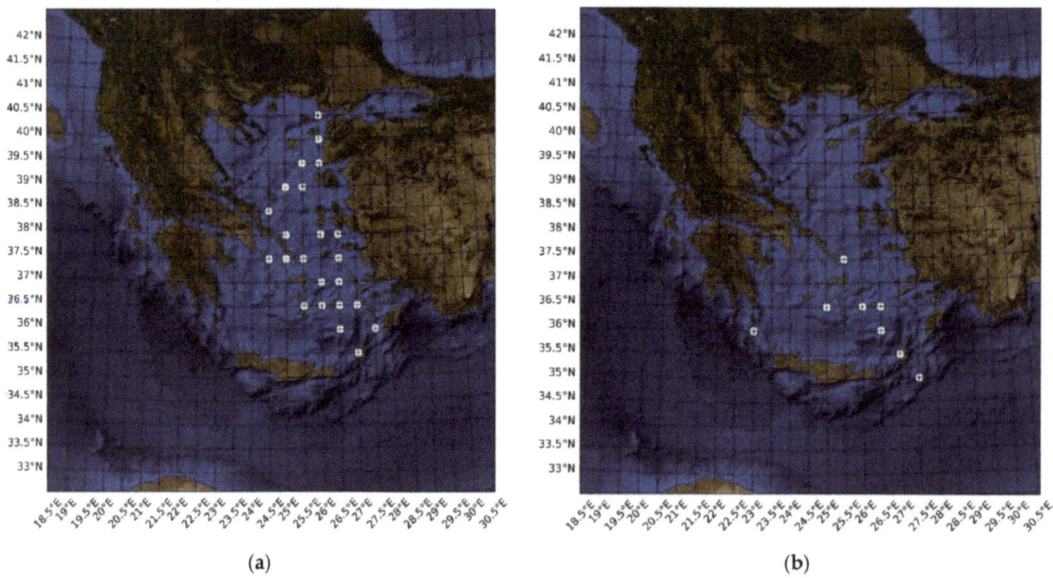

Figure 7. Locations with the highest values of wind to wave complementarity (**a**) and synergy (**b**) indices.

4.3. An Actual Application

For the exploitation of offshore wind energy, the Vestas V164-8.0 (an 8-MW offshore wind turbine specifically designed for offshore wind conditions) was selected. The technical specifications of the wind turbine can be found in https://en.wind-turbine-models.com/turbines/1419-mhi-vestas-offshore-v164-8.0-mw (last accessed 15 September 2021) and the corresponding power curve P_{WT} can be found in [54].

The mean power output of the Vestas V164-8.0 wind turbine at a specific location, where a discrete and sufficiently long time series of wind speeds u_i, $i = 1, 2, \cdots, n$, with a

1 h–time step (as is the time step of the ERA5 wind dataset in our case) is available, can be simply estimated as follows:

$$\overline{P}_W = \frac{1}{N}\sum_{i=1}^{N} P_{WT}(u_i) \tag{17}$$

where $P_{WT}(\cdot)$ is the corresponding power curve and N is the available sample size. The mean annual values of $P_{W,j}$, $j = 1, 2, \cdots, J$ for year j, can be estimated in a straightforward way, while the corresponding annual energy output for year j, $E_{W,j}$, $j = 1, 2, \cdots, J$ can be obtained by integrating $P_{W,j}$ over each year.

For $P_{WT}(\cdot)$ in kW, E_W is expressed in kWh. Although power curve modeling is frequently used in similar applications, see, e.g., [55], in this work, direct use is made of the wind turbine power curve as is provided in [54]. When it was required, linear interpolation was performed for estimating E_W at particular wind speeds.

The mean annual wind energy output $\overline{E}_{W,AN}$ can then be estimated as follows:

$$\overline{E}_{W,AN} = \frac{1}{J}\sum_{j=1}^{J} E_{W,j} \tag{18}$$

where J is the total of years (20 in our case).

For the exploitation of wave energy, the Pelamis wave energy converter was selected. The power characteristics $P_{VP}(H_S, T_e)$ of the Pelamis WEC are described in [56]. Following the same approach as above, the mean wave power output can be estimated as follows:

$$\overline{P}_V = \frac{1}{N}\sum_{i=1}^{N} P_{VP}(H_{S,i}, T_{e,i}) \tag{19}$$

where N is the number of (H_S, T_e)–sea states, and $P_{VP}(H_{S,i}, T_{e,i})$ is the power matrix of the Pelamis WEC. The mean annual values of $P_{V,j}$, $j = 1, 2, \cdots, J$ for year j, can be estimated in a straightforward way, while the corresponding annual energy output for year j, $E_{V,j}$, $j = 1, 2, \cdots, J$ can be obtained by integrating $P_{V,j}$ over each year.

The mean annual wave energy output can then be estimated as follows:

$$\overline{E}_{V,AN} = \frac{1}{J}\sum_{j=1}^{J} E_{V,j} \tag{20}$$

where J is the total of years.

The energy output, using the above methods for the Vestas wind turbine and the Pelamis wave energy converter, respectively, was estimated for the two most favorable locations that are identified in the previous section, in terms of complementarity (A) and synergy (B). The results are summarized in Table 5, including a layout of 1 × 12 Pelamis WECs, in order to achieve a nominal power close to 8 MW, resulting in a farm of a total capacity of 9 MW (12 × 750 kW).

Table 5. Energy outputs of Vestas offshore wind turbine and Pelamis devices in best complementarity (A) and synergy (B) points.

Device	Point A [38° N, 26.5° E]	Point B [35° N, 27.5° E]
1 Vestas (GWh)	33.080	34.406
1 Pelamis (MWh)	64.280	347.321
1 × 12 Array of Pelamis (MWh)	771.360	4167.852

The efficiency of the attenuator's WECs is dependent on the mean wave direction, see [57] and references cited therein. On the other hand, phenomena like diffraction and reflection usually occur inside the area of WEC array installation, as the presence of the wave farms affects the directional spreading and the wave spectral shape while it also reduces the amplitude of the sea waves. For quantification purposes, a coefficient called energy transmission factor, represents the percentage of energy remaining in the sea after the waves pass through the WEC array. Another factor called the "capture width" defines the length of the wave crest that is absorbed by the instrument and varies with both wave height and wave period. Transmission factors for WECs are partially commercially confidential, yet they have a dynamic behavior depending especially on the magnitude of the significant wave height and wave period [58]. The spatial distribution in wave power in the vicinity of a wave farm is strongly dependent on the device-incident wave climate and the device absorption parameters. In WEC locations, large reductions (~25%) in wave power leeward of the devices may occur [59]. Therefore, for an optimized operation, Pelamis arrays are better installed in a single line or in a two-row array. In this respect, as well as taking into account that the contribution of the aforementioned transmission factors is practical under specific basin modeling scenarios, an array of 1 × 12 Pelamis devices is examined (see Table 5), avoiding energy output reductions, due to the attenuated waves at the leeward section of the installation and the direction of the incident, diffracted and reflected ones inside the farm. Moreover, it can be easily seen that the WEC arrays located in point B (i.e., the location with the best synergy value), produce larger amounts of wave energy, while offshore wind energy is similar in both locations.

5. Conclusions

Joint offshore wind and wave energy exploitation seems to be one of the most promising solutions to compensate for the continuously increasing energy demand and high costs of offshore wind energy. The Greek Seas are characterized by a remarkable offshore wind power potential; thus, the main objective of this paper was to identify favorable locations for collocating hybrid offshore wind and wave energy systems. This was performed by assessing the wind and wave energy complementarity of the Greek Seas using ERA5 data. Firstly, the ERA5 reanalysis dataset was evaluated by means of in-situ measurements derived from six Poseidon buoys. It was concluded that there is a strong correlation between the datasets, and that the ERA5 tends to underestimate the actual values of the measured parameters.

In terms of energy potential, offshore wind has maximum wind power density values of the order of 650–800 W/m^2 that are encountered across the central axis of Aegean Sea, an area that is characterized by strong northern winter and summer winds (Etesian winds). In addition, wave energy flux peak values range between 5–7 kW/m in west, southwest and southeast areas of Crete Isl., an area characterized by large fetch lengths that allow the development of larger wind waves and swells. In this respect, the most favorable locations in terms of complementarity and synergy were, respectively, identified as follows: southeast of Chios Isl. (point A: 38° N, 26.5° E—index value 45.143%) and southeast of Karpathos Isl. (point B: 35° N, 27.5° E—index value 29.261%), taking into consideration a threshold value of 300 m water depth.

In the context of an actual application in the aforementioned locations, an 8-MW offshore wind turbine and an array of 1 × 12 Pelamis WECs (750 kW each) could produce at point A an annual energy output of 33.080 GWh and 771.360 MWh, respectively, while the same hybrid system at point B could generate 34.406 GWh of wind energy and 4167.852 MWh of wave energy in the same temporal scale. The difference between the annual energy output from the WEC between locations A and B can be explained by the fact that the grid point with the highest synergy index is located in an area characterized by large fetch lengths and thus higher wave energy potential.

Author Contributions: Conceptualization, T.S.; methodology, T.S.; software, K.K. and I.B.; validation, K.K. and I.B.; formal analysis, T.S., K.K. and I.B.; investigation, K.K. and I.B.; writing—original draft preparation, T.S., K.K. and I.B.; writing—review and editing, T.S., K.K. and I.B.; All authors have read and agreed to the published version of the manuscript.

Funding: This research received no external funding.

Institutional Review Board Statement: Not applicable.

Informed Consent Statement: Not applicable.

Data Availability Statement: In The measured in-situ data have been obtained from the POSEIDON marine monitoring network (https://poseidon.hcmr.gr/) (accessed on 3 March 2016). The ERA5 reanalysis dataset has been obtained from https://cds.climate.copernicus.eu (accessed on 3 March 2021). See also [39]. The provided results have been generated using Copernicus Climate Change Service information (2021).

Acknowledgments: The authors would like to thank F. Karathanasi for her help in data analysis.

Conflicts of Interest: The authors declare no conflict of interest.

References

1. Soukissian, T.H.; Denaxa, D.; Karathanasi, F.; Prospathopoulos, A.; Sarantakos, K.; Iona, A.; Georgantas, K.; Mavrakos, S. Marine Renewable Energy in the Mediterranean Sea: Status and Perspectives. *Energies* **2017**, *10*, 1512. [CrossRef]
2. Soukissian, T. Use of multi-parameter distributions for offshore wind speed modeling: The Johnson SB distribution. *Appl. Energy* **2013**, *111*, 982–1000. [CrossRef]
3. Johnston, B.; Foley, A.; Doran, W.J.; Littler, T. Levelised cost of energy, A challenge for offshore wind. *Renew. Energy* **2020**, *160*, 876–885. [CrossRef]
4. Rubio-Domingo, G.; Linares, P. The future investment costs of offshore wind: An estimation based on auction results. *Renew. Sustain. Energy Rev.* **2021**, *148*, 111324. [CrossRef]
5. Lavidas, G.; Blok, K. Shifting wave energy perceptions: The case for wave energy converter (WEC) feasibility at milder resources. *Renew. Energy* **2021**, *170*, 1143–1155. [CrossRef]
6. Besio, G.; Mentaschi, L.; Mazzino, A. Wave energy resource assessment in the Mediterranean Sea on the basis of a 35-year hindcast. *Energy* **2016**, *94*, 50–63. [CrossRef]
7. Jin, S.; Greaves, D. Wave energy in the UK: Status review and future perspectives. *Renew. Sustain. Energy Rev.* **2021**, *143*, 110932. [CrossRef]
8. Soukissian, T.; Karathanasi, F. Joint Modelling of Wave Energy Flux and Wave Direction. *Processes* **2021**, *9*, 460. [CrossRef]
9. Mazarakos, T.; Konispoliatis, D.; Katsaounis, G.; Polyzos, S.; Manolas, D.; Voutsinas, S.; Soukissian, T.; Mavrakos, S.A. Numerical and experimental studies of a multi-purpose floating TLP structure for combined wind and wave energy exploitation. *Mediterr. Mar. Sci.* **2019**, *20*, 745–763. [CrossRef]
10. Konispoliatis, D.; Katsaounis, G.; Manolas, D.; Soukissian, T.; Polyzos, S.; Mazarakos, T.; Voutsinas, S.; Mavrakos, S. REFOS: A Renewable Energy Multi-Purpose Floating Offshore System. *Energies* **2021**, *14*, 3126. [CrossRef]
11. Pérez-Collazo, C.; Greaves, D.; Iglesias, G. A review of combined wave and offshore wind energy. *Renew. Sustain. Energy Rev.* **2015**, *42*, 141–153. [CrossRef]
12. McTiernan, K.L.; Sharman, K.T. Review of Hybrid Offshore Wind and Wave Energy Systems. *J. Physics Conf. Ser.* **2020**, *1452*, 012016. [CrossRef]
13. Azzellino, A.; Lanfredi, C.; Riefolo, L.; De Santis, V.; Contestabile, P.; Vicinanza, D. Combined Exploitation of Offshore Wind and Wave Energy in the Italian Seas: A Spatial Planning Approach. *Front. Energy Res.* **2019**, *7*, 42. [CrossRef]
14. Veigas, M.; Iglesias, G. Wave and offshore wind potential for the island of Tenerife. *Energy Convers. Manag.* **2013**, *76*, 738–745. [CrossRef]
15. Dalton, G.; Bardócz, T.; Blanch, M.; Campbell, D.; Johnson, K.; Lawrence, G.; Lilas, T.; Friis-Madsen, E.; Neumann, F.; Nikitas, N.; et al. Feasibility of investment in Blue Growth multiple-use of space and multi-use platform projects; results of a novel assessment approach and case studies. *Renew. Sustain. Energy Rev.* **2019**, *107*, 338–359. [CrossRef]
16. Gideon, R.A.; Bou-Zeid, E. Collocating offshore wind and wave generators to reduce power output variability: A Multi-site analysis. *Renew. Energy* **2021**, *163*, 1548–1559. [CrossRef]
17. Tedeschi, E.; Robles, E.; Santos, M.; Duperray, O.; Salcedo, F. Effect of energy storage on a combined wind and wave energy farm. In Proceedings of the 2012 IEEE Energy Conversion Congress and Exposition (ECCE), Raleigh, NC, USA, 15–20 September 2012; pp. 2798–2804.
18. Astariz, S.; Iglesias, G. Selecting optimum locations for co-located wave and wind energy farms. Part I: The Co-Location Feasibility index. *Energy Convers. Manag.* **2016**, *122*, 589–598. [CrossRef]
19. Rusu, L.; Ganea, D.; Mereuta, E. A joint evaluation of wave and wind energy resources in the Black Sea based on 20-year hindcast information. *Energy Explor. Exploit.* **2017**, *36*, 335–351. [CrossRef]
20. Fusco, F.; Nolan, G.; Ringwood, J. Variability reduction through optimal combination of wind/wave resources—An Irish case study. *Energy* **2010**, *35*, 314–325. [CrossRef]
21. Kalogeri, C.; Galanis, G.; Spyrou, C.; Diamantis, D.; Baladima, F.; Koukoula, M.; Kallos, G. Assessing the European offshore wind and wave energy resource for combined exploitation. *Renew. Energy* **2017**, *101*, 244–264. [CrossRef]

22. Astariz, S.; Iglesias, G. Enhancing Wave Energy Competitiveness through Co-Located Wind and Wave Energy Farms. A Review on the Shadow Effect. *Energies* **2015**, *8*, 7344–7366. [CrossRef]
23. Astariz, S.; Iglesias, G. Selecting optimum locations for co-located wave and wind energy farms. Part II: A case study. *Energy Convers. Manag.* **2016**, *122*, 599–608. [CrossRef]
24. Soukissian, T.H.; Karathanasi, F.E.; Zaragkas, D.K. Exploiting offshore wind and solar resources in the Mediterranean using ERA5 reanalysis data. *Energy Convers. Manag.* **2021**, *237*, 114092. [CrossRef]
25. Ferrari, F.; Besio, G.; Cassola, F.; Mazzino, A. Optimized wind and wave energy resource assessment and offshore exploitability in the Mediterranean Sea. *Energy* **2020**, *190*, 116447. [CrossRef]
26. Benassai, G.; Mariani, P.; Stenberg, C.; Christoffersen, M. A Sustainability Index of potential co-location of offshore wind farms and open water aquaculture. *Ocean Coast. Manag.* **2014**, *95*, 213–218. [CrossRef]
27. Weiss, C.V.; Ondiviela, B.; Guinda, X.; Del Jésus, F.; González, J.; Guanche, R.; Juanes, J.A. Co-location opportunities for renewable energies and aquaculture facilities in the Canary Archipelago. *Ocean Coast. Manag.* **2018**, *166*, 62–71. [CrossRef]
28. Monforti, F.; Huld, T.; Bódis, K.; Vitali, L.; D'Isidoro, M.; Arántegui, R.L. Assessing complementarity of wind and solar resources for energy production in Italy. A Monte Carlo approach. *Renew. Energy* **2014**, *63*, 576–586. [CrossRef]
29. Miglietta, M.M.; Huld, T.; Monforti-Ferrario, F. Local Complementarity of Wind and Solar Energy Resources over Europe: An Assessment Study from a Meteorological Perspective. *J. Appl. Meteorol. Clim.* **2017**, *56*, 217–234. [CrossRef]
30. Vega-Sanchez, M.A.; Castaneda-Jimenez, P.D.; Pena-Gallardo, R.; Ruiz-Alonso, A.; Morales-Saldana, J.A.; Palacios-Hernandez, E.R. Evaluation of complementarity of wind and solar energy resources over Mexico using an image processing approach. In Proceedings of the 2017 IEEE International Autumn Meeting on Power, Electronics and Computing (ROPEC), Ixtapa, Mexico, 8–10 November 2017; pp. 1–5. [CrossRef]
31. Prasad, A.; Taylor, R.A.; Kay, M. Assessment of solar and wind resource synergy in Australia. *Appl. Energy* **2017**, *190*, 354–367. [CrossRef]
32. Schindler, D.; Behr, H.D.; Jung, C. On the spatiotemporal variability and potential of complementarity of wind and solar resources. *Energy Convers. Manag.* **2020**, *218*, 113016. [CrossRef]
33. Ren, G.; Wan, J.; Liu, J.; Yu, D. Spatial and temporal assessments of complementarity for renewable energy resources in China. *Energy* **2019**, *177*, 262–275. [CrossRef]
34. Jurasz, J.; Canales, F.; Kies, A.; Guezgouz, M.; Beluco, A. A review on the complementarity of renewable energy sources: Concept, metrics, application and future research directions. *Sol. Energy* **2019**, *195*, 703–724. [CrossRef]
35. Ganea, D.; Amortila, V.; Mereuta, E.; Rusu, E. A Joint Evaluation of the Wind and Wave Energy Resources Close to the Greek Islands. *Sustainability* **2017**, *9*, 1025. [CrossRef]
36. Moschos, E.; Manou, G.; Dimitriadis, P.; Afentoulis, V.; Koutsoyiannis, D.; Tsoukala, V.K. Harnessing wind and wave resources for a Hybrid Renewable Energy System in remote islands: A combined stochastic and deterministic approach. *Energy Procedia* **2017**, *125*, 415–424. [CrossRef]
37. Loukogeorgaki, E.; Vagiona, D.G.; Vasileiou, M. Site Selection of Hybrid Offshore Wind and Wave Energy Systems in Greece Incorporating Environmental Impact Assessment. *Energies* **2020**, *11*, 2095. [CrossRef]
38. Hersbach, H.; Bell, B.; Berrisford, P.; Hirahara, S.; Horanyi, A.; Muñoz-Sabater, J.; Nicolas, J.; Peubey, C.; Radu, R.; Schepers, D.; et al. The ERA5 global reanalysis. *Q. J. R. Meteorol. Soc.* **2020**, *146*, 1999–2049. [CrossRef]
39. Hersbach, H.; Bell, B.; Berrisford, P.; Biavati, G.; Horányi, A.; Muñoz Sabater, J.; Nicolas, J.; Peubey, C.; Radu, R.; Rozum, I.; et al. *ERA5 Hourly Data on Single Levels From 1979 to Present*; Copernicus Climate Change Service (C3S) Climate Data Store (CDS). Available online: https://cds.climate.copernicus.eu/cdsapp#!/dataset/reanalysis-era5-single-levels?tab=overview (accessed on 3 March 2021). [CrossRef]
40. Farjami, H.; Hesari, A.R.E. Assessment of sea surface wind field pattern over the Caspian Sea using EOF analysis. *Reg. Stud. Mar. Sci.* **2020**, *35*, 101254. [CrossRef]
41. Tavares, L.F.D.A.; Shadman, M.; Assad, L.P.D.F.; Silva, C.; Landau, L.; Estefen, S. Assessment of the offshore wind technical potential for the Brazilian Southeast and South regions. *Energy* **2020**, *196*, 117097. [CrossRef]
42. Bruno, M.F.; Molfetta, M.G.; Totaro, V.; Mossa, M. Performance Assessment of ERA5 Wave Data in a Swell Dominated Region. *J. Mar. Sci. Eng.* **2020**, *8*, 214. [CrossRef]
43. Soukissian, T.; Chronis, G. Poseidon: A marine environmental monitoring, forecasting and information system for the Greek seas. *Mediterr. Mar. Sci.* **2000**, *1*, 71. [CrossRef]
44. Soukissian, T.H.; Chronis, G.T.; Nittis, K.; Diamanti, C. Advancement of Operational Oceanography in Greece: The Case of the Poseidon System. *J. Atmos. Ocean Sci.* **2002**, *8*, 93–107. [CrossRef]
45. Gower, J.F.R. Intercalibration of wave and wind data from TOPEX/POSEIDON and moored buoys off the west coast of Canada. *J. Geophys. Res. Space Phys.* **1996**, *101*, 3817–3829. [CrossRef]
46. Soukissian, T.; Kechris, C. About applying linear structural method on ocean data: Adjustment of satellite wave data. *Ocean Eng.* **2007**, *34*, 371–389. [CrossRef]
47. Soukissian, T.H.; Karathanasi, F.E.; Voukouvalas, E.G. Effect of outliers in wind speed assessment. In Proceedings of the Proceedings of the 24th International Ocean and Polar Engineering Conference, Busan, Korea, 15–20 June 2014; pp. 362–369.
48. Karathanasi, F.E.; Soukissian, T.H.; Axaopoulos, P.G. Calibration of wind directions in the Mediterranean Sea. In Proceedings of the International Ocean and Polar Engineering Conference, Rhodes, Greece, 26 June–1 July 2016; pp. 491–497.

49. Soukissian, T.; Papadopoulos, A. Effects of different wind data sources in offshore wind power assessment. *Renew. Energy* **2015**, *77*, 101–114. [CrossRef]
50. Guillou, N. Estimating wave energy flux from significant wave height and peak period. *Renew. Energy* **2020**, *155*, 1383–1393. [CrossRef]
51. Soukissian, T.; Papadopoulos, A.; Skrimizeas, P.; Karathanasi, F.; Axaopoulos, P.; Avgoustoglou, E.; Kyriakidou, H.; Tsalis, C.; Voudouri, A.; Gofa, F.; et al. Assessment of offshore wind power potential in the Aegean and Ionian Seas based on high-resolution hindcast model results. *AIMS Energy* **2017**, *5*, 268–289. [CrossRef]
52. Soukissian, T.H.; Gizari, N.; Chatzinaki, M. Wave potential of the Greek seas. In Proceedings of the WIT Transactions on Ecology and the Environment, Alicante, Spain, 11–13 April 2011; pp. 203–213.
53. Lin, Z.; Liu, X.; Lotfian, S. Impacts of water depth increase on offshore floating wind turbine dynamics. *Ocean Eng.* **2021**, *224*, 108697. [CrossRef]
54. Pandit, R.; Kolios, A. SCADA Data-Based Support Vector Machine Wind Turbine Power Curve Uncertainty Estimation and Its Comparative Studies. *Appl. Sci.* **2020**, *10*, 8685. [CrossRef]
55. Sohoni, V.; Gupta, S.C.; Nema, R.K. A Critical Review on Wind Turbine Power Curve Modelling Techniques and Their Applications in Wind Based Energy Systems. *J. Energy* **2016**, *2016*, 8519785. [CrossRef]
56. Marañon-Ledesma, H.; Tedeschi, E. Energy storage sizing by stochastic optimization for a combined wind-wave-diesel supplied system. In Proceedings of the 2015 International Conference on Renewable Energy Research and Applications (ICRERA), Palermo, Italy, 22–25 November 2015; pp. 426–431.
57. Soukissian, T. Probabilistic modeling of directional and linear characteristics of wind and sea states. *Ocean Eng.* **2014**, *91*, 91–110. [CrossRef]
58. Rusu, E.; Soares, C.G. Coastal impact induced by a Pelamis wave farm operating in the Portuguese nearshore. *Renew. Energy* **2013**, *58*, 34–49. [CrossRef]
59. Greenwood, C. The Impact of Large Scale Wave Energy Converter Farms on the Regional Wave Climate. Ph.D. Thesis, University of the Highlands and Islands and The University of Aberdeen, Aberdeen, UK, 2015.

 MDPI

Article

Warm Island Effect in the Lake Region of the Tengger Desert Based on MODIS and Meteorological Station Data

Nan Meng, Nai'ang Wang *, Liqiang Zhao, Zhenmin Niu, Xiaoyan Liang, Xinran Yu, Penghui Wen and Xianbao Su

Center for Glacier and Desert Research, College of Earth and Environmental Sciences, Lanzhou University, Lanzhou 730000, China; mengn19@lzu.edu.cn (N.M.); zhaolq@lzu.edu.cn (L.Z.); niuzhm09@lzu.edu.cn (Z.N.); liangxy16@lzu.edu.cn (X.L.); yuxr2013@lzu.edu.cn (X.Y.); wenph17@lzu.edu.cn (P.W.); suxb19@lzu.edu.cn (X.S.)

* Correspondence: wangna@lzu.edu.cn; Tel.: +86-1899-316-8377

Abstract: The northeastern part of the Tengger Desert accommodates several lakes. The effect of these lakes on local temperatures is unclear. In this study, the effects of the lakes were investigated using land surface temperature (LST) from MODIS (Moderate Resolution Imaging Spectroradiometer) data from 2003 to 2018 and air temperatures from meteorological stations in 2017. LST and air temperatures are compared between the lake-group region and an area without lakes to the north using statistical methods. Our results show that the lake-group region is found to exhibit a warm island effect in winter on an annual scale and at night on a daily scale. The warm island effect is caused by the differing properties of the land and other surfaces. Groundwater may also be an important heat source. The results of this study will help in understanding the causative factors of warm island effects and other properties of lakes.

Keywords: Tengger Desert; lake; remote-sensing; warm island effect; microclimate

Citation: Meng, N.; Wang, N.; Zhao, L.; Niu, Z.; Liang, X.; Yu, X.; Wen, P.; Su, X. Warm Island Effect in the Lake Region of the Tengger Desert Based on MODIS and Meteorological Station Data. *Atmosphere* **2021**, *12*, 1157. https://doi.org/10.3390/atmos12091157

Academic Editor: Baojie He, Ayyoob Sharifi, Chi Feng and Jun Yang

Received: 10 August 2021
Accepted: 6 September 2021
Published: 8 September 2021

1. Introduction

Deserts account for approximately 1/4 of the global land area [1]. Owing to the extreme arid climate and other special conditions, desert areas have considerably different surface radiation budgets and energy distribution processes than other ecosystems, with unique energy, water, and material circulation processes. These unique exchanges of heat, water, and momentum influence ecosystems at the local, regional, and global scales, consequently exerting strong direct and indirect influences on the global climate [2,3].

One of the most studied phenomena of a city's climate is the urban heat island (UHI) [4]. The UHI means the phenomenon that the urban air temperature is higher than that in the surrounding natural environment. Factors that contribute to the formation of the UHI are the properties of the underlying surface and changes in city pattern [5]. Previous research has shown that due to non-uniform solar heating of oases or lakes in arid areas like the Gobi Desert that are colder than the surrounding environment, a "cold island effect" is formed, different to the UHI [6–8]. However, in the southeastern part of the Badain Jaran Desert (BJD), over 110 perennial lakes are spread among the mega-dunes, referred to as the BJD lake-group region. Zhang et al. (2015) have shown that the winter surface temperature of the lake-group region in the hinterland of the BJD was significantly higher than that of other regions (i.e., a "warm island effect") according to MODIS (Moderate Resolution Imaging Spectroradiometer) data, and its formation mechanism might be related to the underlying surface that the lakes are characterized by, with an elevated heat capacity and thermal inertia, and a small roughness length and albedo but also related to the remote groundwater recharge lakes; the heat is mainly carried by the deep groundwater during the process of recharging lakes [9]. Several subsequent studies have also demonstrated the presence of the warm island effect in the BJD lake-group region. For example, a study of Landsat satellite remote sensing data by Deng et al. (2018) found that the surface temperature in the lake-group region is higher than that of the non-lake area in the northern part of the desert [10]. Liang et al. (2016, 2020) indicated that the annual and seasonal average temperatures in the lake-group region were higher than those of the surrounding area, and believed that the mechanism of the warm island effect was related to two aspects: First, heat is carried by the groundwater recharge to the desert lake groups. Second, due to the sparse vegetation and arid surface of the desert hinterland, most of the net radiation is transferred to the atmosphere through sensible heat flux. Meteorological observations found that accumulated temperatures of ≥0 °C and ≥10 °C and phenological observations also proved the existence of a warm island effect in the lake-group region [11–13]. Zhao et al. (2021) found that the lake region has a warm island effect in winter, and is more intense at night [14]. Like the UHI, the "warm island effect" is a meteorological phenomenon in the atmospheric boundary layer, but their formations are different [14].

The Tengger Desert is located in the southeastern part of arid northwestern China, adjacent to the Badain Jaran Desert. It has an extremely continental climate and is sensitive to climate change [15]. As in the BJD, there is a region covered with numerous lakes of varying sizes and shapes in the northeastern part of the Tengger Desert hinterland. Previous studies have focused on the seasonal changes of the lakes, lake water balance, and lake supply sources [16,17]. There has been little research on the climatic effects of these lakes. Thus, the aim of this paper is to investigate lake effects on temperatures in the Tengger Desert. For this purpose, we used remote sensing data and field observations to compare the land surface temperature (LST) and air temperatures of the lake-group region and the non-lake area located to the north. The results from this study will provide information about the formation and evolution of lake areas and provide a scientific basis for the prevention of desertification and environmental protection.

The remainder of this paper is organized as follows: Section 2 describes the data and methods, including the study area, detailed data processing methods, and research methods. The results are presented in Section 3, including the LST distribution and air temperature differences in the lake and non-lake areas. Section 4 contains a discussion of the results. Section 5 summarizes the major conclusions of this work and plans for future work.

2. Materials and Methods

2.1. Study Area

The Tengger Desert (37°27′–40°00′ N, 102°15′–105°14′ E) is located in the southeastern region of the Alxa Plateau. With an area of 4.27×10^4 km^2, it is the fourth largest desert in China, and is bordered by the Helan Mountains to the east, the Yabulai Mountains to the northwest, and the Qilian Mountains to the southwest. The average altitude is 1200–1400 m [18,19]. The local weather is controlled by westerly circulation. The desert has a typical continental climate and is sensitive to climate change. The average annual temperature is 7.0–9.7 °C, and the average annual precipitation is 100–200 mm. Approximately 80% of the total precipitation occurs from June to September. The average annual wind speed is 2.9–3.7 m/s and is prevalently northwesterly year-round [20]. There are more than 400 lake basins in the desert, approximately 250 of which currently contain water. The lake basins in the northeast are relatively small (generally below 3 km^2) and shallow; Most are located in the inter-hill depressions between the northeast—southwest compound sand dune chains, arranged in a repeated landform pattern [21].

2.2. Data Sources and Processing

2.2.1. LST Data

The LST data were obtained from the 8-day synthetic product (MOD11A2) of the MODIS surface temperature data, available from the USGS website. The product includes daytime and nighttime surface temperatures. The time period of the data was from January 2003 to December 2018, and the data were mosaiced and projected through the MRT (MODIS Reprojection Tool) software, then cropped using the vector boundary map of the Tengger Desert. Clouds can negatively impact the acquisition of remote sensing images;

thus, to ensure the accuracy and reliability of the research results, quality control software (Quality Control) was used to perform data preprocessing to remove areas affected by clouds [22]. The satellite provided images at 11:00 and 22:00 local solar time, which represented day and night surface temperatures, respectively, in the area for 1 d. More information about the data processing can be found in Zhang [9].

2.2.2. Meteorological Data

The air temperature data were extracted from the MAWS301 automatic weather station (Vaisala, Finland). The installation height of the station is 3 m above ground, and the data collection interval is 10 min. The selected time period was from 1 January 2017 to 1 January 2018. Barunjilang Station (38°40′ N, 104°31′ E, H = 1344.0 m) is located in the lake area in the northeastern area of the desert hinterland, and Baxinggaole Station (39°58′ N, 104°10′ E, H = 1215.0 m) is located in the non-lake area to the north of the desert (see Figure 1). The observation field is relatively empty and the surface is bare sand. Data processing conformed to the regulations of the ground meteorological observation standards and has strict uniformity and accuracy after quality control.

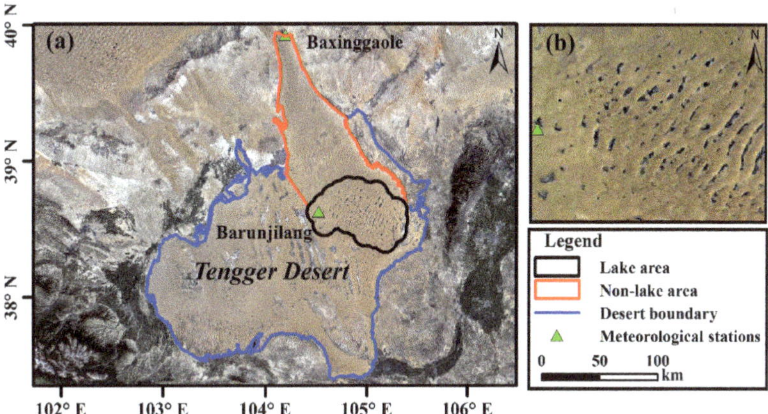

Figure 1. (**a**) Study area and distribution of meteorological stations in Tengger Desert and (**b**) lake-group region located in the northeastern region.

2.2.3. ASTER GDEM Data

Advanced Spaceborne Thermal Emission and Reflection Radiometers (ASTERs) onboard Terra have a unique combination of wide spectral coverage and high spatial resolution in the visible near-infrared, through shortwave infrared, to thermal infrared regions. Japan's Ministry of Economy, Trade and Industry (METI) and NASA announced the release of the ASTER Global Digital Elevation Model (GDEM) on 29 June 2009. ASTER data contributes to a wide array of global change-related application areas including geology and soils, vegetation and ecosystem dynamics, and land cover change. The horizontal resolution of this GDEM is arcsecond (approximately 30 m resolution) and has Z accuracies generally between 10 m and 25 m root mean square error (RMSE) [23]. The ASTER GDEM data was derived from the geospatial data cloud website (http://www.gscloud.cn/, accessed on 4 January 2021).

2.3. Methods

There are many lakes of various sizes and shapes in the Tengger Desert. In addition to freshwater lakes, there are also many saltwater lakes with high total dissolved solids and a certain number of dry alkali-salt lake basins. According to its distribution characteristics, the Tengger Desert can be roughly divided into the following three types of area: (1) In the middle and south of the desert, there are lake basins arranged in a regular north—south direction; most of them may be residual lakes formed when ancient lakes dried up and retreated in the Late Quaternary [24]; (2) Irregularly distributed lake basins in the west and south of the desert; (3) In the northeastern part of the desert, lakes are located in inter-dune depressions between the northeast—southwest trending compound dune chain. This distribution comprises a regular arrangement of small lake basins [21]. The lakes are relatively concentrated and typical in the north-eastern part of the desert. Thus, this is regarded as the study area in the lake area, and the lake area is defined with a 10 km buffer zone from the lakes. Furthermore, the lake area is a blend of surfaces (both water and land) [9,12]. The northern part of the desert has fewer lakes than other areas [25]; Thus, the non-lake area is defined as the northern desert area adjacent to the lake-group region (Figure 1). We computed the mean LST per pixel by aggregating the available 8-day mean in different regions. The seasonal average LST in the daytime and nighttime were compared separately. The geographic parameters of longitude, latitude, and altitude can influence temperature distributions. However, the difference in longitude between the two stations is small, so this study only corrected for the latitudes and altitudes of the two stations. Temperature is assumed to decrease by 0.0065 °C/m for altitude, so that temperatures at different altitudes can be corrected to a consistent altitude [26]; The latitude correction considers one degree of latitude to have the same effect as 100 m of elevation [27]. January, April, July, and October were chosen to represent winter, spring, summer, and autumn, respectively. Remote sensing combined with data from the meteorological stations were used to compare the differences in LST and air temperatures on annual, seasonal, and daily time scales.

3. Results

3.1. Surface Temperature Differences

The average LST distributions of the lake and non-lake areas during the daytime and nighttime from 2003 to 2018 are shown in Figure 2. The figure shows that the average LST during the day and night exhibit different characteristics. LST distributions during the daytime and nighttime in different seasons are shown in Figures 3 and 4. These distributions show that different LST patterns exist in different seasons. This is likely due to seasonal changes in the intensity of solar radiation. The results show that the LST of the lake area was significantly higher than that of the non-lake area in winter on an annual scale and at night on a daily scale. In order to obtain a more reliable conclusion, the average LST of the two regions was calculated as mentioned below.

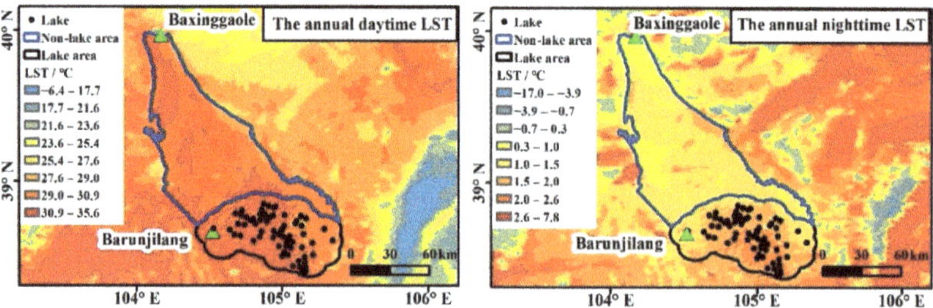

Figure 2. Distribution of average LST during the daytime and nighttime in the lake and non-lake areas from 2003 to 2018.

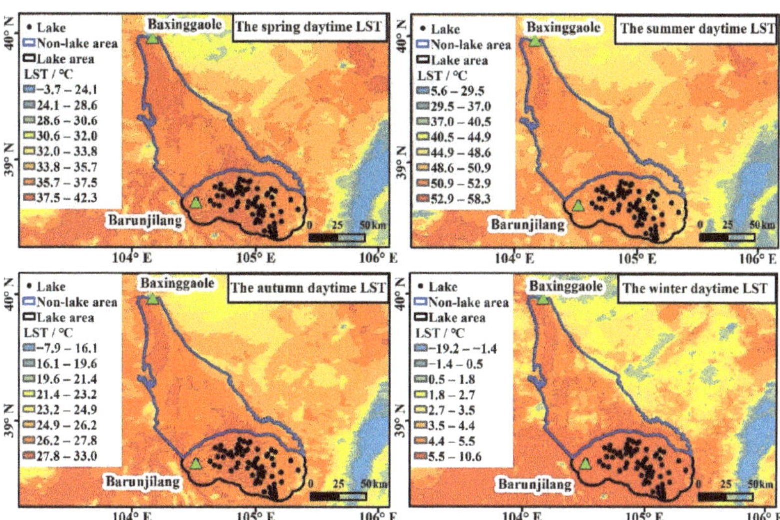

Figure 3. Distributions of average LST during the daytime in the lake and non-lake areas.

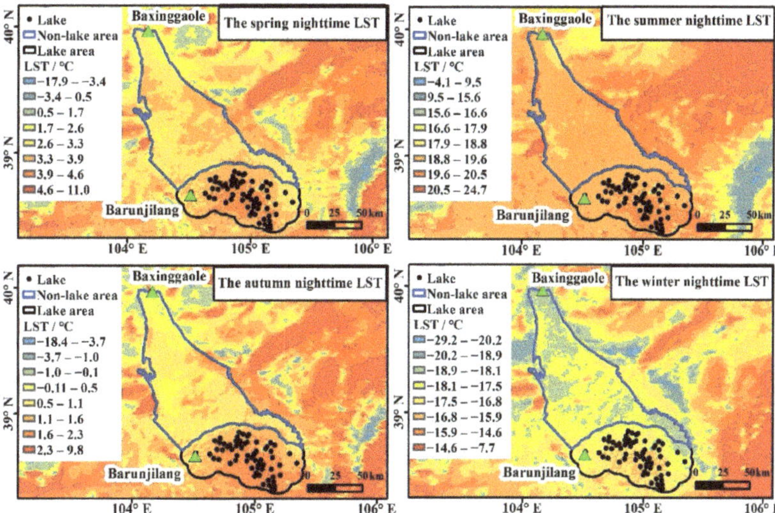

Figure 4. Distributions of average LST during the nighttime in the lake and non-lake areas.

The basis of LST represents the effects of stable factors such as topography, latitude, and longtime energy balance on the land surface. The study concludes that in any study related with spatial distribution of LST over a large area, the effect of changes in elevation at different locations shall also be considered, and LSTs at different locations shall be rationalized on the basis of their comparative elevations [28,29]. The average LST of the two regions was corrected by elevation and latitude. Figure 5a shows the digital elevation model of the area produced from ASTER data.

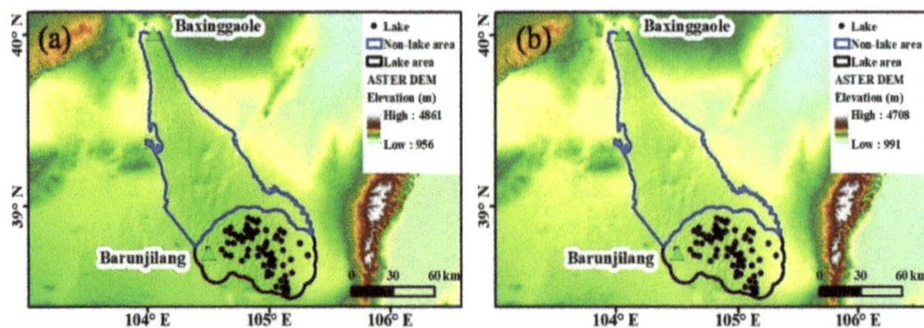

Figure 5. Elevation of the area as derived from ASTER data (**a**), resampled ASTER image of area (**b**).

When the elevation and latitude of LST in the lake and non-lake areas are revised, both LST and elevation shall be at the same spatial resolution. The MODIS and ASTER products have different spatial resolutions; therefore, the elevation data and LST data were resampled to a resolution of 1000 m using the nearest-neighbor sampling approach in ArcMap 10.5 software. Once registered to the same projection system with the same spatial resolution, it is possible to retrieve corresponding values of both LST and elevation for each pixel. Figure 5b shows the resampled ASTER image of the area. After resampling, the highest elevation reduced to 4708 m from 4861 m, and lowest elevation increased to 991 m from 956 m due to averaging of neighborhood pixels.

Table 1 shows the average LST and differences (ΔT) between the two regions after correction. The average LST of the lake area was higher than the non-lake area during the winter and nighttime. Thus, the lake region was found to exhibit a warm island effect in winter and at night. Due to the limited area of the lake, it cannot affect the large-scale surface thermal field.

Table 1. Comparison of average LST of the lake and non-lake areas (unit: °C).

Region	Annual		Spring		Summer		Autumn		Winter	
	Day	Night	Day	Night	Day	Night	Day	Night	Day	Night
Lake area	33.1	4.3	39.4	6.2	53.3	22.1	29.9	4.1	9.2	−14.7
Non-area lake	33.0	4.0	39.9	6.0	54.1	22.0	30.2	3.6	7.9	−15.1
ΔT	0.1	0.3	−0.5	0.2	−0.8	0.1	−0.3	0.5	1.3	0.4

3.2. *Air Temperature Differences*

The Barunjilang station is approximately 1.9 km away from the nearest lake. In addition, the LST change at this station was consistent with the lake group from the above LST distribution, and Baxinggaole station is approximately 135.9 km away from the lake area. Therefore, the temperature characteristics of the Barunjilang weather station can be studied to represent the lake area, and the Baxinggaole station can represent the non-lake area. Figure 6a shows the annual temperature changes collected by the two stations; the annual temperature trends of the two stations were consistent. The standard deviation of the temperatures calculated at the Barunjilang station (11.7) was smaller than the standard deviation of the temperatures calculated at the Baxinggaole station (12.8); thus, the annual temperature variation at the Barunjilang station was smaller than at the Baxinggaole station. Summer temperatures were cooler and winter temperatures warmer in the lake area than in the non-lake area. Figure 6b illustrates that the temperature difference between the two stations was relatively small during the spring and summer, and large during the autumn and winter. Table 2 shows air temperature differences between the lake area and non-lake area. Average air temperatures in the lake area were higher than those of the non-lake area during the spring, autumn, and winter on an annual scale. Lakes generally moderate daily maximum and minimum temperatures. The data show that the average minimum temperature and the daily minimum temperature in the lake group area were both higher than those of the non-lake area; however, the average maximum temperature and the extreme maximum temperature were both lower than those of the non-lake areas, as observed by another study [30]. Additionally, the daily and annual air temperature ranges at Barunjilang station wereredued due to the presence of lakes.

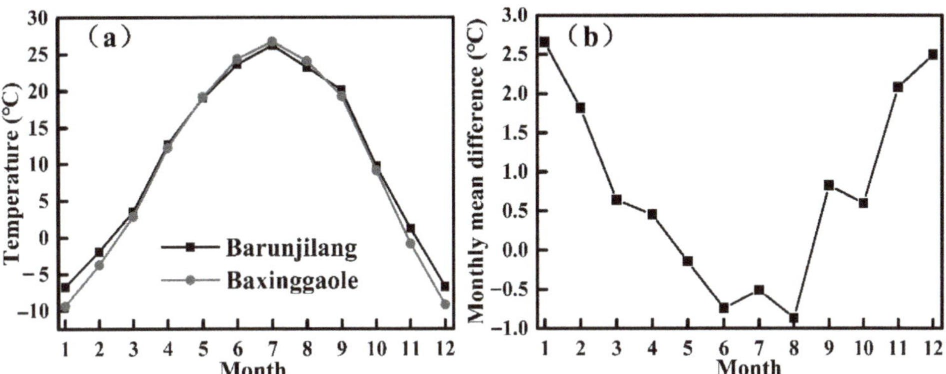

Figure 6. Annual temperature changes (**a**), monthly mean differences (**b**) in the lake and non-lake regions.

Table 2. Comparison of temperature characteristics between lake and non-lake areas (unit: °C).

Site	T_{ave}					T_{min} (January)		T_{max} (July)		Annual Temperature Range	Diurnal Temperature Range
	January	April	July	October	Annual	Average	Daily	Average	Daily		
Barunjilang	−6.7	12.5	26.2	9.7	10.4	−13.9	−21.6	32.1	41.0	32.9	13.2
Baxinggaole	−9.4	12.2	26.7	9.1	9.6	−17.6	−25.1	32.9	42.2	36.1	15.8
ΔT	2.7	0.3	−0.5	0.6	0.8	3.7	3.5	−0.8	−1.2	−3.2	−2.6

The diurnal air temperature variations at the two weather stations are listed in Figure 7; the daily temperature trends at the two stations are consistent. The daily minimum temperature at the Barunjilang station in the lake area occurred slightly later than at Baxinggaole in the non-lake area. The standard deviation of the temperature at Barunjilang Station (3.9) was smaller than the standard deviation of the temperature at Baxinggaole Station (4.7). Thus, the diurnal temperature variations in the lake-group region were smaller than those in the non-lake area. This is typical diurnal behavior that results from the lake effect.

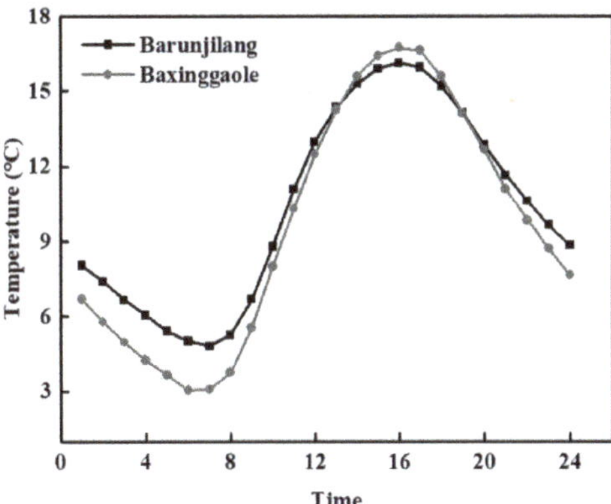

Figure 7. Daily temperature changes in the lake and non-lake areas.

Table 3 compares the annual and seasonal air temperatures during the daytime and nighttime. The results demonstrate that, in terms of the annual averages, the daytime and nighttime temperatures at Barunjilang Station were higher than those at Baxinggaole

Station. The difference between the two stations was largest at night, indicating that the warm island effect was strongest at night. Seasonally, there is a clear warm island effect in the winter. The differences between the air temperature and surface temperature results were related to the distance of the station from the central lake area, the wind direction, and the local environment around the station [31].

Table 3. Comparison of day and night temperatures in the year and in winter between lake and non-lake areas (unit: °C).

Site	Annual		Spring		Summer		Autumn		Winter	
	Day	Night	Day	Night	Day	Night	Day	Night	Day	Night
Barunjilang	12.9	7.8	15.4	10.3	28.7	23.6	11.7	7.5	−4.3	−9.1
Baxinggaole	12.8	6.5	15.7	9.1	29.4	24.0	11.8	6.0	−6.3	−12.4
ΔT	0.1	1.3	−0.3	1.2	−0.7	−0.4	−0.1	1.5	2.0	3.3

4. Discussion

4.1. Temperature Effects of the Lake-Group Region

The average LST and air temperature in the lake-group region were lower than those in the non-lake area, which is a cold island effect. The warm island effect is defined as higher temperatures occurring in a region with lakes than in a region without lakes. The evidence for the occurrence of the warm island effect in the study area is as follows: (1) On a seasonal scale, the average LST and air temperature in the lake-group region were higher than those in the non-lake area in winter; (2) On a daily scale, the average LST and air temperature in the lake-group region were higher than those in the non-lake area during the nighttime; (3) In the lake area, the average minimum temperature and daily minimum temperature during the coldest month were higher, though the average maximum and daily maximum temperatures of the hottest month were lower. The following discussion will focus on the warm island effect.

4.2. Reasons for the Warm Island Effect

The difference in microclimates between the lake and the surrounding land is the result of different underlying surfaces. Lakes have very different radiative and thermal properties than soil [31–33]. The lake surface reflectance is smaller than that of the land surface; the radiation absorbed by the water surface is more than the net radiation absorbed by the land surface which is confirmed. Liang et al. (2020) analysed the annual variation characteristics of monthly net radiation from both land and lake stations in the hinterland of the BJD (Figure 8); they found that the net radiation at the lake station (E1) was much higher than that of the land station (E2). The total annual net radiation of the E1 station (2937.5 MJ/m^2) was more than twice that of the E2 station (1340.3 MJ/m^2); the difference was larger in summer and smaller in winter [11]. The Tengger Desert is adjacent to the BJD; thus, its radiation status should be similar. The absolute value of the radiation difference must be supplemented with field observations in future studies. During the warmer seasons, lakes absorb more net radiation than the land. However, due to the greater depth of solar radiation penetrating into the water, the strong turbulent-mixing effect of the water body makes the heat spread more widely throughout the water, and the water has a large heat storage capacity due to its elevated heat capacity. In addition, lake evaporation consumes a lot of heat energy, so lake surfaces warm more slowly than land surfaces–whereas in the cooler seasons, the heat energy is released from the water. Lake surfaces cool more slowly than land surfaces [30]; the lakes act as heat sinks that absorb and accumulate heat flux from the atmosphere during warm seasons, then act as heat sources that release heat through underwater turbulent exchange and heat exchange between the lake surface and the atmosphere during cold seasons. This leads to cooler summer and warmer winter temperatures in the areas around lakes [34]. In addition, the daily and annual temperature range is reduced; the maximum temperature is reduced and the minimum temperature is increased.

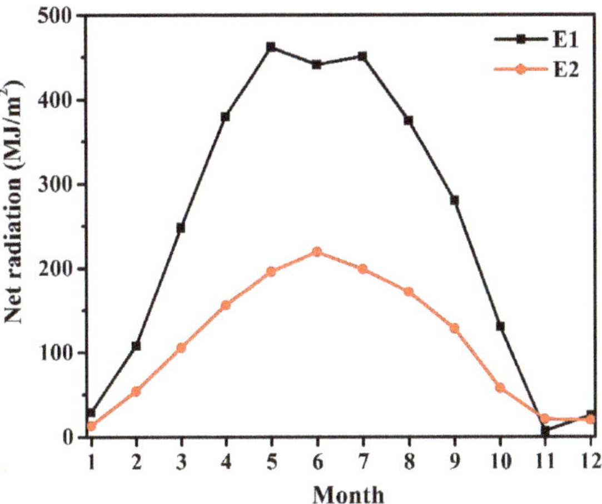

Figure 8. Seasonal net radiation of E1and E2 stations in BJD (Liang et al. 2020).

The warm island effect is not only produced by surface effects, but also may be related to the heat source contribution of deep groundwater. Travertine can be formed by groundwater; Dong et al. (2011) found that extensively distributed travertine deposits were found near a large number of ascending springs, on the shores of lakes, and even at the bottom of some lakes in the southeastern lake area of the BJD [16]. Hu et al. (2015) and Ma et al. (2015) found that the net radiation was less than the sum of the sensible heat flux and latent heat flux on the lake surface in the hinterland of the BJD [3,35]. This implies that other energy supplies to the lake cannot be ignored. Zhang et al. (2016) found that the spring water temperature of the BJD in January 2014 was much higher than the soil and atmospheric temperatures [9]. Groundwater has a warming effect on lakes. The Tengger Desert and BJD are located adjacent to each other in the arid desert area of northwestern China and are only separated by the Yabulai Mountains. As in the BJD, most lakes in the northeastern Tengger Desert are supplied by spring water [17,21,25,36]. Furthermore, Dong et al. (2011) also found widely distributed travertine in the Tengger Desert, the appearance and texture of which are very similar to those of the travertine observed in the BJD [16]. Thus, groundwater recharge of the lakes could import heat from groundwater, substantially contributing to the warm island effect in the Tengger Desert. Owing to a lack of data, the influence of groundwater in this region should to be investigated in future studies.

4.3. Universality and Particularity of the Warm Island Effect

The thermal effect of lakes does not only exist in arid regions. For example, the Great Lakes provide a significant source of warmth to the atmosphere during colder seasons [37–39]. Saaroni et al. (2000) also showed that a water body preserves heat in the winter and at night, and reduces the temperature in hot seasons and during the daytime [4]. Zhao et al. (2010) used a non-hydrostatic mesoscale model (RAMS) to show that a warm island effect occurs at the Jiefangcun reservoir [40]. Near lakes, the daily and annual ranges of near-surface air temperatures are reduced [38,41]. Lakes generally reduce the maximum temperature and increase the minimum temperature throughout the year [42,43]. Thus, the thermal effect of lakes exists globally, caused by the universal nature of the material properties of lake surfaces. For the lake area in the Tengger Desert, the warm island effect is not only affected by the lake surfaces, but also may exhibit large heat contributions from the groundwater. This provides a scientific basis for studying the sources of lakes in this desert area.

4.4. Difference between the Warm Island Effects of the Two Deserts

The temperature difference between the lake-group region and the non-lake area is known as the intensity of the warm island effect. The intensity of the warm island effect varies with the areal extents, depths, and shapes of the lakes. In addition, the velocity and direction of the winds, whether winter ice forms on the lakes, the surrounding topography, and the regional climate of the lakes potentially affect the intensity of the warm island effect. Studies have shown that the larger the lake surface and the deeper the lake water, the stronger the heating effect and range of influence [44,45]. Lakes in the Tengger Desert and the BJD differ greatly in area, water depth, and morphological characteristics. The Tengger Desert contains mostly irregularly shaped, shallow, grassy lakes with no or small areas of water accumulation, while the lakes in the BJD are wider, deeper, and have more regular shapes [17]. Thus, the two desert lakes regions have different heat storage capacities. In addition, studies have shown that the topography around the lake affects wind circulation through thermal and dynamic effects, which in turn affects the temperature [46]. The BJD lake area is surrounded by tall compound sand hills with relative heights of 200–300 m—the highest of which exceeds 430 m [47]—while the height of the tall compound sand dune chain around the Tengger desert lake area is only 50–100 m, which is much lower [21]. Thus the intensity of the warm island effect differs between the two deserts' lake areas.

4.5. Implications

The Tengger Desert has a unique landscape that contains hundreds of lakes and sand dunes. In the past several decades, the formation of the lake group has received continuous attention from many researchers. An explanation for the source of the water in this desert is still controversial, and the origin and flow of groundwater are unclear. Current hypotheses regarding the source of lake water in this region include: (1) residual paleowater, i.e., residual water from precipitation and snowmelt in past humid periods (e.g., the mid-Holocene or the late Pleistocene) [48,49]; (2) percolation of local precipitation, i.e., the mega-dunes can collect sufficient rainfall to fill the lakes in the depressions [20]; (3) subsurface runoff from precipitation in nearby mountains, e.g., the Yabulai and Beida Mountains [50,51]; (4) faults and igneous rock outcrops may contain a large amount of deep groundwater that is free-flowing and becomes the main source of recharge for lakes and shallow groundwater [16].

The warm island effect provides a scientific basis for determining the mechanisms controlling groundwater-recharged lakes. Investigating the source of this water is of importance, as it aids in understanding the hydrologic cycle in the Tengger Desert and in the rational allocation of these water resources. Further investigation of the warm island effect is necessary in order to understand the water sources that sustain the aquifer and the lakes in the Tengger Desert.

5. Conclusions

In this study, we used remote sensing and conventional meteorological data to investigate the effect of lakes on temperature in the northeastern region of the Tengger Desert. The spatial contrasts in land surface temperatures in the lake-group region and the non-lake area located to the north, as well as the seasonal and diurnal variations in air temperature were analysed. The following conclusions were drawn: (1) On a seasonal scale, the warm island effect occurs during the winter; on a daily scale, the warm island effect occurs during the nighttime; (2) The mechanism of the warm island effect may be related to the properties of the lake surfaces, as well as heat contributions from groundwater; (3) There is a difference in the intensities of the warm island effect between the Tengger Desert and the BJD lake areas, which is closely related to lake size, depth, shape, and the terrain surrounding the lake area.

This study has important implications for the water source that recharges the lakes. However, the air temperature study period was only one year. Due to a lack of data, the magnitude of the effect of groundwater on the warm island effect is unclear. Further hydrometeorological observations should be conducted to investigate the effect of groundwater on the warm island effect, and to understand the water cycle and recharge mechanisms of the lake groups in the Tengger Desert.

Author Contributions: N.W. proposed the idea and designed the study. L.Z. contributed to the field work. Z.N. and X.L. implemented the methods. N.M. analyzed the results and wrote the main manuscript text. X.Y., P.W. and X.S. helped analyze the results. All authors have read and agreed to the published version of the manuscript.

Funding: This research was funded by the Project of National Natural Sciences Foundation of China (Grant No. 41871021) and Fundamental Research Funds for the Central Universities (Grant No. lzujbky-2021-sp16).

Institutional Review Board Statement: Not applicable.

Informed Consent Statement: Not applicable.

Data Availability Statement: Not applicable.

Conflicts of Interest: The authors declare no conflict of interest.

References

1. Huo, W.; Zhi, X.F.; Yang, L.M.; Ali, M.; Zhou, C.L.; Yang, F.; Yang, X.H.; Meng, L.; He, Q. Research progress on several problems of desert meteorology. *Trans. Atmos. Sci.* **2019**, *42*, 469–480.
2. Labraga, J.C.; Villalba, R. Climate in the monte desert: Past trends, present conditions, and future projections. *J. Arid Environ.* **2009**, *73*, 154–163. [CrossRef]
3. Ma, L.; Wang, N.A.; Huang, Y.Z.; Li, H.Y.; Lu, J.W. Characteristics of Radiation Budget and Energy Partitioning on Land and Lake Surface under Different Summer Weather Conditions in the Hinterland of Badain Jaran Desert. *J. Nat. Res.* **2015**, *30*, 796–809.
4. Saaroni, H.; Ben-Dor, E.; Bitan, A.; Potchter, O. Spatial distribution and microscale characteristics of the urban heat island in Tel-Aviv, Israel. *Landsc. Urban Plan.* **2000**, *48*, 1–18. [CrossRef]
5. Memon, R.A.; Leung, D.; Liu, C. A review on the generation, determination and mitigation of urban heat island. *J. Environ. Sci.* **2008**, *20*, 120–128.
6. Su, C.X.; Hu, Y.Q.; Zhang, Y.F.; Wei, G.A. The microclimate character and "cold island effect" over the oasis in Hexi region. *Chin. J. Atmos. Sci.* **1987**, *11*, 390–396.
7. Su, C.X.; Hu, Y.Q. Cold island effect over oasis and lake. *Chin. Sci. Bull.* **1987**, 33, 1023–1026.
8. Hu, Y.Q.; Su, C.X.; Zhang, Y.F. Research on the microclimate characteristics and cold island effect over a reservoir in the Hexi Region. *Adv. Atmos. Sci.* **1988**, *5*, 117–126. [CrossRef]
9. Zhang, X.H.; Wang, N.A.; Li, Z.L.; Wu, Y.; Liang, X.Y. Spatial distribution of winter warm islands in Badain Jaran Desert based on MODIS data. *J. Lanzhou Univ. (Nat. Sci.)* **2015**, *51*, 180–185.
10. Deng, X.B. *Spatiotemporal Changes of Warm Island in Badain Jaran Desert and Its Influencing Factors*; Lan Zhou University: Lanzhou, China, 2018; pp. 21–35.
11. Liang, X.Y. *Research on Warm Island Effect in Badain Jaran Desert Based on Observation Data*; Lan Zhou University: Lanzhou, China, 2016; pp. 13–21.
12. Liang, X.Y.; Zhao, L.Q.; Niu, Z.M.; Meng, N.; Wang, N.A. Warm Island Effect in the Badain Jaran Desert Lake Group Region Inferred from the Accumulated Temperature. *Atmosphere* **2020**, *11*, 153. [CrossRef]
13. Liang, X.Y.; Zhao, L.Q.; Xu, X.B.; Niu, Z.M.; Wang, N.A. Plant phenological responses to the warm island effect in the lake group region of the Badain Jaran Desert, northwestern China. *Ecol. Inform.* **2020**, *57*, 101066. [CrossRef]
14. Zhao, L.Q.; Yu, X.R.; Zhang, W.J.; Liang, X.Y.; Wang, N.A.; Cai, W.J. Warm island effect observed in lake areas of the Badain Jaran Desert, China. *Weather* **2021**, in press.
15. Zhang, H.C.; Ma, Y.Z.; Li, J.J.; Cao, J.X. Ancient lake and environment in Tengger Desert, 42–18 ka ago. *Chin. Sci. Bull.* **2002**, *47*, 1847–1857.
16. Dong, C.Y. *Observation Experiment of the Water Cycle and Lake Water Balance in Alxa Desert*; Lan Zhou University: Lanzhou, China, 2011; pp. 46–52.
17. Lai, T.T.; Wang, N.N.; Huang, Y.Z.; Zhang, J.M.; Zhao, L.Q. Seasonal changes of lakes in Tengger Desert of 2002. *J. Lake Sci.* **2012**, *24*, 957–964.
18. Dong, Z.B.; Zhang, Z.C.; Zhao, G.A. Characteristics of wind velocity, Temperature and Humidity Profiles of Near-surface Layer in Tengger Desert. *J. Desert. Res.* **2009**, *29*, 977–981.
19. Zhang, Z.C.; Dong, Z.B.; Wen, Q.; Jiang, C.W. Wind regimes and aeolian geomorphology in the western and southwestern Tengger Desert, NW China. *Geol. J.* **2015**, *50*, 707–719. [CrossRef]
20. Zhao, J.B.; Yu, K.K.; Shao, T.J. A Preliminary Study on the Water Status in Sand Layers and Its Sources in the Tengger Desert. *Resour. Sci.* **2011**, *33*, 259–264.

21. Wu, Z. *Desert and Its Control in China*; Science Press: Beijing, China, 2009; pp. 562–568.
22. Wan, Z. New refinements and validation of the collection-6 modis land-surface temperature/emissivity product. *Remote Sens. Environ.* **2014**, *140*, 36–45. [CrossRef]
23. Wang, W.C.; Yang, X.X.; Yao, T.D. Evaluation of ASTER GDEM and SRTM and their suitability in hydraulic modelling of a glacial lake outburst flood in southeast Tibet. *Hydrol. Process.* **2012**, *26*, 213–225. [CrossRef]
24. Wang, N.A.; Li, Z.L.; Cheng, H.Y.; Li, Y.; Huang, Y.Z. Re-discussion on the Late Quaternary High Lake Surface and Great Lake Period of the Alxa Plateau. Chinese. *Sci. Bull.* **2011**, *56*, 1367–1377.
25. Yan, C.Z.; Li, S.; Lu, J.F. Lake number and area in the Tengger Desert during 1975–2015. *J. Desert Res.* **2020**, *40*, 183–189.
26. Li, M.; Wang, X.L.; Ding, Y.Y. Comparison of several daily temperature interpolation methods. *J. Anhui Agric. Sci.* **2014**, *42*, 8670–8674.
27. Fu, B.P. *Mountain Climate*; Science Press: Beijing, China, 1983; p. 1.
28. Khandelwal, S.; Goyal, R.; Kaul, N.; Mathew, A. Assessment of land surface temperature variation due to change in elevation of area surrounding Jaipur, India. Egypt. *J. Remote Sens.* **2017**, *21*, 87–94.
29. Thanh, P.; Martin, K.; Trong, T. Land surface temperature variation due to changes in elevation in northwest Vietnam. *Climate* **2018**, *6*, 28.
30. Bonan, G.B. Sensitivity of a GCM Simulation to Inclusion of Inland Water Surfaces. *J. Chem. Ecol.* **1995**, *8*, 2691–2704. [CrossRef]
31. Anyah, R.O.; Semazzi, F. Idealized simulation of hydrodynamic characteristics of Lake Victoria that potentially modulate regional climate. *Int. J. Climatol.* **2009**, *29*, 971–981. [CrossRef]
32. Mackay, M.D.; Neale, P.J.; Arp, C.D.; Domis, L.N.D.S.; Fang, X.; Gal, G. Modeling lakes and reservoirs in the climate system. *Limnol. Oceanogr.* **2009**, *54*, 2315–2329. [CrossRef]
33. Sharma, A.; Hamlet, A.F.; Fernando, F. Lessons from Inter-Comparison of Decadal Climate Simulations and Observations for the Midwest U.S. and Great Lakes Region. *Atmosphere* **2019**, *10*, 266. [CrossRef]
34. Long, Z.; Perrie, W.; Gyakum, J.; Caya, D.; Laprise, R. Northern Lake Impacts on Local Seasonal Climate. *J. Hydrometeorol.* **2007**, *8*, 881–896. [CrossRef]
35. Hu, W.F. *Research on Water-Heat Exchange between Land and Air in Badain Jaran Desert Based on Observation*; LanZhou University: LanZhou, China, 2015; pp. 79–83.
36. Zhang, W.W.; Cheng, G.E.; Gao, Y. Characteristics of desert lakes and their protection in Inner Mongolia. *J. Inner Mong. Agric. Univ.* **2006**, *27*, 11–14.
37. Scott, R.W.; Huff, F.A. *Lake Effects on Climatic Conditions in the Great Lakes Basin*; Illinois State Water Survey: Champaign, IL, USA, 1997; p. 73.
38. Notaro, M.; Holman, K.D.; Zarrin, A.; Vavrus, V.; Fluck, E. Influence of the Laurentian Great Lakes on Regional Climate. *J. Clim.* **2013**, *26*, 789–804. [CrossRef]
39. Dobson, K.C.; Beaty, L.E.; Rutter, M.A.; Hed, B.; Campbell, M.A. The influence of Lake Erie on changes in temperature and frost dates. *Int. J. Climatol.* **2020**, *40*, 5590–5598. [CrossRef]
40. Zhao, L.; Chen, Y.C.; Lv, S.F.; Meng, X.H.; Li, W.L.; Li, J.L. Simulation of Hydrometeorological Effect of Jiefangcun Reservoir in Jinta Oasis Summer. *Plateau Meteorol.* **2010**, *29*, 1414–1422.
41. Nikoo, E.; Susanne, G.C.; Hagen, K.; Reik, D.; Jan, V. Effects of the Lak Sobradinho Reservoir (Northeastern Brazil) on the Regional Climate. *Climate* **2017**, *5*, 50.
42. Bates, G.T.; Giorgi, F.; Hostetler, S.W. Toward the Simulation of the Effects of the Great Lakes on Regional Climate. *Mon. Weather Rev.* **1993**, *121*, 1373–1387. [CrossRef]
43. Wen, L.J.; Lv, S.; Li, Z.G.; Zhao, L.; Nagabhatla, N. Impacts of the Two Biggest Lakes on Local Temperature and Precipitation in the Yellow River Source Region of the Tibetan Plateau. *Adv. Meteorol.* **2015**, *2015*, 248031. [CrossRef]
44. Momii, K.; Ito, Y. Heat budget estimates for Lake Ikeda, Japan. *J. Hydrol.* **2008**, *361*, 362–370. [CrossRef]
45. Dutra, E.; Stepanenko, V.A.; Balsamo, G.; Viterbo, P.; Miranda, P.M.A.; Mironov, D. An offline study of the impact of lakes on the performance of the ECMWF surface scheme. *Boreal Environ. Res.* **2010**, *15*, 100–112.
46. Stivari, S.M.S.; Oliveira, A.P.; Karam, H.A.; Soares, J. Patterns of local circulation in the Itaipu Lake area: Numerical simulations of lake breeze. *J. Appl. Meteorol.* **2003**, *42*, 3750. [CrossRef]
47. Shao, T.J.; Zhao, J.B.; Dong, Z.B. Water Chemistry of the Lakes and Groundwater in the Badain Jaran Desert. *Acta Geogr. Sin.* **2011**, *66*, 662–672.
48. Edmunds, W.; Ma, J.; Aeschbachhertig, W.; Kipfer, R.; Darbyshire, D. Groundwater recharge history and hydrogeochemical evolution in the minqin basin, north west china. *Appl. Geochem.* **2006**, *21*, 2148–2170. [CrossRef]
49. Ding, Z.; Ma, J.Z.; Zhao, W.; Jiang, Y.; Love, A.J. Profiles of geochemical and isotopic signatures from the Helan Mountains to the eastern Tengger Desert, northwestern China. *J. Arid Environ.* **2013**, *90*, 77–87. [CrossRef]
50. Ma, J.Z.; Wang, X.S.; Edmunds, W.M. The characteristics of ground-water resources and their changes under the impacts of human activity in the arid Northwest China—A case study of the Shiyang River Basin. *J. Arid Environ.* **2005**, *61*, 277–295. [CrossRef]
51. Gates, J.B.; Edmunds, W.M.; Ma, J.; Scanlon, B.R. Estimating groundwater recharge in a cold desert environment in northern China using chloride. *Hydrogeol. J.* **2008**, *16*, 893–910. [CrossRef]

Article

Surface Urban Heat Island Assessment of a Cold Desert City: A Case Study over the Isfahan Metropolitan Area of Iran

Alireza Karimi [1], Pir Mohammad [2], Sadaf Gachkar [3], Darya Gachkar [4], Antonio García-Martínez [1], David Moreno-Rangel [1] and Robert D. Brown [5,*]

1 Instituto Universitario de Arquitectura y Ciencias de la Construcción, Escuela Técnica Superior de Arquitectura, Universidad de Sevilla, 41012 Sevilla, Spain; alikar1@alum.us.es (A.K.); agarcia6@us.es (A.G.-M.); davidmoreno@us.es (D.M.-R.)
2 Department of Earth Sciences, Indian Institute of Technology, Roorkee, Uttarakhand 247667, India; pmohammad@es.iitr.ac.in
3 Department of Restoration of Historical Heritages, Shahid Beheshti University, Tehran 1983969411, Iran; sadafgachkar74@gmail.com
4 Department of Landscape Architecture, Shahid Beheshti University, Tehran 1983969411, Iran; daryagachkar74@gmail.com
5 Department of Landscape and Architecture, Texas A&M University, College Station, TX 77843, USA
* Correspondence: rbrown@arch.tamu.edu

Abstract: This study investigates the diurnal, seasonal, monthly and temporal variation of land surface temperature (LST) and surface urban heat island intensity (SUHII) over the Isfahan metropolitan area, Iran, during 2003–2019 using MODIS data. It also examines the driving factors of SUHII like cropland, built-up areas (BI), the urban–rural difference in enhanced vegetation index (ΔEVI), evapotranspiration (ΔET), and white sky albedo (ΔWSA). The results reveal the presence of urban cool islands during the daytime and urban heat islands at night. The maximum SUHII was observed at 22:30 p.m., while the minimum was at 10:30 a.m. The summer months (June to September) show higher SUHII compared to the winter months (February to May). The daytime SUHII demonstrates a robust positive correlation with cropland and ΔWSA, and a negative correlation with ΔET, ΔEVI, and BI. The nighttime SUHII displays a negative correlation with ΔET and ΔEVI.

Keywords: surface urban heat island; LST; semi-arid city; MODIS; trend; seasonal; monthly

Citation: Karimi, A.; Mohammad, P.; Gachkar, S.; Gachkar, D.; García-Martínez, A.; Moreno-Rangel, D.; Brown, R.D. Surface Urban Heat Island Assessment of a Cold Desert City: A Case Study over the Isfahan Metropolitan Area of Iran. *Atmosphere* **2021**, *12*, 1368. https://doi.org/10.3390/atmos12101368

Academic Editors: Baojie He, Ayyoob Sharifi, Chi Feng and Jun Yang

Received: 15 September 2021
Accepted: 14 October 2021
Published: 19 October 2021

Publisher's Note: MDPI stays neutral with regard to jurisdictional claims in published maps and institutional affiliations.

1. Introduction

More than half the world's population lives in urban environments [1], and it is predicted that by 2050 the population of the world's cities will reach about 75% [2,3]. Two out of every three people are expected to live in urban areas by that time [4], so the 21st century will be the century of cities [5]. However, the rising urban population has brought many problems, such as stagnation and high concentrations of air pollutants [6,7], changes in rainfall patterns [8], land erosion [9], urban flooding [10], streams [11], the creation of new habitats [12], and urban heat island effects [13,14].

With increasing urbanization and changing natural landscapes to impervious surfaces, cities are more inclined towards the absorption of solar radiation [15], decrease in evapotranspiration [16], increase in runoff [17], augment surface friction [18], and release of anthropogenic heat [19]. Urbanization causes an increase in temperatures around urban areas and affects the urban thermal environment, thereby increasing the risk to human health. During heatwave conditions, when there is a lot of sunshine and a lack of wind to provide ventilation and disperse the warm air, urban environments can enhance the thermal condition and cause severe heat-related morbidity and mortality. The most well-known climate reform index of cities in environmental science is urban heat island (UHI) [20]. It forms when the urban environment's temperature is higher than surrounding rural areas and has destructive effects on the ecosystem [21]. Urban heat islands were classified into three broad groups according to their characteristics in various layers of the urban atmosphere [22]: surface urban heat island (SUHI) [23], canopy layer urban heat island for micro scale analysis (CLUHI) [24], and boundary layer urban heat island for mesoscale analysis (BLUHI) [25]. Surface urban heat island (SUHI) is dependent on land surface temperature (LST) and has a strong correlation with the orientation of the surface relative to the sun as well as land use and land cover [26,27] which can be measured by radiometers onboard aircraft or satellites [28,29]. The great advantage of SUHI is that it can be conveniently measured across large spatial domains for a large number of cities [30]. This makes it possible to compare UHI between cities in large regions to explore the role of urbanization in affecting SUHI [31,32]. Canopy layer urban heat island (CLUHI) is typically quantified by the air temperature data of urban and rural stations [33,34], which are influenced by building geometry and the nature of pavement materials. The boundary layer urban heat island (BLUHI) is governed by the heated air from the upstream urban areas and the basic canopy layer where the canopy's warm urban island occurs [24].

The long-term trends of SUHI and CLUHI were studied together in 272 cities in the mainland of China using satellite data and station-based air temperature data [35]. The studies reveal that the trend of the nighttime SUHI were strongly related to the CLUHI, whereas the relationship between the trends during daytime were relatively weak. The comparison of SUHI and CLUHI were studied recently for a better understanding of their vertical structure over 366 global cities within various background climates [36]. The findings of the study show that the annual mean SUHI is higher than CLUHI by 1.1 ± 1.9 °C during the daytime and 0.3 ± 1.5 °C at nighttime in equatorial, warm temperate, and snow climates. In contrast, in arid regions SUHI is lower than CLUHI by 0.8 °C. The relationship of SUHI and air temperature were investigated across 145 global large cities for the period 2003–2013, based on MODIS data [37]. The relationship was unstable in space, and the correlation differed significantly in macro-climatic conditions and was affected by the difference in vegetation between urban and rural areas.

SUHI has been widely studied in recent decades by analyzing the land surface temperature (LST) with available thermal remote sensing datasets of various spatial resolutions. The peak in SUHI intensity (urban LST minus rural LST) is generally observed at night [28,38]. The intensity of SUHI is weak during the daytime, but it becomes more pronounced after sunset. Various thermal remote sensing data are used to estimate the SUHI intensity, such as Landsat TM/ETM+/TIRS and Moderate Resolution Imaging Spectroradiometer (MODIS), which play an important role in SUHI research [14]. Generally, SUHI has been studied at a fine-scale using Landsat data because of its higher spatial resolution [39–41]. Meanwhile, MODIS LST data are used in regional and global SUHI studies due to their broad spatial coverage and high temporal resolution [31,42–44]. SUHI has been extensively studied over the different cities of the world situated over different climatic conditions. For example, [45] studied the seasonal SUHI effect over Siberian cities and found the strongest relationships between the UHI and population (*log P*). Additionally, [46] explored the indicators for quantifying SUHI over European cities using MODIS, and their study revealed that the temporal aspects and indicator selection were important factors in determining the SUHI intensity. Moreover, [47] studied the SUHI phenomenon over Ahmedabad, India, and found the impact of rural land cover dynamics in estimating the SUHI variation over semi-arid cities. Additionally, some studies also analyzed the underlying influencing factors of the SUHI, such as vegetation, albedo, evapotranspiration, meteorological condition, and anthropogenic heat emissions [48–52]. For example, [53] studied the SUHI phenomenon over 419 global cities and found that the intensity of SUHI was significantly negative with the vegetation during the daytime, while an insignificant relationship was observed during the nighttime. Also, [43] studied the SUHI footprint over 302 Chinese cities, and their findings indicated an increase in anthropogenic heat emission and decrease of vegetation activities, and they asserted that surface albedos should take lead responsibility for the expansion of SUHI footprint area.

However, little attention has been paid to SUHI studies over the cities of cold desert climates. The SUHI over desert cities shows contrasting behavior to that of the cities of temperate and humid zones. An inversion of SUHI during daytime is seen, with the city core appearing cooling than the surrounding suburban area during daytime. The cooling effect over the arid and desert cities has been found over cities around the world [54–56]. For example, the findings of [57] suggest that the nearby green areas play an important role in the control of atmospheric temperatures and in the promotion of citizens' health and fog in the metropolis of Isfahan. In [58] found that SUHI-influenced areas mostly lie in the northern and southern parts of the city, where the vegetation cover is very low. A recent study estimated that the Isfahan metropolis area is cooler than the suburbs during the daytime, but at night it is around 2 K warmer than its surroundings [59]. With reference to the temporal and spatial actions of the SUHI metropolis of Isfahan, it seems that changes in the moisture, albedo, and composition of the atmosphere of the urban environment play a major role in the creation of SUHI. The impermeable surface expansion in the city of Isfahan was studied with Landsat data over the last three decades in the hot and cold months. The findings revealed that impermeable surfaces have increased by 2.8 times during 1985–2015 and observed a negative association with normalized difference vegetation index (NDVI) during hot months, while in the cold months, the urban region was colder than the surrounding rural area [60]. However, all the existing previous studies were focused on the diurnal analysis of SUHI and its relationship with vegetation using a short span of data, and they neglect the influence of other potential drivers of SUHI. Moreover, a detailed diurnal, seasonal, monthly, and temporal study is necessary along with the important driving factors to comprehend the SUHI phenomenon over Isfahan city, using a larger time series of data. In spite of the detailed SUHI investigation, the present study has some limitations, as it does not include the CLUHI study due to the non-availability of air temperature data. The combine study of SUHI and CLUHI can provide a better understanding of the vertical distribution of UHI and more in depth understanding of its dynamics, which are sometimes responsible for the formation of deadly heat wave phenomenon.

Therefore, this research is a comprehensive study of the SUHI phenomenon over the cold desert city of Isfahan, Iran. The main objective of the present research includes (i) revealing the annual, seasonal, monthly, and temporally spatial and temporal variation of LST from 2003 to 2019, (ii) estimating the annual, seasonal, monthly, and temporal variation of SUHI intensity during 2003–2019, and (iii) examining the relationship between the SUHI intensity and its associated influencing factors, including cropland, built-up area, vegetation, evapotranspiration, and albedo over the study period.

2. Study Area

The study was located in Iran in the cold desert city of Isfahan, with a latitude and longitude of 51°40′ E, 32°39′ N (Figure 1), and altitude of 1590 m [58]. The southern and western regions of Isfahan are surrounded by the Zagros mountains, while the northern and eastern regions are full of fertile plains [61]. Thus, this physiography creates a significant difference in terms of climate between the eastern and western regions of Isfahan [46]. According to Koppen–Geiger's climate classification, Isfahan has a characteristic dry climate [62]. Dry weather and very low rainfall are the prominent features of this classification, where the minimum and maximum temperatures are −10.6 °C and 40.6 °C, respectively. The mean annual precipitation over the city is 116.9 mm [63].

According to national censuses, after industrial development and the establishment of different factories around it, the population of Isfahan rose from 255,000 in 1956 to 1,791,000 in 2010 and experienced remarkable urbanization. The urbanization ratio rose in 1956 and 2006, respectively, from 44.2 to 83.5 [64] and the proportion of farmland decreased, while the proportion of urban areas increased significantly during this period, primarily due to the decrease in the agricultural region of Isfahan [65].

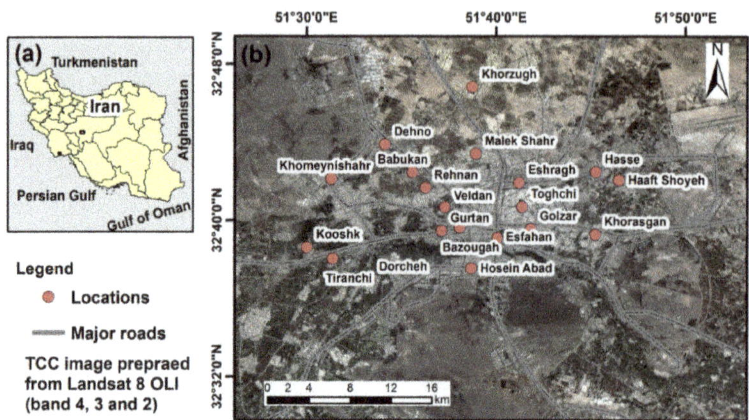

Figure 1. Study area map of (**a**) Isfahan city in Iran map (**b**) Isfahan city and its surrounding area, prepared from true color composite image of Landsat 8 OLI bands (Path/Row: 164/037).

3. Materials and Methods

3.1. Data Sources

The land surface temperature (LST) product obtained from MODIS sensor onboard Terra and Aqua satellite is used in the study. The MODIS sensor consists of 36 spectral bands ranging from 0.4 to 14.4 μm, providing data of different spatial resolutions (250 m, 500 m, and 1 km). The overpass time of the Terra and Aqua is between 10:30 a.m. and 13:30 p.m. local solar time during daytime, while during nighttime, it is between 22:30 p.m. and 01:30 a.m. local solar time.

First, the MODIS derived an 8-day LST product from Terra (MOD11A2) and Aqua (MYD11A2) satellite (https://modis.gsfc.nasa.gov/data/dataprod/mod11.php, accessed on 27 March 2021), available at 1 km × 1 km is used in this study. This LST product is derived using the two thermal infrared channels, band 31 (10.78–11.28 μm) and band 32 (11.77–12.27 μm) of the MODIS sensor. The 8-day LST product is estimated using a generalized split-window algorithm under clear-sky conditions [66], and the accuracy of the product is also checked with in-situ measurements, proving a bias less than 0.5 K [67]. We used the latest version (006) that removed all the cloud-contaminated pixels to enhance the accuracy. Moreover, the coefficients look up table (LUT) in this version is also updated, which considers a wide range of surface and atmospheric conditions, especially extending the upper boundary for (LST-Ts-air) in arid and semi-arid regions. Based on the quality control (QC) flag value, only pixels with high-quality LST were selected to eliminate the effect of retrieval algorithm and cloud cover. Data from 2003 to 2019 (17 years) from both Terra and Aqua satellite during both daytime and nighttime were used in this present study. The seasonal LST of Isfahan city is prepared separately for the summer (April, May, and June) and winter (December, January, and February) period. Besides, the monthly LST images were also prepared for the city to analyze the monthly variation.

LULC maps were extracted from the yearly MODIS land use land cover product, MCD12Q1 (https://lpdaac.usgs.gov/products/mcd12q1v006/, accessed on 27 March 2021), at a spatial resolution of 500 m from 2003 to 2019 and classified the Earth's surface according to the International Geosphere-Biosphere Project (IGBP). The land cover classification scheme identifies a total of 17 classes, which include 11 classes of natural vegetation, 3 classes of mosaic use, and 3 classes of non-vegetated land. Based on the official urban limits of the city metropolitan areas studied, all pixels of the urban category inserted in the respective boundaries were defined as the urban core of the metropolitan area. The LULC maps were utilized for two main purposes: first, for urban and rural boundary area delineation and second, to extract the two important parameters (cropland land and built-up area), which were dominating in the study region for considering as SUHI drivers.

Besides, three more SUHI affective drivers were also analyzed in this present study: vegetation, evapotranspiration, and albedo. As for LST, vegetation and albedo data were extracted from the MODIS product. The enhanced vegetation index (EVI) was obtained from the MOD13A1 (16-day composite) at a spatial resolution of 500 m. The main algorithm of EVI generation is based on the constrained view angle and maximum value composite. In this approach, the number of observations with the highest vegetation value are compared and the observation with the smallest view angle, i.e., closest to nadir view, is chosen to represent the 16-day composite cycle. These methods inevitably result in spatial discontinuities because disparate days can always be chosen for adjacent pixels over the 16-day period. Thus, adjacent selected pixels may originate from different days, with different sun-pixel sensor viewing geometries and different atmospheric and residual cloud/smoke contamination. The white sky albedo (WSA) data are extracted from the daily MCD43A3 bi-hemispherical reflectance product at a spatial resolution of 500 m. The MCD43A3 albedo model parameters product supplies the weighting parameters associated with the Ross Thick Li Sparse Reciprocal BRDF model that best describe the anisotropy of each pixel. These parameters can be used in a forward version of the model to reconstruct the surface anisotropy effects and thus correct directional reflectance's to a common view geometry or to compute the integrated black-sky and white-sky albedos. Alternatively, these parameters can be used with a simple polynomial to easily estimate the black-sky albedo with good accuracy for any desired solar zenith angle. Data with the best (QC = 0) and good (QC = 1) quality flags were used. The evapotranspiration (ET) data used in this study are model data obtained from the operational simplified surface energy balance (SSEB) model [68]. This model provides monthly ET data at a spatial resolution of 1 km. The SSEB model has a unique parameterization for operational applications with the pre-defined, seasonally dynamic boundary condition that is unique to each pixel. The original formulation of this SSEB model is based on the hot/cold pixel principles of the SEBAL [69] and METRIC [70] models. SEBAL (Surface Energy Balance Algorithm for Land) uses surface temperature, hemispherical surface reflectance, and Normalized Difference Vegetation Index (NDVI), as well as their interrelationships to infer surface fluxes for a wide spectrum of land types. METRIC (Mapping Evapotranspiration at High Resolution using Internalized Calibration) uses the instantaneous to daily ET extrapolation method called alfalfa reference ET fraction, which employs wind speed and air temperature that according to better incorporate local/regional surface/environmental conditions than the evaporative fraction of other remote sensing ET algorithms.

3.2. *Methods*

3.2.1. Definitions of Urban and Rural Regions

The MODIS yearly land use land cover (LULC) data product (MCD12Q1) at a spatial resolution of 500 m is used in this study to delineate the urban and rural regions from 2003 to 2019. Firstly, a square box around the city of Isfahan is chosen in such a way that the urban and surrounding non-urban pixels are nearly in an equal ratio. The MODIS LULC data consist of five different classes like open shrubland, grassland, cropland, urban or built-up land, and barren land. Urban or built-up land is used to delineate the urban pixels, while all other four classes were used to delineate the non-urban pixels [71], as shown in Figure 2. As urban area changes continuously, in order to avoid biases in our result by using the same boundary, we changed our urban and rural boundaries in the years 2003, 2007, 2011 and 2015. The mean LULC area for the above years is shown in Table A1 of Appendix A. Whereas, the temporal LULC maps from 2003 to 2019 is shown in Figure A1 of Appendix A.

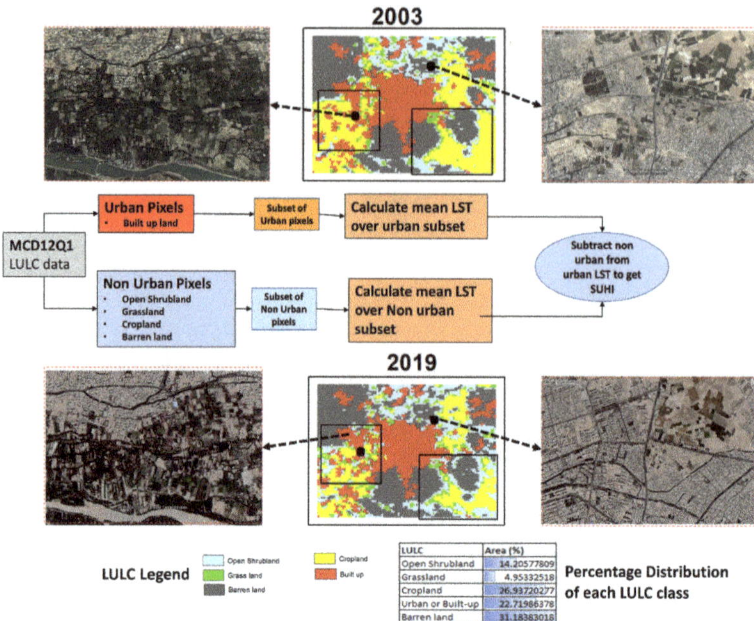

Figure 2. Procedure for estimating surface urban heat island intensity from MODIS.

3.2.2. Surface Urban Heat Island Assessment

The urban and non-urban (rural) boundary extracted from MCD12Q1 is used to calculate surface urban heat island (SUHI) intensity. The Simplified Urban-Extent (SUE) algorithm is used in this study to calculate to intensity of SUHI [71,72]. As per this algorithm, SUHI defines as the difference in the LST of the urban pixels and the non-urban pixels (Equation (1)).

$$\text{SUHI} = \text{meanLST}_{\text{urban pixels}} - \text{mean LST}_{\text{non-urban pixels}} \quad (1)$$

Figure 2 shows the steps for estimating the SUHI using 500 m MCD12Q1 LULC from a 1 km MODIS LST product. Firstly, the urban and non-urban extent are clipped from land cover data for the years 2003, 2007, 2011, and 2015 as defined in Section 3.2.1. After sub setting, the mean LST of both subsets are calculated for each year, and their difference is the surface UHI (SUHI) for that particular year.

3.2.3. LST Trend Analysis Method

We used the Mann–Kendall test to detect the temporal trend in the LST datasets [73]. The test has been extensively used in the hydro-climatic time series data and has always proven to be an efficient tool in comparison to other available tools for detecting trends [74–78]. This test has numerous advantages as it analyzes time series that are not required to follow a specific linear or nonlinear trend. We calculated the standardized Mann–Kendall statistics, which display whether there is a significant trend is present in the datasets or not at a specific significance level, p. A significance level of $p = 0.5$ is considered in this study. Besides, the magnitude of the trend is also determined by Sen's slope estimator test [79]. A positive value indicates an increasing trend and a negative value indicates a falling trend.

3.2.4. Exploring the SUHI Drivers

It is very well known that the changes in land use land cover have significantly affected the seasonal and diurnal variation of SUHI [14]. In the present study, five different potential driving variables of SUHI are considered to understand the reason for SUHI variability over the study period. Two of them, cropland and built-up index, are extracted from MODIS LULC product, which predominated in the study area. The vegetation, evapotranspiration, and albedo are the other three important factors (extracted from MODIS) that can affect SUHI. The differences between an urban and rural area in enhanced vegetation idea (ΔEVI), evapotranspiration (ΔET), and white sky albedo (ΔWSA) were considered to investigate the SUHI variability. This difference was estimated exactly the same way as we did for SUHI estimation in Equation (1). ΔEVI represents the difference in mean EVI over rural pixels and mean EVI over non-rural pixels, and ΔET and ΔWSA are estimated in the same manner. Then, a Pearson's correlation coefficient between the potential driving variables and SUHII across the city and the years was computed.

4. Results

4.1. Spatial Variation and Temporal Trend of LST

The diurnal, seasonal, annual, monthly, and temporal variation of LST from 2003 to 2019 are analyzed in this section. The daytime (10:30 a.m. and 13:30 p.m.) and nighttime (22:30 p.m. and 01:30 a.m.) variation was shown in this study.

4.1.1. Annual Spatial Pattern of LST

The spatial variation (first row) and temporal trend (second row) of the mean annual LST during both daytime and nighttime from 2003 to 2019 are shown in Figure 3. It is evident that the SUHI effect over the city was more pronounced during the nighttime compared to the daytime. The mean LST over urban (U) and non-urban (NU) pixels (as define in Section 3.2.1) was also calculated as in Table 1. During the daytime, the mean annual LST was higher in non-urban pixels compared to urban pixels at both 10:30 a.m. and 13:30 p.m. The observed LST is higher at 13:30 p.m. (33.4 $\pm$ 2.5 °C for NU and 32.6 $\pm$ 2 °C for U pixels), than 10:30 a.m. (29.9 $\pm$ 2.3 °C for NU and 28.9 $\pm$ 1.9 °C for U pixels). In contrast, during nighttime, a reverse phenomenon is observed, witnessing a higher LST over urban pixels compared to non-urban pixels at both 22:30 p.m. and 01:30 a.m. The mean LST was lower at 01:30 a.m. (8 $\pm$ 1.4 °C for NU and 9.2 $\pm$ 1.9 °C for U pixels) in comparison to 22:30 p.m. (10.1 $\pm$ 1.3 °C for NU and 11.4 $\pm$ 1.7 °C for U pixels). The surrounding land cover is dominated by barren land, which can absorb heat much more rapidly than built-up land after sunrise, explaining the higher LST in non-urban pixels. Meanwhile, urban built-up land can retain heat for longer time than barren land cover due to the higher thermal inertia, resulting in the slow emission of heat from built-up lands. Due to this reason, the urban pixels witness a higher LST than the non-urban pixels in nighttime.

The second row in Figure 3 shows Sen's slope trend of mean annual LST during 2003–2019. During the daytime, the central urban area has a lower trend in mean LST compared to the surrounding rural area. However, a significantly dispersed increasing trend is observed during the nighttime compared to the cluster trend in the daytime. The magnitude of Sen's slope was higher in the daytime (0.49 to −0.26 °C/year) compared to the nighttime (0.14 to −0.05 °C/year) trend.

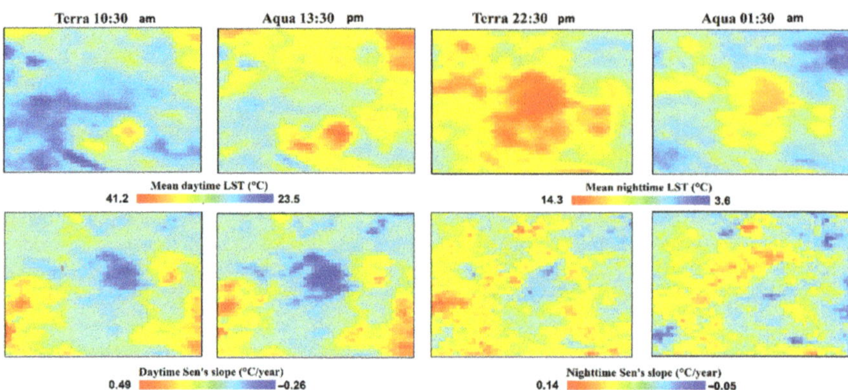

Figure 3. Spatial variation of mean annual LST (average over 2003–2019) during daytime (Terra 10:30 a.m. and Aqua 13:30 p.m.) and nighttime (Terra 22:30 p.m. and Aqua 01:30 a.m.) in first row and the value of Sen's slop for the corresponding LST images in the second row.

Table 1. Annual, seasonal (summer and winter), and monthly (January-December) mean LST from 2003 to 2019 for urban and non-urban pixels at daytime (10:30 a.m. and 13:30 p.m.) and nighttime (22:30 p.m. and 01:30 a.m.).

		Terra 10:30 a.m.	Aqua 13:30 p.m.	Terra 22:30 p.m.	Aqua 01:30 a.m.
Annual	Non-Urban Pixels	29.9 ± 2.3	33.4 ± 2.5	10.1 ± 1.3	8 ± 1.4
	Urban Pixels	28.9 ± 1.9	32.6 ± 2	11.4 ± 1.7	9.2 ± 1.9
Summer	Non-Urban Pixels	37.4 ± 3	40.5 ± 3.2	15.4 ± 1.4	12.6 ± 1.6
	Urban Pixels	36.6 ± 2.3	40.2 ± 2.5	16.6 ± 1.7	13.8 ± 1.9
Winter	Non-Urban Pixels	14.5 ± 1.6	18.5 ± 1.8	−0.5 ± 1.1	−1.8 ± 1.1
	Urban Pixels	13.6 ± 1.3	17.7 ± 1.5	0.7 ± 1.6	−0.5 ± 1.8
Jane	Non-Urban Pixels	11.9 ± 1.6	16.1 ± 1.8	−1.9 ± 1.2	−3 ± 1.2
	Urban Pixels	11.1 ± 1.2	15.4 ± 1.4	−0.5 ± 1.7	−1.6 ± 1.8
February	Non-Urban Pixels	18.4 ± 1.9	23.1 ± 1.9	1 ± 1.1	−0.7 ± 1.1
	Urban Pixels	17.5 ± 1.6	22.2 ± 1.6	2.2 ± 1.6	0.5 ± 1.7
March	Non-Urban Pixels	25.9 ± 2	29.6 ± 2.1	5.7 ± 1.1	3.4 ± 1.2
	Urban Pixels	25.3 ± 1.5	29 ± 1.6	6.9 ± 1.6	4.5 ± 1.7
April	Non-Urban Pixels	30.1 ± 2.4	33.1 ± 2.5	9.9 ± 1.2	7.4 ± 1.3
	Urban Pixels	29.6 ± 1.8	32.9 ± 1.9	11 ± 1.5	8.4 ± 1.7
May	Non-Urban Pixels	36.3 ± 3.2	39.2 ± 3.3	14.7 ± 1.4	12 ± 1.5
	Urban Pixels	35.7 ± 2.4	39.1 ± 2.6	15.9 ± 1.6	13.2 ± 1.8
June	Non-Urban Pixels	43.9 ± 3.5	47.3 ± 3.6	20.1 ± 1.7	17 ± 1.9
	Urban Pixels	42.8 ± 2.8	46.6 ± 2.9	21.5 ± 1.8	18.3 ± 2.1
July	Non-Urban Pixels	45.8 ± 3.5	49.6 ± 3.7	22.2 ± 1.7	19 ± 1.9
	Urban Pixels	44.4 ± 3	48.5 ± 3.1	23.6 ± 1.8	20.2 ± 2.2
August	Non-Urban Pixels	43.6 ± 3.6	47.5 ± 3.9	20.1 ± 1.9	17 ± 2.1
	Urban Pixels	42.3 ± 3	46.4 ± 3.2	21.4 ± 2	18.2 ± 2.3
September	Non-Urban Pixels	38.4 ± 3.2	41.8 ± 3.4	15.9 ± 1.9	13.2 ± 2.2
	Urban Pixels	37 ± 2.7	40.6 ± 2.8	17.2 ± 2.1	14.4 ± 2.3
October	Non-Urban Pixels	30.7 ± 2.5	33.3 ± 2.5	10.3 ± 1.7	8 ± 1.9
	Urban Pixels	29.3 ± 2.2	32.1 ± 2.2	11.4 ± 1.9	9 ± 2.1
November	Non-Urban Pixels	20.3 ± 1.7	22.7 ± 1.9	3.2 ± 1.3	1.8 ± 1.2
	Urban Pixels	19.2 ± 1.6	21.4 ± 1.7	4.3 ± 1.7	2.7 ± 1.7
December	Non-Urban Pixels	13.1 ± 1.5	16.2 ± 1.7	−0.8 ± 1.2	−1.7 ± 1.2
	Urban Pixels	12.3 ± 1.2	15.3 ± 1.5	0.5 ± 1.7	−0.4 ± 1.8

4.1.2. Seasonal Spatial Pattern of LST

The seasonal variation of mean LST during 2003–2019 is shown in Figure 4a,b, where the first row represents the mean LST, and the second-row represents Sen's slope trend during the study period. The daytime represents a clear heterogeneity in the LST distribution compared to the nighttime imagery during both the season. The summer season shows higher land surface temperature (50.2 to 31.2 °C during daytime and 19.5 to 9.2 °C during the night) compared to the winter season (24.1 to 7.3 °C during daytime and 3.6 to −6.9 °C during the night) as shown in Figure 4a,b respectively. More clear evidence of the surface urban heat island phenomenon is visible in the nighttime, while in the daytime, a cool island effect is seen, where the surrounding non-urban regions show a higher LST than the central urban region. The mean LST over urban and non-urban pixels were also calculated as shown in Table 1. During daytime, the mean LST over urban pixels (36.6 ± 2.3 °C at 10:30 a.m. and 40.2 ± 2.5 °C at 13:30 p.m. during summer and 13.6 ± 1.3 °C at 10:30 a.m. and 17.7 ± 1.5 °C at 13:30 p.m.) is comparatively lower than that of non-urban pixels (37.4 ± 3 °C at 10:30 a.m. and 40.5 ± 3.2 °C at 13:30 p.m. during summer and 14.5 ± 1.6 °C at 10:30 a.m. and 18.5 ± 1.8 °C at 13:30 p.m.). Meanwhile, at night, the urban pixels show a higher LST (16.6 ± 1.7 °C in summer and 0.7 ± 1.6 °C in winter at 22:30 p.m., and 13.8 ± 1.9 °C in summer and −0.5 ± 1.8 °C in winter at 1:30 a.m.) compared to non-urban pixels (15.4 ± 1.4 °C in summer and −0.5 ± 1.1 °C in winter and 22:30 p.m., and 12.6 ± 1.6 °C in summer and −1.8 ± 1.1 °C in winter at 1:30 a.m.), witnessing the presence of the urban heat island phenomenon.

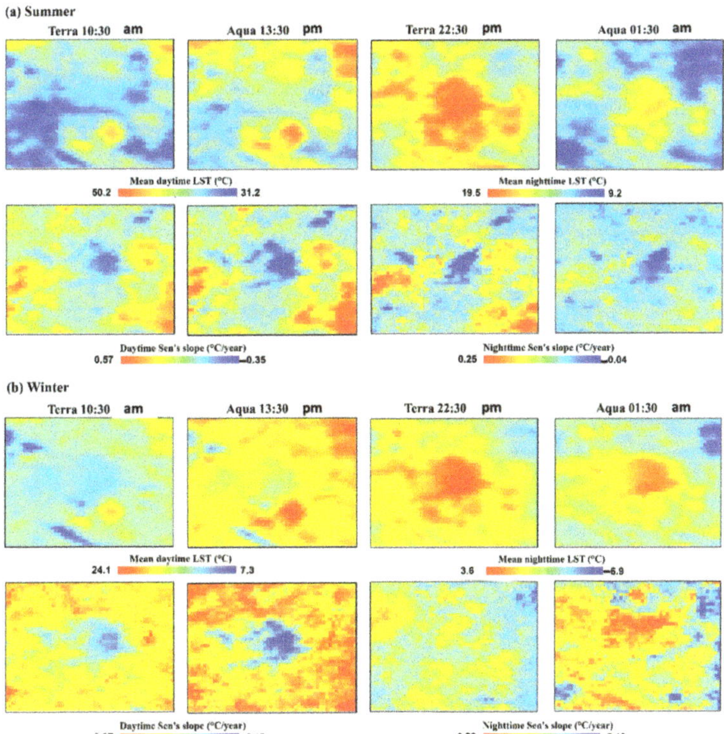

Figure 4. (**a**) Spatial variation of mean summer LST (average over 2003–2019) during daytime (Terra 10:30 a.m. and Aqua 13:30 p.m.) and nighttime (Terra 22:30 p.m. and Aqua 01:30 a.m.) in first row and the value of Sen's slop for the corresponding LST images in the second row, and (**b**) for the winter season.

The result of Sen's slope suggests a higher increasing trend in the summer season compared to the winter season which is also higher in the daytime than the nighttime within the same season. The center urban area witnesses a lower trend compared to the surrounding rural area in the summer season (second row of Figure 4a), while the winter season shows a relatively different trend pattern than summer and annual. In winter nighttime, a positive LST trend is observed in an urban area (more pronounced at 1:30 a.m.) as compared to daytime (Figure 4b). The trend varies from 0.57 to −0.35 °C/year in the daytime to 0.25 to −0.04 °C/year in the nighttime in the summer season. By contrast, in the winter season, it varies from 0.37 to −0.19 °C/year in the daytime to 0.23 to −0.10 °C/year in the nighttime.

4.1.3. Monthly Spatial Pattern of LST

The monthly spatial variation of mean LST during 2003–2019 over the city is also analyzed in this study during both daytime and nighttime (Figure 5). The result obtained from monthly variation also suggests that the SUHI was intensified during the nighttime (more at 22:30 p.m. as compared to 1:30 a.m.) compared to during daytime, in which an urban cool island effect can be seen. The mean LST variation is plotted as a box and whisker plot in Figure 6. It can be seen that the maximum mean LST was observed during the month of July, while the minimum was observed in January. Noticeably, the mean LST at 13:30 p.m. was always higher than at 10:30 a.m. during the daytime, while during nighttime, it was always lower at 1:30 a.m. compared to 22:30 p.m. This also suggests an occurrence of the urban cool island during daytime as non-urban pixels had a higher mean LST than urban pixels at 10:30 a.m. and 13:30 p.m. However, during the nighttime (both 22:30 p.m. and 1:30 a.m.), the mean LST over urban pixels was always higher than non-urban pixels suggesting a positive urban heat island over the city (Table 1).

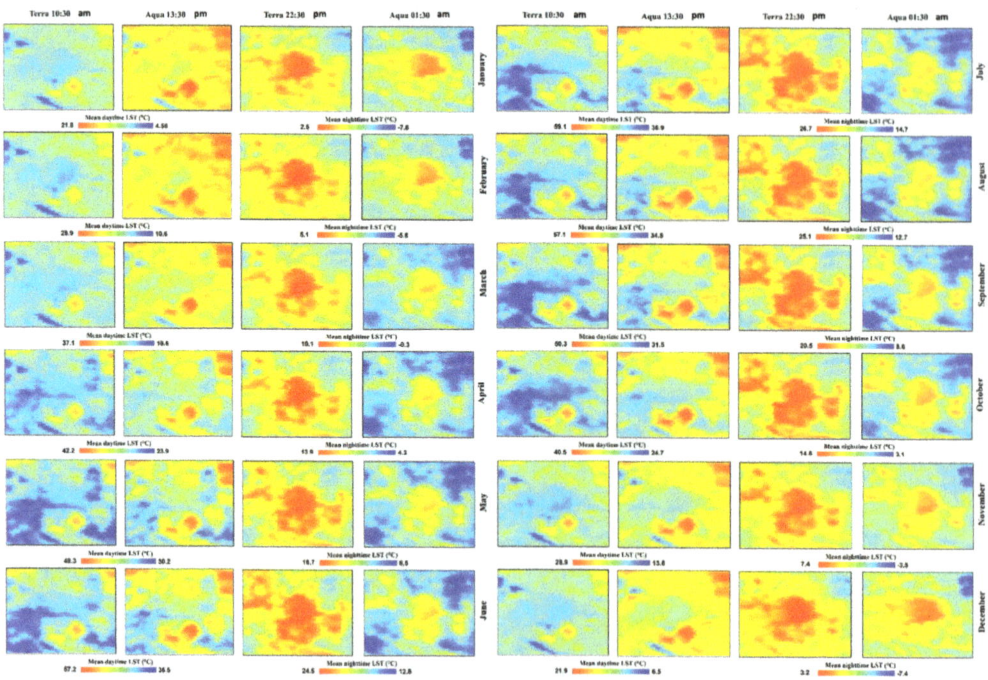

Figure 5. Monthly spatial variation of mean LST (average over 2003–2019) during daytime (Terra 10:30 a.m. and Aqua 13:30 p.m.) and nighttime (Terra 22:30 p.m. and Aqua 01:30 a.m.) from January to June.

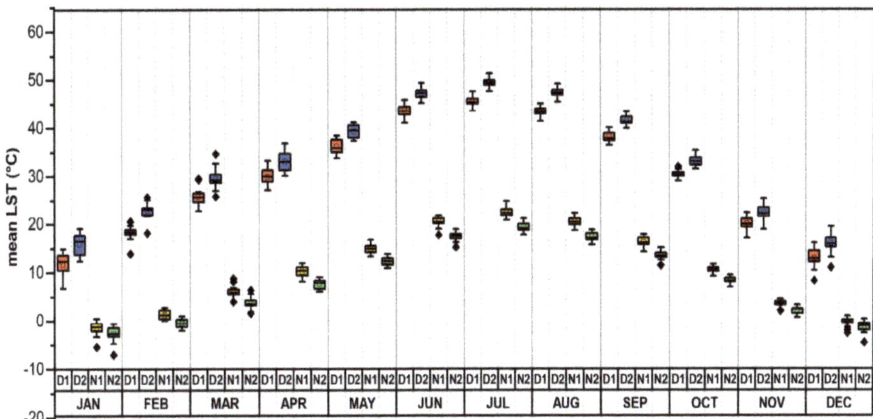

Figure 6. Monthly variation of (average over 2003–2019) during daytime (Terra 10:30 a.m. (D1) and Aqua 13:30 p.m. (D2)) and nighttime (Terra 22:30 p.m. (N1) and Aqua 01:30 a.m. (N2)).

Further, the pixel-based Sen's slope trend of mean LST is also calculated at a monthly level (Figure 7) from 2003 to 2019 during both the daytime and nighttime. The trend was comparatively higher in the daytime hours compared to nighttime. During the daytime, the maximum trend was observed from June to September (in the range of 0.70 to 0.87 °C/year), while the minimum trend was observed in the months of November to February (in the range of 0.19 to 0.36 °C/year. At nighttime, the maximum trend was only 0.33 °C/year in September and the least in February with 0.16 °C/year.

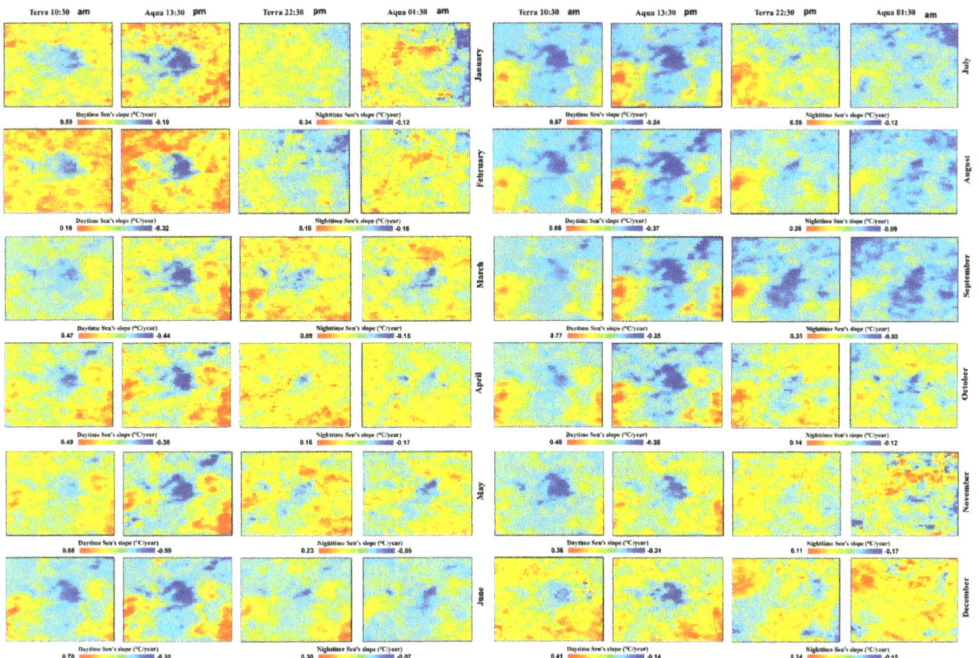

Figure 7. Monthly spatial variation of Sen's slope trend of LST (average over 2003–2019) during daytime (Terra 10:30 a.m. and Aqua 13:30 p.m.) and nighttime (Terra 22:30 p.m. and Aqua 01:30 a.m.) from January to June.

4.2. SUHI Intensity Variation

The annual, seasonal, and monthly variations of SUHI intensity during daytime and nighttime hours are shown in Figure 8. The results clearly indicate the presence of urban cool islands during the daytime and urban heat islands in the nighttime. During the daytime, the SUHI intensity was much lower at 10:30 a.m. compared to at 13:30 p.m. The months from July to October show minimum SUHI intensity values compared to other months. In the nighttime, there was not much variation in SUHI values thought out the year. The intensity of SUHI was higher at 22:30 p.m. compared to that at 1:30 a.m.

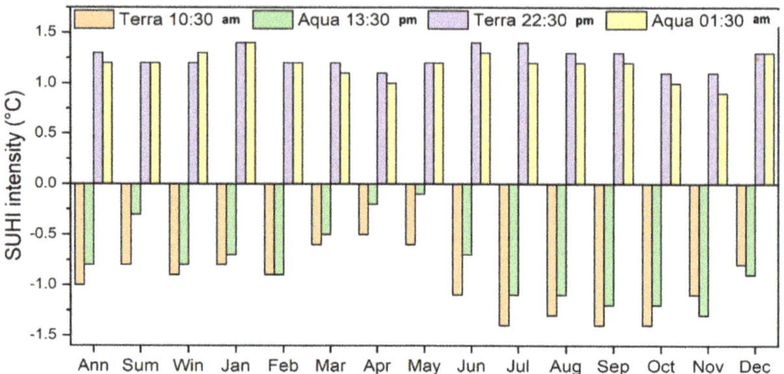

Figure 8. Annual, summer, winter, and monthly variation of mean SUHI intensity (°C) during the period of 2003–2019 at daytime (Terra 10:30 a.m. and Aqua 13:30 p.m.) and nighttime (Terra 22:30 p.m. and Aqua 01:30 a.m.).

4.3. Relationships between the SUHI Intensity and Its Potential Influencing Factors

The relationship between SUHI intensity (SUHII) and its potential influencing factors was examined, which contribute to understanding the specific cause of the SUHI effect. To explore the cause of SUHI variability, five different variables were analyzed and correlated with SUHII: Cropland area (Crop), built-up area (BI), ΔET, ΔEVI, and ΔWSA (i.e., ET, EVI, and WSA urban–rural difference). Figure 9a shows the annual, seasonal (summer and winter), and monthly (January to December) Pearson's correlation coefficient plot of SUHII with the different driving parameters, average to yearly level during daytime (Terra 10:30 a.m. and Aqua 13:30 p.m.), while Figure 9b shows during nighttime (Terra 22:30 p.m. and Aqua 01:30 a.m.). The daytime SUHII shows a significant strong positive correlation with Crop and ΔWSA, while negative with ΔET, ΔEVI, and to some extent BI as well. The daytime correlation was more significant at 13:30 p.m. than 10:30 a.m. The nighttime SUHII shows very less correlation except with the negative correlation of ΔET and ΔEVI. For seasonal correlation, the summer has a much more significant correlation with the driving parameters compared to the winter.

The results of monthly variation reveal that the daytime SUHII was significantly positively correlated with Crop irrespective of the months. The BI shows a significant correlation with SUHII during daytime (more so in the months of July, August, and September), while in the nighttime, the correlation was very less and insignificant. ΔET shows a significant negative correlation in the daytime compared to the nighttime, especially in the months from March to September. The nighttime correlation of ΔET with SUHII is less negative than daytime, where a slight positive correlation is seen from January to March. The correlation of ΔEVI was also negative with SUHII and more in the daytime than nighttime, and higher in the summer months (March–May) than in the winter months. The correlation of ΔWSA is the opposite, as it shows a positive correlation during daytime (higher in summer months than winter months) and negative in the nighttime (significant in winter months, especially in November and December).

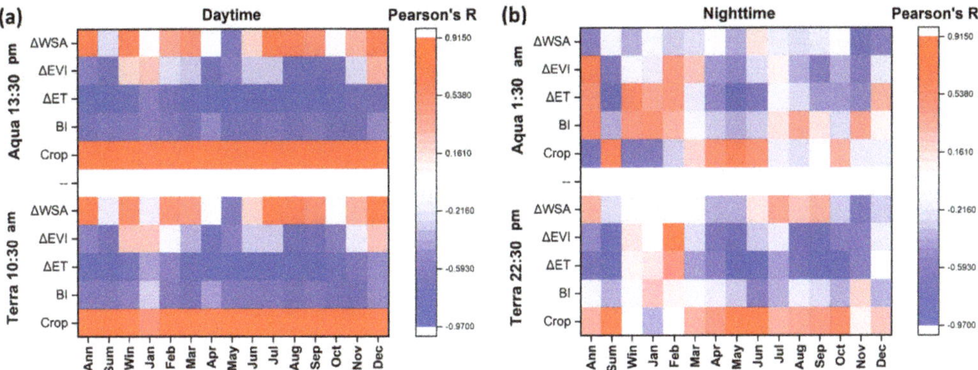

Figure 9. Pearson's correlation coefficient plot of SUHII with the different driving parameters during the (**a**) daytime (Terra 10:30 a.m. and Aqua 13:30 p.m.) and (**b**) nighttime (Terra 22:30 p.m. and Aqua 01:30 a.m.).

5. Discussion

5.1. Variations in the SUHI Intensity

The present study examines the diurnal, annual, seasonal, and monthly behavior of LST and SUHI intensity over Isfahan city from 2003 to 2019 using MODIS datasets. The diurnal and monthly variation of the SUHI index can be found in the previous study [59]. Their findings suggest that the Isfahan metropolitan area is 3–4 K colder than the surrounding suburban area during the daytime, while at nighttime, it is 2 K warmer than its surroundings. The result of the present study also found evidence of the occurrence of the urban cool island during daytime 10:30 a.m. and 13:30 p.m.), while at nighttime (22:30 p.m. and 01:30 a.m.), a positive heat island phenomenon is profoundly visible over the city. The SUHI intensity was found to be minimum at 10:30 a.m. and maximum at 22:30 p.m. The lower land surface temperature over the Isfahan city as compared to surrounding rural area during daytime is also found in previous studies using Landsat data [57,58,60,80]. Moreover, another study [57] suggests that the central portion of the Isfahan metropolis exhibits the highest portion of cold and very cold classes of LST. The findings also suggest that the northern, southern, and eastern portions of the city show much warmer LST, which is also seen in the present study. Considering the seasonal behavior of SUHI, it is evident that the SUHI intensity is weaker in the summer season and stronger in the winter season [59]. The maximum SUHI intensity was found during the January month while the minimum in the months of the summer season.

The phenomenon of a higher LST over the urban area compared to the surrounding rural areas pertaining to negative SUHI during daytime is a typical characteristic of cities situated over arid regions [14,54,55,81]. Moreover, during the daytime, there was an obvious seasonal variation in SUHI, except for the city of the equatorial zone. The amount of rural vegetation cover, direct solar radiation, and longer sunshine duration in summer may contribute to the seasonal variation in SUHI [82]. A negative SUHI during daytime is also reported in various urban areas of the world, including northwestern China [83], central Asia [53], western United States [84], and cities in India [85–87] and South America [88]. All these urban areas are situated over arid or semi-arid regions, a similar climatic characteristic prevailing over the present study area of Isfahan city. Cities situated over arid and semi-arid climatic zones such as Abu Dhabi, Kuwait, Riyadh, Las Vegas, and Phoenix experienced lower LST than the suburban area during daytime. The data suggest that this negative SUHI is due to the existence of a large number of bare lands present in suburban areas, which mostly absorb sunlight rather than reflecting it, leading to a higher temperature in suburban areas than in city cores [59,89,90]. The present study area's dominating land cover is bare land in the suburban region, which leads to a higher land surface temperature than the urban area of the city.

5.2. The Effects of Each Factor on SUHI Intensity

The second objective of the present research was to explain the potential driving variables affecting the diurnal, seasonal, and monthly variation of SUHI intensity. The SUHI phenomenon is driven by various factors that can be broadly categorized as LULC distribution and its spatial pattern, urban site characteristics, and landscape configuration. During the daytime, SUHII was negatively correlated with ΔET, meaning that an increase of SUHI intensity is related to a reduction of the urban ET or a rise of the rural ET. The evaporation from the water surfaces and transpiration from green leaves provides cooling to the rural area, resulting in positive SUHI over the cities. The evaporative cooling is a much more dominant controlling parameter in the daytime for cities over dry climatic regions [55]. The results of the study conducted by [86] suggest that the ET increase in the city area due to higher water use and gardening with irrigating resulted in negative SUHII during the daytime. The relationship between the SUHI intensity and vegetation was significant during the daytime due to the evaporative cooling affect by vegetation [53,55]. The correlation for summer days was significant and negative between the SUHII and ΔEVI in this study compared to winter days, which was similar to the results of the other previous studies [47,53,91]. The vegetation over the city is less affected by seasons, and so a decreasing ΔEVI can intensify the SUHII, even in the summer. The lack of vegetation transpiration in the absence of sunlight during the nighttime leads to an insignificant relation of SUHII with ΔEVI [91,92]. The rural area with higher vegetation cover can increase the latent heat flux via the process of transpiration, which provides cooling, thus lowering the LST of the urban area and resulting in a cool urban heat island during the daytime. It is inferred from the previous studies that vegetation plays an important role in the mitigation of the SUHI effect [93]. The study in Guangzhou, China, found that a decrease in 16% of urban vegetation from 1990–2007 caused an increase in LST by 2.5 °C [94]. Except for ΔEVI and ΔET, there is no significant negative correlation of SUHI intensity with other parameters. Cropland and ΔWSA are found to be effective SUHI controlling variables during the nighttime with a significant positive correlation. In the winter, the LST over cropland is lower than the urban LST due to irrigation and plantation growth, and the rural LST is lower during nighttime, which causes an insignificant correlation with SUHI compared to the summer season. Urbanization generally decreases surface albedo and emissivity, largely due to the urban canopy effect and the nature of the newly added urban material [95,96]. They are responsible for reducing heat loss through increased heat storage and net all-wave radiation in urban areas. The night SUHII is mainly driven by the surface heat fluxes that originate from heat storage during daytime, resulting in the positive correlation of SUHII with the urban–rural difference in albedo (ΔWSA) in the nighttime as seen in our study. This result is also supported by the previous research that suggests urban regions with lower surface albedo tend to increase the diurnal LST and have more energy for releasing at night [97]. Moreover, the recent study on SUHI study also supports a strong diurnal control of albedo in SUHI variation [55].

6. Conclusions

The present study investigated the diurnal, seasonal, annual, and monthly spatiotemporal variation of the LST and SUHII of Isfahan city, Iran, during the last 17 years (2003–2019) based on the MODIS datasets of 1 km resolution. Correlation analyses were also conducted to reveal the relationship between the SUHII and its driving factors, including cropland, built-up area, ΔEVI, ΔET, and ΔWSA.

The results show the presence of an urban cool island during the daytime and a positive heat island during the nighttime. Considering the diurnal variation of SUHI, the maximum SUHII was observed at 22:30 p.m., while the minimum was observed at 10:30 a.m. The SUHII values ranged from −1.5 to +1.5 °C over the metropolis area of

Isfahan. This seasonal variation suggests that the SUHI was much more dominating in the winter season compared to the summer season, with both higher and lower SUHII value during daytime and nighttime in the winter season. The SUHII were higher from June to August months, while minimum from February to May months. The SUHI intensity was significantly and negatively correlated with ΔEVI and ΔET during the daytime. This negative correlation suggests that the negative SUHI during daytime is ascribed to the low vegetation activity in the rural area, dominated by croplands and bare land, which absorb much more heat and warm the surrounding rural surfaces. Furthermore, the nighttime SUHII was related to cropland and ΔWSA. The lower albedo material in the urban area due to urbanization is responsible for reducing heat loss through increased heat storage and net all-wave radiation in urban areas. The nighttime SUHI mainly drivees the surface heat flux driven by heat storage. Therefore, a strong positive correlation of WSA is observed with the SUHII over Isfahan during the nighttime.

Due to the significant SUHI effects observed over Isfahan city in recent years, variation in the urban thermal environment and its related ecological responses should be comprehensively analyzed. The result of present study can be an important reference for understanding the spatiotemporal variations in the SUHI effect, and the interaction between human activities and land-surface ecosystems. However, some uncertainties remain, and other factors (e.g., soil moisture, thermal inertia, landscape configuration) that are associated with the SUHI phenomenon should be thoroughly studied in the future work.

Author Contributions: Conceptualization, methodology, software, writing—original draft preparation, data curation, investigation: A.K.; conceptualization, methodology, software, data curation, original draft preparation, reviewing and editing: P.M.; writing-original draft preparation, visualization: S.G.; data curation, visualization, investigation: D.G.; methodology, investigation, supervision, reviewing and editing: A.G.-M.; methodology, investigation, supervision, reviewing and editing: D.M.-R.; conceptualization, methodology, investigation, supervision, writing—review & editing: R.D.B. All authors have read and agreed to the published version of the manuscript.

Funding: This research received no external funding.

Institutional Review Board Statement: Not applicable.

Informed Consent Statement: Not applicable.

Data Availability Statement: Data are available on request due to privacy/ethical restrictions.

Conflicts of Interest: The authors declare no conflict of interest.

Appendix A

Table A1. Mean area (km^2) of the different LULC classes.

LULC	2003	2007	2011	2015
Open Shrub land	218.75	243.58	285.73	280.77
Grassland	76.27	75.10	80.81	102.34
Cropland	414.81	404.67	344.75	316.01
Urban or Built-up	319.86	326.46	339.53	350.55
Barren land	480.20	467.09	479.08	490.28

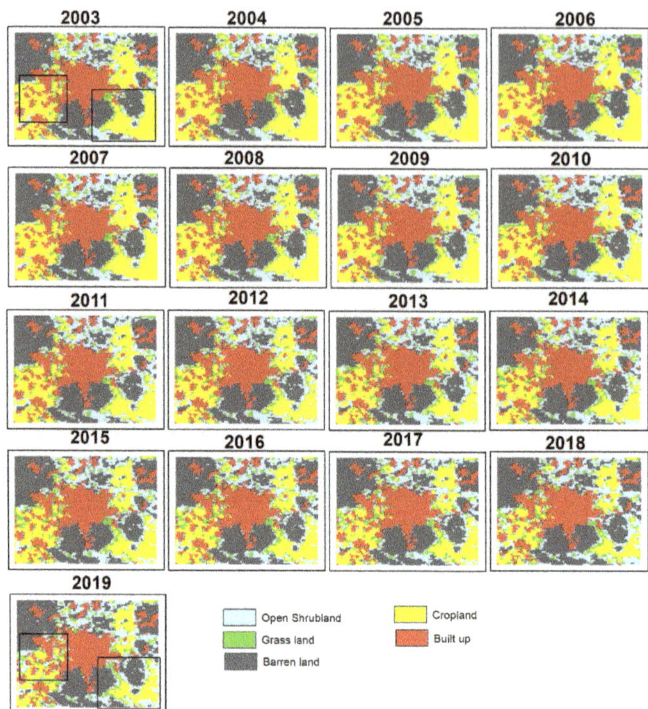

Figure A1. Annual LULC maps of the Isfahan metropolitan area from 2003 to 2019 extracted from MCD12Q1.

References

1. Shen, H.; Huang, L.; Zhang, L.; Wu, P.; Zeng, C. Long-term and fi ne-scale satellite monitoring of the urban heat island effect by the fusion of multi-temporal and multi-sensor remote sensed data: A 26-year case study of the city of Wuhan in China. *Remote Sens. Environ.* **2016**, *172*, 109–125. [CrossRef]
2. Sun, Y.; Zhao, S. Spatiotemporal dynamics of urban expansion in 13 cities across the Jing-Jin-Ji urban agglomeration from 1978 to 2015. *Ecol. Indic.* **2018**, *87*, 302–313. [CrossRef]
3. Weng, Q.; Firozjaei, M.K.; Sedighi, A.; Kiavarz, M.; Alavipanah, S.K. Statistical analysis of surface urban heat island intensity variations: A case study of Babol city, Iran. *GIScience Remote Sens.* **2019**, *56*, 576–604. [CrossRef]
4. Moonen, P.; Defraeye, T.; Dorer, V.; Blocken, B.; Carmeliet, J. Urban Physics: Effect of the micro-climate on comfort, health and energy demand. *Front. Archit. Res.* **2012**, *1*, 197–228. [CrossRef]
5. Kikegawa, Y.; Genchi, Y.; Kondo, H.; Hanaki, K. Impacts of city-block-scale countermeasures against urban heat-island phenomena upon a building's energy-consumption for air-conditioning. *Appl. Energy* **2006**, *83*, 649–668. [CrossRef]
6. Li, Z.; Zhou, Y.; Wan, B.; Chen, Q.; Huang, B.; Cui, Y.; Chung, H. The impact of urbanization on air stagnation: Shenzhen as case study. *Sci. Total Environ.* **2019**, *664*, 347–362. [CrossRef] [PubMed]
7. Liang, L.; Wang, Z.; Li, J. The effect of urbanization on environmental pollution in rapidly developing urban agglomerations. *J. Clean. Prod.* **2019**, *237*, 117649. [CrossRef]
8. Uttara, S.; Bhuvandas, N.; Aggarwal, V. Impacts of urbanization on environment. *Int. J. Res. Eng. Appl. Sci.* **2012**, *2*, 1637–1645.
9. Jeong, A.; Dorn, R.I. Soil erosion from urbanization processes in the Sonoran Desert, Arizona, USA. *Land Degrad. Dev.* **2019**, *30*, 226–238. [CrossRef]
10. Malik, S.; Pal, S.C.; Sattar, A.; Singh, S.K.; Das, B.; Chakrabortty, R.; Mohammad, P. Trend of extreme rainfall events using suitable Global Circulation Model to combat the water logging condition in Kolkata Metropolitan Area. *Urban Clim.* **2020**, *32*, 100599. [CrossRef]
11. Mbao, E.O.; Gao, J.; Wang, Y.; Sitoki, L.; Pan, Y.; Wang, B. Sensitivity and reliability of diatom metrics and guilds in detecting the impact of urbanization on streams. *Ecol. Indic.* **2020**, *116*, 106506. [CrossRef]
12. Blair, R.B. Birds and butterflies along urban gradients in two ecoregions of the United States: Is urbanization creating a homogeneous fauna? In *Biotic Homogenization*; Springer: Berlin/Heidelberg, Germany, 2001; pp. 33–56.

13. Fan, C.; Myint, S.W.; Kaplan, S.; Middel, A.; Zheng, B.; Rahman, A.; Huang, H.-P.; Brazel, A.; Blumberg, D.G. Understanding the impact of urbanization on surface urban heat islands—A longitudinal analysis of the oasis effect in subtropical desert cities. *Remote Sens.* **2017**, *9*, 672. [CrossRef]
14. Zhou, D.; Xiao, J.; Bonafoni, S.; Berger, C.; Deilami, K.; Zhou, Y.; Frolking, S.; Yao, R.; Qiao, Z.; Sobrino, J.A. Satellite remote sensing of surface urban heat islands: Progress, challenges, and perspectives. *Remote Sens.* **2019**, *11*, 48. [CrossRef]
15. Kim, H.H. Urban heat island. *Int. J. Remote Sens.* **1992**, *13*, 2319–2336. [CrossRef]
16. Qiu, G.Y.; Zou, Z.; Li, X.; Li, H.; Guo, Q.; Yan, C.; Tan, S. Experimental studies on the effects of green space and evapotranspiration on urban heat island in a subtropical megacity in China. *Habitat Int.* **2017**, *68*, 30–42. [CrossRef]
17. Qin, Y.; He, Y.; Hiller, J.E.; Mei, G. A new water-retaining paver block for reducing runoff and cooling pavement. *J. Clean. Prod.* **2018**, *199*, 948–956. [CrossRef]
18. Mohajerani, A.; Bakaric, J.; Jeffrey-Bailey, T. The urban heat island effect, its causes, and mitigation, with reference to the thermal properties of asphalt concrete. *J. Environ. Manag.* **2017**, *197*, 522–538. [CrossRef]
19. Du, H.; Wang, D.; Wang, Y.; Zhao, X.; Qin, F.; Jiang, H.; Cai, Y. Influences of land cover types, meteorological conditions, anthropogenic heat and urban area on surface urban heat island in the Yangtze River Delta Urban Agglomeration. *Sci. Total Environ.* **2016**, *571*, 461–470. [CrossRef] [PubMed]
20. Stewart, I.; Oke, T.R. Newly developed "thermal climate zones" for defining and measuring urban heat island magnitude in the canopy layer. In Proceedings of the Eighth Symposium on Urban Environment, Phoenix, AZ, USA, 12 January 2009.
21. Marando, F.; Salvatori, E.; Sebastiani, A.; Fusaro, L.; Manes, F. Regulating ecosystem services and green infrastructure: Assessment of urban heat island effect mitigation in the municipality of Rome, Italy. *Ecol. Modell.* **2019**, *392*, 92–102. [CrossRef]
22. Hu, Y.; Hou, M.; Jia, G.; Zhao, C.; Zhen, X.; Xu, Y. Comparison of surface and canopy urban heat islands within megacities of eastern China. ISPRS J. Photogramm. *Remote Sens.* **2019**, *156*, 160–168.
23. Zhang, X.; Zhong, T.; Feng, X.; Wang, K. Estimation of the relationship between vegetation patches and urban land surface temperature with remote sensing. *Int. J. Remote Sens.* **2009**, *30*, 2105–2118. [CrossRef]
24. Kato, S.; Yamaguchi, Y. Estimation of storage heat flux in an urban area using ASTER data. *Remote Sens. Environ.* **2007**, *110*, 1–17. [CrossRef]
25. Mirzaei, P.A.; Haghighat, F. Approaches to study urban heat island–abilities and limitations. *Build. Environ.* **2010**, *45*, 2192–2201. [CrossRef]
26. Pichierri, M.; Bonafoni, S.; Biondi, R. Satellite air temperature estimation for monitoring the canopy layer heat island of Milan. *Remote Sens. Environ.* **2012**, *127*, 130–138. [CrossRef]
27. Gaur, A.; Eichenbaum, M.K.; Simonovic, S.P. Analysis and modelling of surface Urban Heat Island in 20 Canadian cities under climate and land-cover change. *J. Environ. Manag.* **2018**, *206*, 145–157. [CrossRef] [PubMed]
28. Voogt, J.A.; Oke, T.R. Thermal remote sensing of urban climates. *Remote Sens. Environ.* **2003**, *86*, 370–384. [CrossRef]
29. Li, X.; Zhou, W.; Ouyang, Z. Relationship between land surface temperature and spatial pattern of greenspace: What are the effects of spatial resolution? *Landsc. Urban Plan.* **2013**, *114*, 1–8. [CrossRef]
30. Li, X.; Zhou, Y.; Asrar, G.R.; Imhoff, M.; Li, X. The surface urban heat island response to urban expansion: A panel analysis for the conterminous United States. *Sci. Total Environ.* **2017**, *605*, 426–435. [CrossRef]
31. Clinton, N.; Gong, P. MODIS detected surface urban heat islands and sinks: Global locations and controls. *Remote Sens. Environ.* **2013**, *134*, 294–304. [CrossRef]
32. Heinl, M.; Hammerle, A.; Tappeiner, U.; Leitinger, G. Determinants of urban–rural land surface temperature differences—A landscape scale perspective. *Landsc. Urban Plan.* **2015**, *134*, 33–42. [CrossRef]
33. Camilloni, I.; Barros, V. On the urban heat island effect dependence on temperature trends. *Clim. Change* **1997**, *37*, 665–681. [CrossRef]
34. Li, L.; Zha, Y. Satellite-Based Spatiotemporal Trends of Canopy Urban Heat Islands and Associated Drivers in China's 32 Major Cities. *Remote Sens.* **2019**, *11*, 102. [CrossRef]
35. Yao, R.; Wang, L.; Huang, X.; Liu, Y.; Niu, Z.; Wang, S.; Wang, L. Long-term trends of surface and canopy layer urban heat island intensity in 272 cities in the mainland of China. *Sci. Total Environ.* **2021**, *772*, 145607. [CrossRef] [PubMed]
36. Du, H.; Zhan, W.; Liu, Z.; Li, J.; Li, L.; Lai, J.; Miao, S.; Huang, F.; Wang, C.; Wang, C.; et al. Simultaneous investigation of surface and canopy urban heat islands over global cities. *ISPRS J. Photogramm. Remote Sens.* **2021**, *181*, 67–83. [CrossRef]
37. Li, L.; Zha, Y.; Wang, R. Relationship of surface urban heat island with air temperature and precipitation in global large cities. *Ecol. Indic.* **2020**, *117*, 106683. [CrossRef]
38. Oke, T.R. Canyon geometry and the nocturnal urban heat island: Comparison of scale model and field observations. *J. Climatol.* **1981**, *1*, 237–254. [CrossRef]
39. Rizvi, S.H.; Fatima, H.; Iqbal, M.J.; Alam, K. The effect of urbanization on the intensification of SUHIs: Analysis by LULC on Karachi. *J. Atmos. Solar-Terr. Phys.* **2020**, *207*, 105374. [CrossRef]
40. Tsou, J.; Zhuang, J.; Li, Y.; Zhang, Y. Urban Heat Island Assessment Using the Landsat 8 Data: A Case Study in Shenzhen and Hong Kong. *Urban Sci.* **2017**, *1*, 10. [CrossRef]
41. Gohain, K.J.; Mohammad, P.; Goswami, A. Assessing the impact of land use land cover changes on land surface temperature over Pune city, India. *Quat. Int.* **2021**, *575–576*, 259–269. [CrossRef]

42. Hung, T.; Uchihama, D.; Ochi, S.; Yasuoka, Y. Assessment with satellite data of the urban heat island effects in Asian mega cities. *Int. J. Appl. Earth Obs. Geoinf.* **2006**, *8*, 34–48. [CrossRef]
43. Yang, Q.; Huang, X.; Tang, Q. The footprint of urban heat island effect in 302 Chinese cities: Temporal trends and associated factors. *Sci. Total Environ.* **2019**, *655*, 652–662. [CrossRef]
44. Mohammad, P.; Goswami, A. Surface urban heat island variation over major Indian cities across different climatic zone. In Proceedings of the EGU General Assembly Conference Abstracts, 22nd EGU General Assembly, Online. 4–8 May 2020; p. 6444. Available online: https://ui.adsabs.harvard.edu/abs/2020EGUGA..22.6444M/abstract (accessed on 13 October 2021).
45. Miles, V.; Esau, I. Seasonal and spatial characteristics of Urban Heat Islands (UHIs) in northern West Siberian cities. *Remote Sens.* **2017**, *9*, 989. [CrossRef]
46. Schwarz, N.; Lautenbach, S.; Seppelt, R. Exploring indicators for quantifying surface urban heat islands of European cities with MODIS land surface temperatures. *Remote Sens. Environ.* **2011**, *115*, 3175–3186. [CrossRef]
47. Mohammad, P.; Goswami, A.; Bonafoni, S. The Impact of the Land Cover Dynamics on Surface Urban Heat Island Variations in Semi-Arid Cities: A Case Study in Ahmedabad City, India, Using Multi-Sensor/Source Data. *Sensors* **2019**, *19*, 3701. [CrossRef]
48. Yao, R.; Wang, L.; Huang, X.; Niu, Z.; Liu, F.; Wang, Q. Temporal trends of surface urban heat islands and associated determinants in major Chinese cities. *Sci. Total Environ.* **2017**, *609*, 742–754. [CrossRef] [PubMed]
49. Ramamurthy, P.; Sangobanwo, M. Inter-annual variability in urban heat island intensity over 10 major cities in the United States. *Sustain. Cities Soc.* **2016**, *26*, 65–75. [CrossRef]
50. Cao, C.; Lee, X.; Liu, S.; Schultz, N.; Xiao, W.; Zhang, M.; Zhao, L. Urban heat islands in China enhanced by haze pollution. *Nat. Commun.* **2016**, *7*, 1–7. [CrossRef]
51. Yue, W.; Liu, X.; Zhou, Y.; Liu, Y. Impacts of urban configuration on urban heat island: An empirical study in China mega-cities. *Sci. Total Environ.* **2019**, *671*, 1036–1046. [CrossRef]
52. Manoli, G.; Fatichi, S.; Schläpfer, M.; Yu, K.; Crowther, T.W.; Meili, N.; Burlando, P.; Katul, G.G.; Bou-Zeid, E. Magnitude of urban heat islands largely explained by climate and population. *Nature* **2019**, *573*, 55–60. [CrossRef] [PubMed]
53. Peng, S.; Piao, S.; Ciais, P.; Friedlingstein, P.; Ottle, C.; Bréon, F.M.; Nan, H.; Zhou, L.; Myneni, R.B. Surface urban heat island across 419 global big cities. *Environ. Sci. Technol.* **2012**, *46*, 696–703. [CrossRef]
54. Rasul, A.; Balzter, H.; Smith, C.; Remedios, J.; Adamu, B.; Sobrino, J.; Srivanit, M.; Weng, Q. A Review on Remote Sensing of Urban Heat and Cool Islands. *Land* **2017**, *6*, 38. [CrossRef]
55. Mohammad, P.; Goswami, A. Quantifying diurnal and seasonal variation of surface urban heat island intensity and its associated determinants across different climatic zones over Indian cities. *GIScience Remote Sens.* **2021**, 1–27. [CrossRef]
56. Lazzarini, M.; Molini, A.; Marpu, P.R.; Ouarda, T.B.M.J.; Ghedira, H. Urban climate modifications in hot desert cities: The role of land cover, local climate, and seasonality. *Geophys. Res. Lett.* **2015**, *42*, 9980–9989. [CrossRef]
57. Mirzaei, M.; Verrelst, J.; Arbabi, M.; Shaklabadi, Z.; Lotfizadeh, M. Urban heat island monitoring and impacts on citizen's general health status in Isfahan metropolis: A remote sensing and field survey approach. *Remote Sens.* **2020**, *12*, 1350. [CrossRef]
58. Shirani-Bidabadi, N.; Nasrabadi, T.; Faryadi, S.; Larijani, A.; Roodposhti, M.S. Evaluating the spatial distribution and the intensity of urban heat island using remote sensing, case study of Isfahan city in Iran. *Sustain. Cities Soc.* **2019**, *45*, 686–692. [CrossRef]
59. Montazeri, M.; Masoodian, S.A. Tempo-Spatial Behavior of Surface Urban Heat Island of Isfahan Metropolitan Area. *J. Indian Soc. Remote Sens.* **2020**, *48*, 263–270. [CrossRef]
60. Madanian, M.; Soffianian, A.R.; Soltani Koupai, S.; Pourmanafi, S.; Momeni, M. The study of thermal pattern changes using Landsat-derived land surface temperature in the central part of Isfahan province. *Sustain. Cities Soc.* **2018**, *39*, 650–661. [CrossRef]
61. Ramezankhani, R.; Sajjadi, N.; Jozi, S.A.; Shirzadi, M.R. Climate and environmental factors affecting the incidence of cutaneous leishmaniasis in Isfahan, Iran. *Environ. Sci. Pollut. Res.* **2018**, *25*, 11516–11526. [CrossRef]
62. Kottek, M.; Grieser, J.; Beck, C.; Rudolf, B.; Rubel, F. World map of the Köppen-Geiger climate classification updated. *Meteorol. Z.* **2006**, *15*, 259–263. [CrossRef]
63. Eslamian, S.S.; Feizi, H. Maximum monthly rainfall analysis using L-moments for an arid region in Isfahan province, Iran. *J. Appl. Meteorol. Climatol.* **2007**, *46*, 494–503. [CrossRef]
64. Bihamta, N.; Soffianian, A.; Fakheran, S.; Gholamalifard, M. Using the SLEUTH urban growth model to simulate future urban expansion of the Isfahan metropolitan area, Iran. *J. Indian Soc. Remote Sens.* **2015**, *43*, 407–414. [CrossRef]
65. Abbasnia, M.; Tavousi, T.; Khosravi, M.; Toros, H. Investigation of interactive effects between temperature trend and urban climate during the last decades: A case study of Isfahan-Iran. *Eur. J. Sci. Technol.* **2016**, *4*, 74–81.
66. Wan, Z. A generalized split-window algorithm for retrieving land-surface temperature from space. *IEEE Trans. Geosci. Remote Sens.* **1996**, *34*, 892–905. [CrossRef]
67. Wan, Z. New refinements and validation of the collection-6 MODIS land-surface temperature/emissivity product. *Remote Sens. Environ.* **2014**, *140*, 36–45. [CrossRef]
68. Senay, G.B.; Budde, M.E.; Verdin, J.P. Enhancing the Simplified Surface Energy Balance (SSEB) approach for estimating landscape ET: Validation with the METRIC model. *Agric. Water Manag.* **2011**, *98*, 606–618. [CrossRef]
69. Bastiaanssen, W.G.M.G.M.; Menenti, M.; Feddes, R.A.; Holtslag, A.A.M.; Pelgrum, H.; Wang, J.; Ma, Y.; Moreno, J.F.; Roerink, G.J.; Van der Wal, T.; et al. A remote sensing surface energy balance algorithm for land (SEBAL): 1. Formulation. *J. Hydrol.* **1998**, *212–213*, 198–212. [CrossRef]

70. Gowda, P.H.; Chävez, J.; Howell, T.A.; Marek, T.H.; New, L.L. Surface energy balance based evapotranspiration mapping in the Texas high plains. *Sensors* **2008**, *8*, 5186–5201. [CrossRef]
71. Chakraborty, T.; Lee, X. A simplified urban-extent algorithm to characterize surface urban heat islands on a global scale and examine vegetation control on their spatiotemporal variability. *Int. J. Appl. Earth Obs. Geoinf.* **2019**, *74*, 269–280. [CrossRef]
72. Chakraborty, T.; Hsu, A.; Manya, D.; Sheriff, G. A spatially explicit surface urban heat island database for the United States: Characterization, uncertainties, and possible applications. *ISPRS J. Photogramm. Remote Sens.* **2020**, *168*, 74–88. [CrossRef]
73. Mann, H.B. Nonparametric Tests Against Trend. *Econometrica* **1945**, *13*, 245–259. [CrossRef]
74. Zhang, T.; Peng, J.; Liang, W.; Yang, Y.; Liu, Y. Spatial-temporal patterns of water use efficiency and climate controls in China's Loess Plateau during 2000–2010. *Sci. Total Environ.* **2016**, *565*, 105–122. [CrossRef]
75. Radhakrishnan, K.; Sivaraman, I.; Jena, S.K.; Sarkar, S.; Adhikari, S. A Climate Trend Analysis of Temperature and Rainfall in India. *Clim. Chang. Environ. Sustain.* **2017**, *5*, 146. [CrossRef]
76. Araghi, A.; Mousavi Baygi, M.; Adamowski, J.; Malard, J.; Nalley, D.; Hasheminia, S.M. Using wavelet transforms to estimate surface temperature trends and dominant periodicities in Iran based on gridded reanalysis data. *Atmos. Res.* **2015**, *155*, 52–72. [CrossRef]
77. Zhao, W.; He, J.; Wu, Y.; Xiong, D.; Wen, F.; Li, A. An Analysis of Land Surface Temperature Trends in the Central Himalayan Region Based on MODIS Products. *Remote Sens.* **2019**, *11*, 900. [CrossRef]
78. Mohammad, P.; Goswami, A. Temperature and precipitation trend over 139 major Indian cities: An assessment over a century. *Model. Earth Syst. Environ.* **2019**, *5*, 1481–1493. [CrossRef]
79. Sen, K.P. Estimates of the Regression Coefficient Based on Kendall' s Tau Pranab Kumar Sen. *J. Am. Stat. Assoc.* **1968**, *63*, 1379–1389. [CrossRef]
80. Madanian, M.; Soffianian, A.R.; Koupai, S.S.; Pourmanafi, S.; Momeni, M. Analyzing the effects of urban expansion on land surface temperature patterns by landscape metrics: A case study of Isfahan city, Iran. *Environ. Monit. Assess.* **2018**, *190*, 189. [CrossRef]
81. Wheeler, S.M.; Abunnasr, Y.; Dialesandro, J.; Assaf, E.; Agopian, S.; Gamberini, V.C. Mitigating Urban Heating in Dryland Cities: A Literature Review. *J. Plan. Lit.* **2019**, *34*, 434–446. [CrossRef]
82. Peel, M.C.; Finlayson, B.L.; McMahon, T.A. Updated world map of the Koppen-Geiger climate classification. *Hydrol. Earth Syst. Sci.* **2007**, *11*, 1633–1644. [CrossRef]
83. Zhou, D.; Zhao, S.; Zhang, L.; Sun, G.; Liu, Y. The footprint of urban heat island effect in China. *Sci. Rep.* **2015**, *5*, 2–12. [CrossRef]
84. Hawkins, T.W.; Brazel, A.J.; Stefanov, W.L.; Bigler, W.; Saffell, E.M. The Role of Rural Variability in Urban Heat Island Determination for Phoenix, Arizona. *J. Appl. Meteorol.* **2004**, *43*, 476–486. [CrossRef]
85. Mohammad, P.; Goswami, A. Spatial variation of surface urban heat island magnitude along the urban-rural gradient of four rapidly growing Indian cities. *Geocarto Int.* **2021**, 1–23. [CrossRef]
86. Shastri, H.; Barik, B.; Ghosh, S.; Venkataraman, C.; Sadavarte, P. Flip flop of Day-night and Summer-Winter Surface Urban Heat Island Intensity in India. *Sci. Rep.* **2017**, *7*, 1–8. [CrossRef]
87. Kumar, R.; Mishra, V.; Buzan, J.; Kumar, R.; Shindell, D.; Huber, M. Dominant control of agriculture and irrigation on urban heat island in India. *Sci. Rep.* **2017**, *7*, 1–10. [CrossRef]
88. Wu, X.; Wang, G.; Yao, R.; Wang, L.; Yu, D.; Gui, X. Investigating Surface Urban Heat Islands in South America Based on MODIS Data from 2003–2016. *Remote Sens.* **2019**, *11*, 1212. [CrossRef]
89. Lazzarini, M.; Marpu, P.R.; Ghedira, H. Temperature-land cover interactions: The inversion of urban heat island phenomenon in desert city areas. *Remote Sens. Environ.* **2013**, *130*, 136–152. [CrossRef]
90. Georgescu, M.; Moustaoui, M.; Mahalov, A.; Dudhia, J. An alternative explanation of the semiarid urban area "oasis effect". *J. Geophys. Res. Atmos.* **2011**, *116*, 1–13. [CrossRef]
91. Zhou, D.; Zhao, S.; Liu, S.; Zhang, L.; Zhu, C. Surface urban heat island in China's 32 major cities: Spatial patterns and drivers. *Remote Sens. Environ.* **2014**, *152*, 51–61. [CrossRef]
92. Arnfield, A.J. Two decades of urban climate research: A review of turbulence, exchanges of energy and water, and the urban heat island. *Int. J. Climatol.* **2003**, *23*, 1–26. [CrossRef]
93. Mohammad, P.; Aghlmand, S.; Fadaei, A.; Gachkar, S.; Gachkar, D.; Karimi, A. Evaluating the role of the albedo of material and vegetation scenarios along the urban street canyon for improving pedestrian thermal comfort outdoors. *Urban Clim.* **2021**, *40*, 100993. [CrossRef]
94. Hu, Y.; Jia, G. Influence of land use change on urban heat island derivedfrom multi-sensor data. *Int. J. Climatol.* **2010**, *30*, 1382–1395. [CrossRef]
95. Gachkar, D.; Taghvaei, S.H.; Norouzian-Maleki, S. Outdoor Thermal Comfort Enhancement using Various Vegetation Species and Materials (Case study: Delgosha Garden, Iran). *Sustain. Cities Soc.* **2021**, *75*, 103309. [CrossRef]
96. Karimi, A.; Sanaieian, H.; Farhadi, H.; Norouzian-Maleki, S. Evaluation of the thermal indices and thermal comfort improvement by different vegetation species and materials in a medium-sized urban park. *Energy Rep.* **2020**, *6*, 1670–1684. [CrossRef]
97. Zhou, D.; Bonafoni, S.; Zhang, L.; Wang, R. Remote sensing of the urban heat island effect in a highly populated urban agglomeration area in East China. *Sci. Total Environ.* **2018**, *628–629*, 415–429. [CrossRef] [PubMed]

Article

Environmental Impact of District Heating System Retrofitting

Aleksandrs Zajacs [1,*], Anatolijs Borodinecs [1] and Nikolai Vatin [2]

1 Department of Heat Engineering and Technology, Riga Technical University, LV-1048 Riga, Latvia; anatolijs.borodinecs@rtu.lv
2 Institute of Civil Engineering, Peter the Great Saint Petersburg Polytechnic University, 195251 Saint Petersburg, Russia; vatin_ni@spbstu.ru
* Correspondence: aleksandrs.zajacs@rtu.lv

Abstract: Retrofitting of district heating systems is a comprehensive process which covers all stages of district heating (DH) systems: production, distribution and consumption. This study quantitatively shows the effect of retrofitting measures and represents strengths and weaknesses of different development scenarios. Improvements in production units show improvements in fuel use efficiency and thus indirectly reduce CO_2 emissions due to unburned fuel. For this purpose, validated district planning tools have been used. Tool uses mathematical model for calculation and evaluation of all three main components of the DH system. For the quantitative evaluation, nine efficiency and balance indicators were used. For each indicator, recommended boundary values were proposed. In total, six simulation scenarios were simulated, and the last scenario have shown significant reduction in CO_2 emissions by 40% (from 3376 to 2000 t CO_2 compared to the actual state), while share of biomass has reached 47%.

Keywords: environmental impact; district heating; retrofitting

Citation: Zajacs, A.; Borodinecs, A.; Vatin, N. Environmental Impact of District Heating System Retrofitting. *Atmosphere* **2021**, *12*, 1110. https://doi.org/10.3390/atmos12091110

Academic Editors: Baojie He, Ayyoob Sharifi, Chi Feng and Jun Yang

Received: 29 July 2021
Accepted: 26 August 2021
Published: 29 August 2021

1. Introduction

The world is experiencing a surge in the use of renewable energy for generating electricity, and for heating and cooling. This can be explained by the fact that many respected information sources declare that fossil fuel resources are diminishing, especially in the last decade, and clear evidence of a climate change, which stems in large part from the use of fossil oil and gas [1,2].

The Paris Protocol set new ambitious goals regarding climate change, and since December 2015 the EU is committed to greenhouse gas reduction targets; a 40% cut compared to 1990 levels by 2030, and 60% cut below 2010 levels by 2050. The overarching goal of this agreement is to limit the global average temperature rise by 2030 to below 2 °C compared with pre-industrial levels. Additionally, more ambitious goals are set within EU Green Deal, where a central objective is to set out the trajectory for the EU to be climate neutral by 2050 [3]. While renewable energy is critical for achievement of these set goals, the shift to renewables is also driving a surge in demand for metals and minerals used in renewable energy sources and new infrastructure [4]. Most of the economic, social and environmental risks were addressed in [5], in which the authors concluded that international efforts to promote renewable energy appear to be reinforcing patterns of displacement that exacerbate vulnerability and inequality among and within states. The evaluation of the transition towards renewable energy by [6] led to the conclusion that it effectively creates a "decarbonization divide" between those who benefit from RE and those who are harmed.

Energy demand in the European Union building sector is responsible for about 40% of energy consumption. The reduction in the energy use for heating and cooling in buildings and the introduction of renewables in district heating and cooling sectors are top priorities to achieve reduction in fossil fuel consumption and CO_2 emissions [7]. Significant energy savings are only possible if the energy sectors are prioritized and capital investments are mobilized and available for modernization and refurbishment in those sectors. It has been found that in the North European climate, human behavior can lead to 50% higher heating demand and 60% higher heating power then the standard reference values for energy-efficient buildings [8].

The development of (4th generation district heating) 4GDH involves meeting the challenge of constructing more energy-efficient buildings, as well as the integration of district heating into a future smart energy system based on renewable energy sources [9]. To respond adequately and meet challenges of developing the district heating sector, it is necessary to have a district heating (DH) system planning tool to evaluate different development pathways.

Energy balances performed annually by EUROSTAT show that the amount of energy wasted in conventional thermal power plants located in the European Union (EU), when they are working in condensing mode, is greater than the amount of energy that residential and commercial buildings use for heating. If this waste heat could be used in district heating networks, it would be possible to significantly decrease the amount of imported fossil fuels and CO_2 emissions into the atmosphere, and as result increase the security of economic energy supply, whilst being more environmental friendly [10]. The economic and environmental benefits from district heating are obvious in places where infrastructure has historically been developed and waste heat from combined heat and power (CHP) and industry is available for further utilization for heating needs. However, the volatility of prices for imported fuels, as well as global economic instability, can be a reason for volatility in thermal energy prices in DH networks. As a response to what is happening in the global market, heating operators are forced to adjust tariffs for their services. For example, the planned very sharp tariff increase of 27% in Riga city [11]. Cogeneration and district heating networks should experience significant growth in coming years. The study [12] indicates that the market shares for district heating for buildings may increase to 30% in 2030 and 50% in 2050. However, the penetration of larger scale district heating networks to 80% and higher may take decades; for example, in Copenhagen it took more than 40 years [13]. Study [14] provide some estimations of possible district heating sector development scenarios leading up to 2050 (an overview of scenarios is provided in Energy Roadmap 2050 by the European Commission, 2011). An energy-efficiency scenario indicates a strong reduction in thermal energy demand by about 30% as compared with the reference scenario. The widespread use of energy in the European construction sector creates opportunities for the implementation of energy conservation measures (ECM) in residential buildings. If ECM is implemented in buildings connected to a DH system, it can affect the operation of DH stations, which in turn can change both revenue and electricity production in CHP plants [15]. Taking this into account, before the decision-making step, it is vital to make a careful assessment of development scenarios/projects in terms of environmental impact as well. In order to investigate in detail the feasibility of the selected energy system development option, it is important to make a detailed assessment of it.

During the production of heat or electricity at industrial process plants, combustion of fuel is one of the main origins of greenhouse gas emissions. Some economically reasonable actions that would reduce the emissions include changing fuels and retrofitting the plant's energy system in order to increase fuel use efficiency and lower CO_2 and NO_x emissions [16,17]. It would be advantageous to plan for power production plants located in the vicinity where there are significant heat demands, in order to facilitate the development of district heating networks [18,19]. Among the solutions for the achievement of environmental sustainability in the energy sector, DH with CHP systems is increasingly being used. Study [20] on the development of the DH network addressed the issue of reducing the presence of pollutants in a city (especially in terms of nitrogen oxides (NO_x) and particulate matter (PM) emissions), showing that it is possible to achieve a reduction in ground level average NO_x concentration ranging between 0.2 and 4 $\mu g/m^3$.

The configuration and operational features of a DH system significantly affect its environmental performance. Relevant environmental performance indicators, such as total emissions, pollutant concentration (NO_x, CO, PM), and health damage external costs, were defined within study [21]. Results show that lower pollutant emissions are associated with the installation of a DH system compared to autonomous residential boilers. While NO_x emissions are mostly related to natural gas-fired boilers, emissions of PM2.5 are mostly produced by coal/biomass burning, according to the study [22].

2. Materials and Methods

Energy-efficiency evaluation of district heating systems is comprehensive and extended process, which can be performed in very different ways. The evaluation and assessment of district heating systems always consider different assumptions and simplifications. If these assumptions are not accurate, this will lead to uncertainties and probable errors in the expected results, and this cannot be avoided since any theoretical-mathematical model can simulate different scenarios only for correctly predetermined processes. In the case of district heating system, there are plenty of socio-economic factors, governmental limitations and geopolitical interests, which may be accepted provisionally with some degree of probability. Mathematical description of certain physical processes such as heat exchange in pipelines, heat losses in buildings, combustion of fuel, etc., can be described quite accurately and then validated by real data from physical experiments, while multi-elemental district heating systems require another approach to solve and predict the expected result [23]. Precise and accurate prediction of heat demand is vital for DH operators. While short term heat load numerical prediction can be carried out using available statistic data [24], the current study is focused on long-term planning, where goals of national and EU energy policy and different socio-economic factors are taken into account. Planning of DH systems can be performed with different techniques, which are broadly presented in scientific literature [25–27].

For the detailed assessment of a district heating system, a validated district planning tool has been used. Validation of the tool was performed by comparing calculated values with the actual state of the system with real data provided from the district heating operator's data storage and management system [28]. The tool uses a mathematical model for calculation and evaluation of the three main components of the DH system:

- Heat production;
- Heat distribution;
- Heat consumption.

There are some uncertainties which can impact the accuracy of the presented results or parameters, which cannot be defined at the time of planning:

- Compatibility of technology (low temperature district heating vs. high energy demand of non-renovated buildings);
- Rates of the city growth (new customers);
- Changes in the demand profiles, etc.

This planning tool is based on the mathematically described heat energy production, distribution and consumption processes within one district heating system. All the relationships consider changing heat energy consumption, which was carefully studied and validated within previous studies [28].

The district heating planning tool includes nine separate blocks. Each block has input and output parameters and they are interconnected to tie up all processes in a common calculation algorithm. The scheme of the planning tool with all nine blocks (Introduction, Production, Production Several Heat Sources, Distribution, Consumption, Living Stock, Emissions, Economics, Visualization-Criteria) including the main parameters is presented in Figure 1. Any spreadsheet software is considered an appropriate software for the planning tool.

Figure 1. Smart District Heating Planning Tool overview.

For the simulations of development scenarios a real district heating system was considered, which is located in Riga. There are the following main characteristics of the existing DH system:

Multiapartment buildings—51,000 m^2 of heated area;

Distribution pipeline network—2 km (buried in the ground 1.1 m, overground transit through the building basement—0.9 km);

Heat capacity installed—3 MW natural gas water heating boilers 2 pc (total 6 MW).

Six different development scenarios for this particular district heating system were proposed and simulated, which are presented in Table 1. Additionally, the sixth scenario is assumed to be a combination of the most efficient and perspective development solutions.

Table 1. Description of simulation scenarios.

Scenario / Changing Parameters	Reference Scenario (Actual State)	Low Refurbishment Rate	Moderate Refurbishment Rate	High Refurbishment Rate and Low Temperature Supply	Low Refurbishment Rate and Low Temperature Supply
Distribution network	Not refurbished	100% refurbished	100% refurbished	100% Refurbished Temperatures in distribution network 60/35 °C	100% Refurbished Temperatures in distribution network 60/35 °C
Production units and fuel	2 natural gas boilers (2 × 3 MW)	2 natural gas boilers (2 × 3 MW), natural gas CHP unit (600 kW)	2 natural gas boilers (2 × 3 MW), natural gas CHP unit (600 kW), wooden biomass water boiler 2 MW	2 natural gas boilers (2 × 3 MW), natural gas CHP unit (600 kW), local solar collectors	1 natural gas boiler (3 MW), natural gas CHP unit (600 kW), wooden biomass water boiler 3 MW
Building stock refurbishment rate	Not refurbished	3% per year	5% per year	7% per year	3% per year

Thermal energy production units were prioritized by their efficiency and sustainability. Base load is covered by co-generation units and biomass fired boilers and solar collectors if they are available, and fossil energy units are used to cover peak demand.

Trends in national and European energy policy were used to justify the chosen values for different scenarios. Directive 2012/27/EU of the EU on energy efficiency defines that from the year 2014, 3% of the public buildings owned and occupied by its central government should be renovated each year to meet at least the minimum energy performance requirements. A refurbishment rate of 3% per year is considered low, since it corresponds to the minimum required rate applicable only for buildings owned by government, while the private sector could also attract European co-funding programs to speed up the renovation process [29].

Regulation No. 222 of the Cabinet of Ministers of Latvia states that: "Methods for calculating the energy performance of buildings and rules for energy certification of buildings" [30], which determines the minimum levels of energy performance and efficiency for heating in new buildings. Starting from 1 January 2021, the minimal requirement for new single-family houses is less than 50 kWh/m^2/year, and for multiapartment residential buildings, less than 40 kWh/m^2/year. A renovation rate of 7% is considered very ambitious but possible, if renovation measures are prioritized by national and EU co-financing programs. Higher values are unlikely to be possible, and are actually limited by the availability qualified workforce and citizen activity. While 5% per year is chosen as a moderate value for the evaluation of intermediate values, more results will be observed in order to estimate possible trends and for extrapolation purposes. Based on previous studies on average energy efficiency improvements among refurbished buildings, it was possible to predict at least a 30% reduction in heat energy consumption after the building undergoes the refurbishment process [28]. Evaluation and simulations were made for a 15-year period. Reduced temperature regimes of 60/35 °C were proposed as the minimum possible supply temperatures that can be used for preparation of domestic hot water. Reducing supply temperatures below 60 °C can jeopardize hygienic requirements and cause additional electricity consumption for domestic hot water preparation [31,32]. It is important to highlight that reducing the supply temperatures in the district heating network is not possible without additional renovation measures in the multiapartment buildings; otherwise, the existing building heat supply system will be undersized and will not provide the required comfort level during peak cold periods during the seasons where heating is required.

3. Results

For each scenario production program or each production unit, demand profiles on a monthly basis were developed. In Table 2, the example for a "Low refurbishment rate" scenario is presented:

For each scenario, results were presented in a form of a table with 12 efficiency criteria, where all values are marked in a color from green to red to indicate whether the parameter meets certain requirements. An example of the output from the planning tool in Figure 2 is given as a reference for the "Low refurbishment rate and low temperature supply" scenario.

When the distribution network is 100% renovated, a co-generation unit is installed, wooden biomass fired water boiler 3MW is installed, and flow/return temperature in the distribution network is 60/35 °C, residential sector energy efficiency increases by 30% in 3% of the buildings every year (15-year perspective).

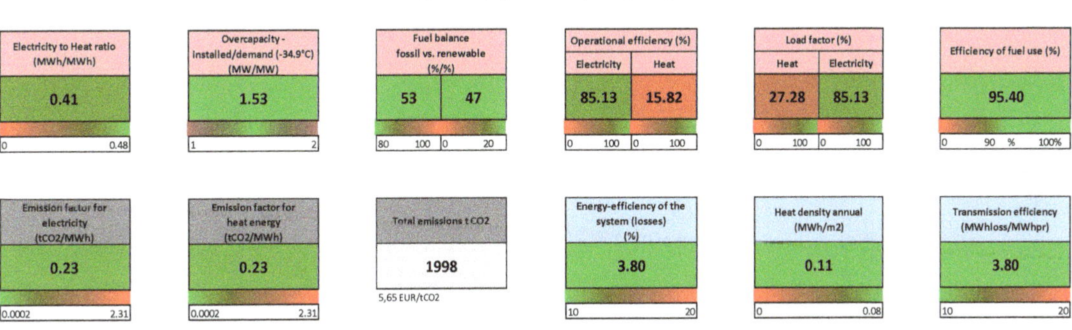

Figure 2. "Low refurbishment rate and low temperature supply" scenario criteria evaluation results.

Additionally, all results for all scenarios are presented in Table 3.

Table 2. Production program for "Low refurbishment rate" scenario.

Installed Heat Capacity	**6.97**	MW		Total Emissions		**2990.447tCO2**			*Production programs*					
Capacity of installed cogeneration unit	**0.50**	MW el		Emissions for heat		**2112.16 tCO2**								
	0.64	MW th		Emissions for electr		**878.29 tCO2**								
Peak load for heating period (heat)	**3.85**	MW												
Peak load for heating period (electr)	**0.50**	MW												
Cogeneration heating plants														
	Units	Jan	Feb	Mar	Apr	May	Jun	Jul	Aug	Sep	Oct	Nov	Dec	**Total**
Hours year		744	672	744	720	744	720	744	744	720	744	720	744	
Working hours of cogeneration unit	h	744	672	744	720	744	720	744	744	720	744	720	744	**8760**
Working hours of natural gas boilers	h	744	672	744	720	0	0	0	0	0	744	720	744	**5088**
Heat energy to the network	**MWh**	**1443**	**1284**	**1237**	**874**	**339**	**307**	**311**	**313**	**307**	**623**	**1017**	**1335**	**9392**
Average heating load	MW	1.94	1.91	1.66	1.21	0.46	0.43	0.42	0.42	0.43	0.84	1.41	1.79	
Cogeneration unit el.load	MW	0.50	0.50	0.50	0.50	0.36	0.33	0.33	0.33	0.33	0.50	0.50	0.50	
alfa (electr/heat)		0.785	0.785	0.785	0.785	0.785	0.785	0.785	0.785	0.785	0.785	0.785	0.785	
Cogeneration unit produced electr.	MWh	372	336	372	360	266	241	245	246	241	372	360	372	**3783**
El. self-consumption of cog.unit	MWh	46	46	50	27	15	14	15	13	14	17	47	39	**343**
Electricity to the grid	MWh	326	290	322	333	251	227	230	233	227	355	313	333	**3440**
Heat load of cogeneration unit	MW	0.64	0.64	0.64	0.64	0.46	0.43	0.42	0.42	0.43	0.64	0.64	0.64	
Heat energy produced by cogeneration unit	**MWh**	**474**	**428**	**474**	**459**	**339**	**307**	**311**	**313**	**307**	**474**	**459**	**474**	**4819**
Heat energy poduced by heating boiler (ng)	**MWh**	**969**	**856**	**764**	**416**	**0**	**0**	**0**	**0**	**0**	**149**	**558**	**861**	**4573**
Natural gas consumption	1000 m^3	219	196	195	151	76	69	70	70	69	123	168	207	**1611**
Efficiency water boiler		0.93	0.93	0.93	0.93	0.93	0.93	0.93	0.93	0.93	0.93	0.93	0.93	
Efficiency cogeneration unit		0.87	0.87	0.87	0.87	0.87	0.87	0.87	0.87	0.87	0.87	0.87	0.87	
Total efficiency		0.90	0.90	0.90	0.89	0.87	0.87	0.87	0.87	0.87	0.88	0.89	0.90	**0.89**
Heat energy produced with renewables	MWh													0

Table 3. Summary of results for simulation of all scenarios.

		Reference Scenario (Actual State)	Low Refurbishment Rate	Moderate Refurbishment Rate	High Refurbishment Rate and Low Temperature Supply	Low Refurbishment Rate and Low Temperature Supply
Emissions, tCO2		2372	2990	2102	1737	1998
Loss in the network (% from consumed energy)		10.61	5.77	6.40	2.71	4.15
Consumption, MWh		10,230.79	8849.63	7928.86	7161.55	8849.63
Produced from renewable sources, MWh		0.00	0.00	3036.79	2078.62	3701.51
Electricity to Heat ratio (MWh/MWh)		0.00	0.40	0.43	0.45	0.43
Overcapacity - installed/demand		1.39	1.61	2.31	1.99	1.53
Fuel balance fossil vs. renewable (%/%)		100/0	100/0	64/36	72/28	56/44
Operational efficiency (%)	**Electricity**	0.00	86.36	83.78	53.65	82.55
	Heat	17.89	15.38	10.78	8.64	14.49
Load factor (%)	**Heat**	28.02	27.85	28.04	19.34	24.98
	Electricity	0.00	86.36	83.78	53.65	82.55
Efficiency of fuel use (%)		93.00	89.00	93.61	88.71	95.40
Emission factor for electricity (tCO2/MWh)		N/A	0.23	0.23	0.23	0.23
Emission factor for heat energy (tCO2/MWh)		0.22	0.22	0.23	0.23	0.23
Total emissions tCO2		2372	2990	2102	1737	1934
Energy-efficiency of the system (losses) (%)		10.61	5.77	6.40	2.71	4.15
Heat density annual (MWh/m2)		0.13	0.11	0.10	0.06	0.10
Transmission efficiency (MWhloss/MWhpr)		10.61	5.77	6.40	3.82	4.15

4. Discussion

Various renovation rates of 3%, 5%, and 7% in the residential sector have shown a reduction in energy consumption in a 15-year perspective by 14%, 23% and 30%, respectively. The estimated reduction shows the overall trend in the energy sector, especially energy for domestic needs, and shows that existing DH network operators will face a situation when the capacity of the existing infrastructure will not be used at full potential. This situation may lead to a reduction in DH company revenues, which, in turn, will raise the question of tariff increase. However, there is always a possibility that growing cities and towns will attract new customers to increase the amount of heat energy sold. The attraction of new customers and decommissioning of decentralized fossil fuel heating plants will provide great benefits for future development perspectives, from both an economic and an environmental perspective.

The last three scenarios include renewable energy sources, a wooden biomass fired water boiler equipped with a flue gas condensing unit with total efficiency 112% (by lower heating value—LHV efficiency) and solar collectors. In these scenarios, overall efficiency has increased and fossil vs. renewable fuel balance has noticeably shifted towards renewable energy, reaching a maximum of 44% in the last "Low refurbishment rate and low temperature supply" scenario. Additionally, CO_2 emissions were reduced in the last three scenarios, even compared with the "Reference scenario" (Actual state), where a CHP unit is not installed. Compared with the "Low refurbishment rate" scenario, where CO_2 emissions are the highest, the last three scenarios show a reduction from 30% to 42%. Within this study, wooden biomass was considered as a CO_2-neutral fuel, with an emission factor equal to "0", according to the Republic of Latvia Cabinet Regulation No. 42 (23.01.2018) "Methodology for Calculating Greenhouse Gas Emissions" [33] and EU DIRECTIVE 2003/87/EC 13.10.2003 [34]. However, the U.S. Environmental Protection Agency [35] and other scientific literature sources report CO_2 factors that are different from "0". It is useful to note, even if wooden biomass is considered to be a CO_2-neutral fuel, without a properly planned reforestation policy, standard CO_2 emissions from burning wooden biomass (0.403 tCO_2/MWh) are twice as high than those from natural gas (0.202 tCO_2/MWh), according to the Joint Research Centre of the European Commission [36]. In combination with increased load on traffic and use of heavy traffic, the environmental impact of biomass fuels needs to be reassessed, taking into account all the embedded energy input before the biomass goes into the furnace of heating plant. Nevertheless, it is important to understand that biomass gives an opportunity for countries, which are highly dependent on energy imports to develop local economy and produce energy from local sources.

Comprehensive assessment of the simulation results shows that the "Low refurbishment rate and low temperature supply" scenario is the most technologically and environmentally acceptable, and meets the targets of national and EU energy policies. For this particular DH system, the payback period of such a development scenario would be 5.78 years (without EU funding). Estimations are made using average equipment and construction costs during the period 2016–2019. Taking into account the extremely high price volatility on steel, wood and services during the last 6 months, there is very high risk that the obtained results are overestimated and the payback period may be considerably increased. However, for equipment in DH systems, payback periods are slightly higher than average, and could be predicted to be up to 10 years, considering an equipment service life up to 35 years.

Author Contributions: Conceptualization, A.B., A.Z.; methodology, A.Z.; validation, A.Z., writing—review and editing, A.B. and A.Z.; visualization, A.Z.; supervision, A.B., N.V. All authors have read and agreed to the published version of the manuscript.

Funding: This work has been supported by the European Regional Development Fund within the Activity 1.1.1.2 "Post-doctoral Research Aid" of the Specific Aid Objective 1.1.1 "To increase the research and innovative capacity of scientific institutions of Latvia and the ability to attract external financing, investing in human resources and infrastructure" of the Operational Programme "Growth and Employment" (1.1.1.2/VIAA/2/18/344).

Conflicts of Interest: The authors declare no conflict of interest.

References

1. Shafiee, S.; Topal, E. When will fossil fuel reserves be diminished? *Energy Policy* **2009**, *37*, 181–189. [CrossRef]
2. Luo, M.; Ji, Y.; Ren, Y.; Gao, F.; Zhang, H.; Zhang, L.; Li, H. Characteristics and health risk assessment of PM2.5-bound pahs during heavy air pollution episodes in winter in urban area of Beijing, China. *Atmosphere* **2021**, *12*, 323. [CrossRef]
3. BP Annual Review 2020, British Petroleum 2020. Available online: https://www.bp.com/content/dam/bp/business-sites/en/global/corporate/pdfs/energy-economics/statistical-review/bp-stats-review-2020-full-report.pdf (accessed on 15 August 2021).
4. European Commission. EU Climate Action and the European Green Deal. 14 July 2021. Available online: https://ec.europa.eu/clima/policies/eu-climate-action_en (accessed on 15 August 2021).
5. The World Bank. The Growing Role of Minerals and Metals for a Low Carbon Future. 2018. Available online: https://www.worldbank.org/en/topic/energy/publication/minerals-and-metals-to-play-significant-role-in-a-low-carbon-future#:~{}:text=The%20rise%20of%20green%20energy,new%20World%20Bank%20report%2C%20%E2%80%9CThe (accessed on 15 August 2021).
6. Kramarz, T.; Park, S.; Johnson, C. Governing the dark side of renewable energy: A typology of global displacements. *Energy Res. Soc. Sci.* **2021**, *74*, 101902. [CrossRef]
7. Sovacool, B.K.; Hook, A.; Martiskainen, M.; Brock, A.; Turnheim, B. The decarbonisation divide: Contextualizing landscapes of low-carbon exploitation and toxicity in Africa Global Environ. *Change* **2020**, *60*, 102028. [CrossRef]
8. Remeikienė, R.; Gasparėnienė, L.; Fedajev, A.; Szarucki, M.; Đekić, M.; Razumienė, J. Evaluation of sustainable energy development progress in EU member states in the context of building renovation. *Energies* **2021**, *14*, 4209. [CrossRef]
9. Dalla Rosa, A.; Christensen, J.E. Low-energy district heating in energy-efficient building areas. *Energy* **2011**, *36*, 6890–6899. [CrossRef]
10. Lund, H.; Østergaard, P.A.; Chang, M.; Werner, S.; Svendsen, S.; Sorknæs, P.; Thorsen, J.E.; Hvelplund, F.; Mortensen, B.O.G.; Mathiesen, B.V.; et al. The status of 4th generation district heating: Research and results. *Energy* **2018**, *164*, 147–159. [CrossRef]

11. Jouhara, H.; Olabi, A.G. Industrial waste heat recovery. *Energy* **2018**, *160*, 1–2. [CrossRef]
12. Submission of Heat Tariffs Set (Offered) by JSC "RĪGAS SILTUMS" to the Public Utilities Commission. Available online: https://www.rs.lv/lv/saturs/rigas-siltums-siltumenergijas-tarifs (accessed on 15 August 2021).
13. David, A.; Mathiesen, B.V.; Averfalk, H.; Werner, S.; Lund, H. Heat roadmap europe: Large-scale electric heat pumps in district heating systems. *Energies* **2017**, *10*, 578. [CrossRef]
14. Thornton, R. *Copenhagen's District Heating System: Recycling Waste Heat Reduces Carbon Emissions and Delivers Energy Security*; International District Energy Association: Westborough, MA, USA, 2009; 9p. Available online: http://www.districtenergy.org/assets/pdfs/White-Papers/Copenhagen-Clean-District-Heating-final-Web4.pdf (accessed on 15 August 2021).
15. Colmenar-Santosa, A.; Rosales-Asensio, E.; Borge-Diez, D.; Blanes-Peiro, J.J. District heating and cogeneration in the EU-28: Current situation, potential and proposed energy strategy for its generalization. *Renew. Sustain. Energy Rev.* **2016**, *62*, 621–639. [CrossRef]
16. Difs, K.; Bennstam, M.; Trygg, L.; Nordenstam, L. Energy conservation measures in buildings heated by district heating—A local energy system perspective. *Energy* **2010**, *35*, 3194–3203. [CrossRef]
17. Kosorukov, D.; Aksenov, A. Use of condensing economizers with developed surfaces to improve the energy efficiency of conventional gas-fired heat generators in boilers. *E3S Web Conf.* **2021**, *263*, 04024. [CrossRef]
18. Dagilis, V.; Vaitkus, L.; Balčius, A.; Gudzinskas, J.; Lukoševičius, V. Low grade heat recovery system for woodfuel cogeneration plant using water vapour regeneration. *Therm. Sci.* **2018**, *22*, 2667–2677. [CrossRef]
19. Chen, J.; Huang, S.; Shahabi, L. Economic and environmental operation of power systems including combined cooling, heating, power and energy storage resources using developed multi-objective grey wolf algorithm. *Appl. Energy* **2021**, *298*, 117257. [CrossRef]
20. Li, X.; Wu, X.; Gui, D.; Hua, Y.; Guo, P. Power system planning based on CSP-CHP system to integrate variable renewable energy. *Energy* **2021**, *232*, 121064. [CrossRef]
21. Ravina, M.; Panepinto, D.; Zanetti, M.C.; Genon, G. Environmental analysis of a potential district heating network powered by a large-scale cogeneration plant. *Environ. Sci. Pollut. Res.* **2017**, *24*, 13424–13436. [CrossRef] [PubMed]
22. Ravina, M.; Panepinto, D.; Zanetti, M. District heating networks: An inter-comparison of environmental indicators. *Environ. Sci. Pollut. Res.* **2021**, *28*, 33809–33827. [CrossRef]
23. Le Truong, N.; Gustavsson, L. Cost and primary energy efficiency of small-scale district heating systems. *Appl. Energy* **2014**, *130*, 419–427. [CrossRef]
24. Rusovs, D.; Jakovleva, L.; Zentins, V.; Baltputnis, K. Heat load numerical prediction for district heating system operational control. *Latv. J. Phys. Tech. Sci.* **2021**, *58*, 121–136. [CrossRef]
25. Kudela, L.; Špiláček, M.; Pospíšil, J. Influence of control strategy on seasonal coefficient of performance for a heat pump with low-temperature heat storage in the geographical conditions of central europe. *Energy* **2021**, *234*, 121276. [CrossRef]
26. Siksnelyte-Butkiene, I.; Streimikiene, D.; Balezentis, T. Multi-criteria analysis of heating sector sustainability in selected north european countries. *Sustain. Cities Soc.* **2021**, *69*, 102826. [CrossRef]
27. Lebedeva, K.; Borodinecs, A.; Krumins, A.; Tamane, A.; Dzelzitis, E. Potential of end-user electricity peak load shift in latvia. *Latv. J. Phys. Tech. Sci.* **2021**, *58*, 32–44. [CrossRef]
28. Zajacs, A.; Borodiņecs, A. Assessment of development scenarios of district heating systems. *Sustain. Cities Soc.* **2019**, *48*, 101540. [CrossRef]
29. Latvian Public Broadcasting, State to Grant Public Loans for Apartment Building Renovation. 2021. Available online: https://eng.lsm.lv/article/economy/economy/state-to-grant-public-loans-for-apartment-building-renovation.a411764/ (accessed on 15 August 2021).
30. Regulations No. 222 of Cabinet of Ministers of Latvia "Methods for Calculating the Energy Performance of Buildings and Rules for Energy Certification of Buildings". Available online: https://likumi.lv/ta/id/322436-eku-energoefektivitates-aprekina-metodes-un-eku-energosertifikacijas-noteikumi (accessed on 15 August 2021).
31. Toffanin, R.; Curti, V.; Barbato, M.C. Impact of legionella regulation on a 4th generation district heating substation energy use and cost: The case of a swiss single-family household. *Energy* **2021**, *228*, 120473. [CrossRef]
32. Hajian, H.; Ahmed, K.; Kurnitski, J. Estimation of energy-saving potential and indoor thermal comfort by the central control of the heating curve in old apartment buildings. *E3S Web Conf.* **2021**, *246*, 09002. [CrossRef]
33. Republic of Latvia Cabinet Regulation No. 42 (23 January 2018). "Methodology for Calculating Greenhouse Gas Emissions". Available online: https://likumi.lv/ta/en/en/id/296651 (accessed on 15 August 2021).

34. Directive 2003/87/EC of the European Parliament and of the Council of 13 October 2003 Establishing a Scheme for Greenhouse Gas Emission Allowance Trading within the Community and Amending Council Directive 96/61/EC. Available online: https://eur-lex.europa.eu/legal-content/EN/TXT/PDF/?uri=CELEX:32003L0087&from=LV (accessed on 15 August 2021).
35. U.S. Environmental Protection Agency, Emission Factors for Greenhouse Gas Inventories. 2014. Available online: https://www.epa.gov/sites/default/files/2015-07/documents/emission-factors_2014.pdf (accessed on 15 August 2021).
36. Koffi, B.; Cerutti, A.; Duerr, M.; Iancu, A.; Kona, A.; Janssens-Maenhout, G. CoM Default Emission Factors for the Member States of the European Union—Version 2017, European Commission, Joint Research Centre (JRC) [Dataset] PID. Available online: http://data.europa.eu/89h/jrc-com-ef-comw-ef-2017 (accessed on 15 August 2021).

 atmosphere

MDPI

Communication

Preliminary Study on Water Bodies' Effects on the Decomposition Rate of Goldenrod Litter

Szabina Simon *, Brigitta Simon-Gáspár, Gábor Soós and Angéla Anda

Hungarian University of Agriculture and Life Sciences (Georgikon Campus), P.O. Box 71, H-8361 Keszthely, Hungary; Simon.Gaspar.Brigitta@uni-mate.hu (B.S.-G.); Soos.Gabor@uni-mate.hu (G.S.); anda.angela@uni-mate.hu (A.A.)

* Correspondence: Simon.Szabina@uni-mate.hu; Tel.: +36-83-545102

Abstract: Leaf-litter input constitutes a major load in natural waters; therefore, to achieve and maintain high water quality, it is important to thoroughly examine and understand the litter decomposition process. The widespread *Solidago canadensis* exerts a negative effect on the composition of the ecosystem, causes extinction of species, and modifies the function of the system. In Hungary, goldenrod constantly spreads to newer areas, which can also be observed around Lake Balaton and at the bank of the Hévíz canal. In our investigation, we examined the decomposition rate of the leaves and stems of the goldenrod with the commonly applied method of leaf litter bags. As water temperature, ranging from 24.0 °C to 13.7 °C, decreases in Hévíz canal away from Lake Hévíz (−0.32 °C/100 m), we chose three different sampling sites with different water temperatures along the canal to determine how water temperature influences the rate of decomposition. For both leaves and stems, the fastest decomposition rate was observed at the first site, closest to the lake. At further sites with lower water temperatures, leaf litter decomposition rates decreased. Results observed through Hévíz canal demonstrated that higher water temperature accelerated the goldenrod decomposition dynamics, while the drift also impacted its efficiency.

Keywords: leaf litter decomposition; *Solidago canadensis*; Lake Hévíz; Hévíz canal

Citation: Simon, S.; Simon-Gáspár, B.; Soós, G.; Anda, A. Preliminary Study on Water Bodies' Effects on the Decomposition Rate of Goldenrod Litter. *Atmosphere* **2021**, *12*, 1394. https://doi.org/10.3390/atmos12111394

Academic Editors: Baojie He, Ayyoob Sharifi, Chi Feng and Jun Yang

Received: 30 September 2021
Accepted: 22 October 2021
Published: 25 October 2021

Publisher's Note: MDPI stays neutral with regard to jurisdictional claims in published maps and institutional affiliations.

1. Introduction

Water is one of the main natural resources of the world and a prime constituent of the biosphere. Fresh water sources are primarily present in forms of lakes, rivers, precipitation, and underground water. In the past century, the increased nutrient load of the wetland due to anthropogenic intervention highly deteriorated the water quality of Lake Balaton. Water coming from Lake Hévíz (LH) through the 10–12 m wide and 13 km long Hévíz canal (Hc) enters Zala River, after nutrients are retained within the natural wetland of Ingói berek.

The Balaton's largest tributary, the River Zala, which enters the lake through its westernmost and smallest basin, the Keszthely Basin, supplies ~50% of the lake's total water input and accounts for 35–40% of the lake's nutrient input [1]. The Hc, as a part of the Kis-Balaton Water Protection System—designed on the lower part of River Zala, contributes to the nutrition cycle of the wetland's ecosystem. The Kis-Balaton Water Protection System (KBWPS) is a wetland restoration project whose purpose is the water quality protection of Lake Balaton [2] retaining elements through a reservoir food chain. Based on local measurements, the nutrient retention of KBWPS at the mouth of River Zala is functioning well.

Water quality parameters are commonly determined in the months between spring and autumn in the case of non-thermal waters [3]; however, the water temperature (T_W) of LH and its canals are also increased in the winter months, as compared to other water sources. LH is the deepest thermal water in Europe originating from two crater springs with different temperatures: 26 °C and 41 °C. The water of the two springs becomes blended in the cave and flows into the lake. The T_W of LH is evenly distributed as a result of a mist layer over its surface which functions as a cover thus preventing heat loss from its surface. The average summer and winter temperatures are about 33–35 °C and 24–28 °C, respectively. As water moves away from the source through the canal, its temperature constantly decreases.

Leaf litter falling into water includes leaves, leaves fragments, stems, twigs, harvest, and other plant components [4]. Depending on the location of the vegetation, the composition of leaf litter may vary to a certain extent; however, usually leaves constitute the biggest proportion of leaf litter with 41–98% [5]. Leaf-litter input is responsible for a major proportion of the background load in natural waters. Allochthon sources deriving from plants in the water or at the banks contributes to the inner nutrient load of water bodies [6]. After a leaf falls into water, it loses one quarter of its original dry leaf mass through the dissolution of its water-soluble components already within the first 24 h [7]. It is followed by microbial decomposition, which causes the most significant change in leaf structure. The next phase is the macroinvertebrate and finally the physical shredding of leaves. These decomposition dynamics are significantly influenced by ecological and chemical parameters and the temperatures; thus, these parameters should be examined. In addition to plant biological characteristics the response of decomposition to other changes in nutrients availability [8], temperatures [9], and vegetation under water (watershed one) [10] may also modify the process [11].

The Solidago genus includes more than one hundred different species, most of which originate from North America, eight species are present in Mexico, four in South America and six–ten species are native species in Europe [12]. In Europe, *Solidago canadensis*—Goldenrod—is the most examined species. The goldenrod—as a widespread species—exerts a negative effect on other native plant species disturbing the ecosystem. In Hungary, the goldenrod constantly spreads to newer areas, and it can also be observed around Lake Balaton and alongside the bank of the Hc. Investigating the natural background load as a vital part of maintaining water quality, our aim was to examine the decomposition rate for both the stem and leaf material of the goldenrod. To date, there have been no studies concerning goldenrod decomposition in water, even of the two plant organs separately. As the River Zala strongly impacts the water quality of Lake Balaton, the watercourses flowing in tributary also determine the water quality of the Lake. The novelty of the investigation was also the study area, with different sampling places connecting Lake Hévíz, the deepest thermal lake in Europe, to Lake Balaton, the largest freshwater lake in Central Europe. The impact of T_W on leaf litter decomposition in such a complex ecosystem was not investigated until now, even in wintertime. This was the reason for our investigation at three sampling sites in the Hc (Hc1–Hc3) and one in the LH with different T_W. This study highlights the importance of considering different T_W aspects of litter decomposition in order to predict the decomposition process in warming environments, including rising T_W.

2. Materials and Methods

The investigation was conducted in the winter period at four different sampling sites (LH, and three sites at the Hc as Hc1, Hc2 and Hc3) with different T_W to determine the decomposition rate for both the stem and leaf material of the goldenrod (*Solidago canadensis*). Four sampling points were designated, one of them in the LH, and three other points from the LH as follows:

- Hc1: 400 m from the LH,
- Hc2: 1562 m from the LH,
- Hc3: 4280 m from the LH.

In the course of our investigation, we applied the commonly used and accepted method of leaf litter bags. Samples were collected at the time of litter fall, then were dried until they reached constant dry weight and finally 10 g of each were placed into 15 cm × 15 cm polyethylene bags with 3 mm mesh sizes. The mesh size ensured access for the macroinvertebrate species to leaf litter. 120 litter bags were used in the study (i.e., 2 plant part, 4 sampling points, 5 temporal sampling events, 3 replicates each). The bags were randomly attached to a plastic compartment with bricks placed in the center of the compartment. Bags were randomized and separated to avoid spatial confounds.

The filled leaf litter bags were placed into the water at about 1 m below the surface in the littoral zone to ensure that the bags will constantly remain under water. Three parallel samples of each plant component—the stem and leaves—were retrieved 7, 21, 42, 70 and 98 days after placing the leaf litter bags into the water. The litter bags were transferred to the laboratory where the foreign material was carefully removed, then samples were air dried until reaching constant weight, and finally the weight of the remaining litter mass was measured. Leaf litter mass (stem and leaf) breakdown rates, k was calculated applying an exponential model of the percent mass, W_t mass loss over time (t) as follows:

$$W_t = W_0 \times e^{-kt} \tag{1}$$

where W_0 was the initial weight [13].

T_W measurements were in situ recorded at 10-min intervals and integrated to daily intervals (Delta Ohm HD-226-1).

Conductivity and pH were measured in situ, using Adwa AD111 and AD310 field equipment. ammonium (NH_4^+) and phosphate (PO_4^{3-}) were determined in the laboratory, using a Lovibond MultiDirect (type 0913462) spectrophotometer. Biological oxygen demand (BOD_5) was measured after the procedure of the following standard: MSZ ISO 5813.

The composition in microorganisms was excluded from the study. There was no significant difference in the density of macroinvertebrate shredders between the sampling points.

The remaining mass was analyzed through a two-way repeated ANOVA in SPSS 25.0 [14] for five sampling dates at four different places in LH and Hc1-3.

3. Results

A gradual decrease in mean T_W of the warmer LH from 24.0 °C to cooler Lake Balaton to 13.7 °C (Hc3) was measured. With the exception for LH and Hc1 ($p = 0.566$), T_W of every other sampling place differed significantly ($p < 0.001$). The average T_W was lower by 22.0 and 29.5% in relation Hc1 to Hc2, and in Hc2 to Hc3, respectively. The farther the distance from LH, the lower the mean T_W was measured in the Hc between December 2019 and March 2020. The largest difference in T_W, above 10 °C, was observed in the case of Hc3. The length of the sampling area is at about 3.2 km from the LH. The highest mean T_W from December to March was measured in the LH. Declines in mean T_W were 1.3, 5.0 and 10.3 °C in Hc1, Hc2 and Hc3, respectively (Table 1).

Table 1. Water temperature (°C) measured in Lake Hévíz (LH) and Hévíz canal (sample sites Hc1, Hc2 and Hc3) from December 2019 to March 2020.

Header	Min	Average	Max
LH	22.2	24.0	25.8
Hc1	18.4	22.7	25.7
Hc2	14.5	19.0	22.4
Hc3	9.8	13.7	16.9

Based on previous study, the conductivity in the water samples ranged from 715 to 867 µS cm^{-1} (Table 2). Conductivity of Hc3 and the other sampling points, moreover LH and Hc2 also differed significantly ($p < 0.001$) in the investigated period [15]. The pH of the study sites occurred in the range of 7.37–8.05, indicating slightly alkaline nature of the water quality. The pH of different sampling sites of the study varied significantly ($p < 0.001$). BOD_5 ranged from 0.2 to 1.7 mg L^{-1} in the studied period. LH differed with 20.9% ($p < 0.001$), 32.9% ($p < 0.001$) and 53.5% ($p < 0.001$) in Hc1, Hc2 and Hc3, respectively. There were no significant differences between the sampling points regarding ammonium.

In case of PO_4^{2-}, Hc3 differed with 34.8% (p = 0.049) and 25.0% (p = 0.026) from LH and Hc2, respectively.

Table 2. The conductivity, pH, BOD_5, NH_4^+ and PO_4^{2-} concentration values of the four sampling points in the studied period.

Sampling Place	Conductivity (μS cm^{-1})	pH	BOD_5 (mg L^{-1})	NH_4^+ (mg L^{-1})	PO_4^{2-} (mg L^{-1})
LH	750.0 ± 45.4	7.5 ± 0.2	0.5 ± 0.2	0.3 ± 0.1	0.5 ± 0.2
Hc1	763.8 ± 19.2	7.6 ± 0.2	0.7 ± 0.4	0.3 ± 0.1	0.5 ± 0.2
Hc2	775.8 ± 27.1	7.8 ± 0.2	0.8 ± 0.4	0.3 ± 0.1	0.6 ± 0.2
Hc3	821.7 ± 25.8	7.9 ± 0.1	1.1 ± 0.3	0.3 ± 0.1	0.7 ± 0.3

The passage time and the sample place significantly influenced the decomposition of both leaf and stem (Figure 1). The highest decomposition rate was observed in Hc1 resulting in the least amount of litter mass at the end of investigation. The farther the distance from LH, the more intense the litter decomposition was. Although, the T_W in the LH and Hc1 was similar, the fall was more intense in Hc1 as compared to LH. At Hc1, the remaining mass increased by 27.6–34.7% in leaf and 5.1–12.1% in stem in comparison to other sampling places. The lowest remaining mass was measured in the coolest Hc3.

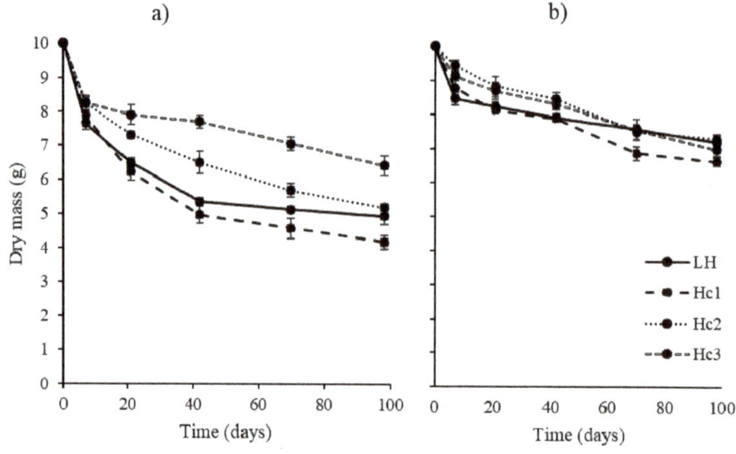

Figure 1. Changes in mean (± SE) leaf (**a**) and stem (**b**) mass loss (g) of goldenrod in Lake Hévíz (LH) and its canal (Hc1–3), n = 3.

The mass loss of leaf litter as a function of time was long ago approximated by an exponential decay model [9]. Exponential decay coefficients (k) were determined according to Bärlocher et al. [16]. In the LH, mean k varied from 0.009 to 0.038 and from 0.003 to 0.024 in leaf and stem (Figure 2), respectively. Somewhat lower values were observed in the canal ranging from 0.034 (leaf, Hc1) to 0.003 (stem, Hc3). Portillo-Estrada et al. [17] also revealed that the decomposition k rates were higher in warmer sites than in colder ones.

As Mauchly's Test of Sphericity was significant, the Greenhouse–Geisser test, one of the alternative univariate tests, was applied. There were significant main effects of place (p = 0.013) and of passage time (p = 0.001) on stem mass decay. In contrast, there was no significant interaction between place x time (p = 0.065) in case of stem mass loss (Table 3).

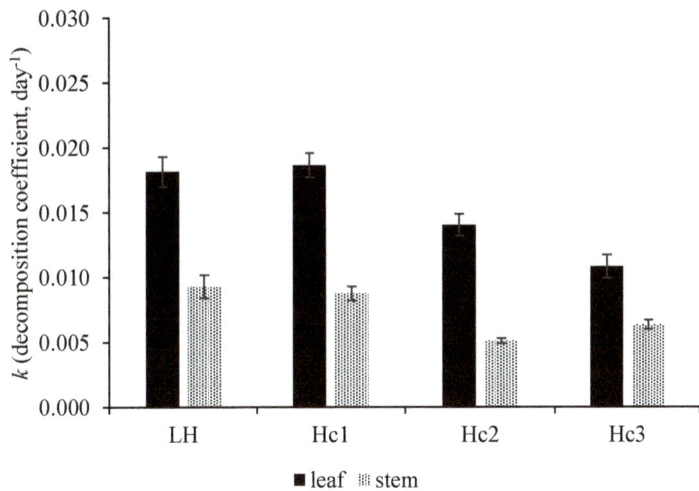

Figure 2. Decomposition coefficients (k) of goldenrod leaf and stem in Lake Hévíz and its canal (Hc1-3) (mean ± SD).

Table 3. Tests of within-subjects effects of goldenrod stem mass loss (g) in Lake Hévíz and through Hévíz canal.

Variable	Type III Sum of Squares	df	Mean Square	F	Sig.	Partial η^2
Place	3.075	1.013	3.036	74.3	0.013	0.974
Error (Place)	0.083	2.026	0.041	-	-	-
Time	69.646	1.025	67.954	983.9	0.001	0.998
Error (Time)	0.142	2.05	0.069	-	-	-
Place X Time	2.754	1.039	2.65	13	0.065	0.867
Error (Place X Time)	0.423	2.078	0.203	-	-	-

Similarly, there were significant main effects of place ($p = 0.000$) and of passage time ($p = 0.000$) on leaf (Table 3).

The results also revealed that there was a significant interaction between place and time ($p = 0.000$) in case of the leaf mass decay (Table 4).

Table 4. Tests of within-subjects effects of leaf mass decay (g) in Lake Hévíz and through Hévíz canal.

Variable	Type III Sum of Squares	df	Mean Square	F	Sig.	Partial η^2
Place	26.464	1.081	24.485	71,003.5	0	1
Error (Place)	0.001	2.162	0	-	-	-
Time	187.715	1.787	105.017	57,642.1	0	1
Error (Time)	0.007	3.575	0.002	-	-	-
Place X Time	15.133	1.799	8.41	2463.5	0	0.999
Error (Place X Time)	0.012	3.599	0.003	-	-	-

The statistical investigation showed that the decomposition rate of stem did not depend significantly on time and place interaction. The probable explanation is that the stem with higher lignin content requires more time for breakdown, and the decomposition rate is less determined by temperature (also by place). The dense nature, hydrophobicity, and nonspecific structure of lignin make it difficult for enzymes to attack [18].

4. Discussion

Previous works demonstrated that T_W modulates litter decomposition by driving activity of decomposers [19–22]. Many other authors investigated synergistic influences with T_Ws and dissolved nutrients [23–25]. As there was minimal difference in flow velocity between sampling sites (data not shown), we assumed that the variation in their decomposition rate might be the changed T_W.

In accordance with the results of this study, Geraldes et al. [22] found faster leaf decomposition under increased T_W (from 16 to 24 °C) in alder (*Alnus glutinosa*) using a microcosmos experiment. The used temperature range of Geraldes et al. [22] was close to the applied T_Ws of this study. Temperature determines organism activity, with increased temperatures stimulating biological processes at least within physiological limits [22,26]. Faster decomposition of the goldenrod leaf litter was probably due to low lignin and polyphenolics and high nutrient concentrations compared to the stem, that might be preferred by microorganisms and invertebrates [27]. In addition to the internal properties of the decomposition, the T_W is also an important factor, with higher temperatures stimulating biological activities [22,26]. Although the higher decomposition dynamics measured in Hc1 as compared to warmer LH was probably due to more intense litter drift in Hc1 [16].

The quality of water should be impaired by processes within the catchment area. The effect of small canals on lake health could be clarified by assessing litter decomposition due to water quality that may be limited [28]. Pressures exerted by a diversity of human activities (application of fertilizers, pesticides, removal of vegetation, ploughing, soil surface sealing) are transmitted, and their effects can potentially accumulate in the water bodies. On the other hand, vegetation cover can modify these processes, adding to or alleviating the pressures [29]. The study of the decomposition of goldenrod on the embankment of Hc and LH is of primary importance as it may impact the water quality of Lake Balaton as well. Migliorini and Romero [30] projected that future temperature rise might accelerate decomposition rates in aquatic systems by stimulating the leaf litter decomposition. This study can be one such example.

5. Conclusions

Temperature is an important factor in litter decomposition and biological processes. Future climate change scenarios predict that rising temperature, mainly in winter, will modify the nutrient cycle, including decomposition processes. In the winter period, we carried out a three-month investigation to examine the decomposition rate of the stem and leaf material of the widespread goldenrod. Except for LH, covered by water lily species, the dominant vegetation of the remaining three sampling points is close to each other; the common surrounding area is herbaceous vegetation, with dominant goldenrod.

In LH and Hc, we selected three sampling sites with different mean T_W ranging from 24.0 to 13.7 °C with the hottest T_W in the lake. On account of the distance from the LH, the T_W of the canal continuously decreased. The T_W gradient was −0.32 °C/100 m on the site of the study. Such an experiment has not been previously conducted at these sampling sites; thus, it can be considered a new, innovative investigation. Furthermore, in contrast to previous studies, the sampling sites were selected in thermal water, which is connected to non-thermal water, thus it can contribute to a more complete and in-depth understanding of the background load of the Keszthely Bay.

The study site was a natural "microcosm", in which litter decomposition rate increased, as the main driver of the process, the T_W rose, regardless of the similarity in water quality parameters and the absence of difference in invertebrates' colonization. The highest decay rate could be observed at the sampling site, which was nearest to LH. Due to possible impacts of climate changes, any modification in decomposition processes will affect the strictly protected Balaton's water quality. Results in this study could help in preparing mitigation of these future challenges.

Author Contributions: Conceptualization, S.S. and B.S.-G.; methodology, S.S. and A.A.; software, G.S.; validation, A.A.; investigation, S.S.; data curation, S.S. and G.S.; writing—original draft preparation, S.S. and A.A.; writing—review and editing, everybody; visualization, B.S.-G.; supervision, A.A. All authors have read and agreed to the published version of the manuscript.

Funding: This research received no external funding.

Institutional Review Board Statement: Not applicable.

Informed Consent Statement: Not applicable.

Data Availability Statement: The data presented in this study are available on request from the corresponding author.

Acknowledgments: This publication is supported by the EFOP-3.6.3-VEKOP-16-2017-00008 project, which is co-financed by the European Union and the European Social Fund.8

Conflicts of Interest: The authors declare no conflict of interest.

References

1. Istvánovics, V.; Clement, A.; Somlyódy, L.; Specziár, A.; Tóth, G.; Padisák, J. Updating water quality targets for shallow Lake Balaton (Hungary), recovering from eutrophication. In *Eutrophication of Shallow Lakes with Special Reference to Lake Taihu, China;* Qin, B., Liu, Z., Havens, K., Eds.; Springer: Dordrecht, The Netherlands, 2007; pp. 305–318.
2. Pomogyi, P. Nutrient retention by the Kis-Balaton Water Protection System. *Hydrobiologia* **1993**, *251*, 309–320. [CrossRef]
3. Hatvani, I.G.; Deganutti de Barros, V.; Tanos, P.; Kovács, J.; Székely Kovács, I.; Clement, A. Spatiotemporal changes and drivers of trophic status over three decades in the largest shallow lake in Central Europe, Lake Balaton. *Ecol. Eng.* **2020**, *151*, 804–811. [CrossRef]
4. Benfield, E.F. Comparison of litterfall input to streams. *J. N. Am. Benthol. Soc.* **1997**, *16*, 104–108. [CrossRef]
5. Oelbermann, M.; Gordon, A.M. Quantity and quality of autumnal litterfall into a rehabilitated agricultural stream. *J. Environ. Qual.* **2000**, *29*, 603–611. [CrossRef]
6. Dobson, M.; Frid, C. *Ecology of Aquatic Systems*, 2nd ed.; Oxford University Press: Oxford, UK, 2009; pp. 10–68.
7. Webster, J.; Benfield, E. Vascular plant breakdown in freshwater ecosystems. *Annu. Rev. Ecol. Syst.* **1986**, *17*, 567–594. [CrossRef]
8. Li, L.J.; Zeng, D.H.; Yu, Z.Y.; Fan, Z.P.; Yang, D.; Liu, Y.X. Impact of litter quality and soil nutrient availability on leaf decomposition rate in a semi-arid grassland of Northeast China. *J. Arid Environ.* **2011**, *75*, 787–792. [CrossRef]
9. Wetterstedt, J.; Persson, T.; Agren, G.I. Temperature sensitivity and substrate quality in soil organic matter decomposition: Results of an incubation study with three substrates. *Glob. Chang. Biol.* **2010**, *16*, 1806–1819. [CrossRef]
10. Casas, J.J.; Larranaga, A.; Menéndez, M.; Pozo, J.; Basaguren, A.; Martinez, A.; Pérez, J.; González, J.M.; Mollá, S.; Casado, C.; et al. Leaf litter decomposition of native and introduced tree species of contrasting quality in headwater streams: How does the regional setting matter? *Sci. Total Environ.* **2013**, *458–460*, 197–208. [CrossRef]
11. Júnior, E.S.A.; Martinez, A.; Goncalves, A.L.; Canhoto, C. Combined effects of freshwater salinization and leaf traits on litter decomposition. *Hydrobiologia* **2020**, *847*, 3427–3435. [CrossRef]
12. Beck, J.B.; Semple, J.C.; Brull, J.M.; Lance, S.L.; Phillips, M.M.; Hoot, S.B.; Meyer, G.A. Genus-wide microsatellite primers for the goldenrods (Solidago; Asteraceae). *Appl. Plant Sci.* **2014**, *2*, 1300093. [CrossRef]
13. Petersen, R.C.; Cummins, K.W. Leaf processing in a woodland stream. *Freshw. Biol.* **1974**, *4*, 343–368. [CrossRef]
14. SPSS 25.0 IBM Corp. *IBM SPSS Statistics for Windows, Version 25.0;* IBM Corp: Armonk, NY, USA, 2017.
15. Simon, S.; Simon, B.; Soós, G.; Kucserka, T.; Anda, A. Some preliminary investigations of water quality parameters in a Hungarian thermal lake, Hévíz. *J. Cent. Eur. Agric.* **2020**, *21*, 896–904. [CrossRef]
16. Bärlocher, F.; Gessner, M.O.; Graca, M.A.S. Leaf Mass Loss Estimated by the Litter Bag Technique. In *Methods to Study Litter Decomposition. A Practical Guide*, 2nd ed.; Springer Nature Switzerland AG: Cham, Switzerland, 2020; Part 1; pp. 43–51. [CrossRef]
17. Portillo-Estrada, M.; Pihlatie, M.; Korhonen, J.F.J.; Levula, J.; Frumau, F.K.A.; Ibrom, A.; Lembrechts, J.J.; Morillas, L.; Horváth, L.; Jones, K.S.; et al. Climatic controls on leaf litter decomposition across European forests and grasslands revealed by reciprocal litter transplantation experiments. *Biogeosciences* **2016**, *13*, 1621–1633. [CrossRef]
18. Horwath, W.H. Carbon Cycling and formation of soil organic matter. In *Soil Microbiology, Ecology and Biochemistry, 3d ed.;* Eldor, A.P., Ed.; Academic Press: Cambridge, MA, USA, 2007; pp. 303–337. [CrossRef]
19. Pazianoto, L.H.R.; Solla, A.; Ferreira, V. Leaf litter decomposition of sweet chestnut is affected more by oomycte infection of trees than by water temperature. *Fungal Ecol.* **2019**, *41*, 269–278. [CrossRef]
20. Friberg, N.; Dybkjaer, J.B.; Olfasson, J.S.; Gislason, G.M.; Larsen, S.E.; Lauridsen, T.L. Relationship between structure and function in streams contrasting in temperature. *Freshw. Biol.* **2009**, *54*, 2051–2068. [CrossRef]

21. Ferreira, V.; Chauvet, E. Future increase in temperature more than decrease in litter quality can affect microbial litter decomposition in streams. *Oecologia* **2011**, *167*, 69–291. [CrossRef]
22. Geraldes, P.; Pascoal, C.; Cássio, F. Effects of increased temperature and aquatic fungal diversityon litter decomposition. *Fungal Ecol.* **2012**, *5*, 734–740. [CrossRef]
23. Ferreira, V.; Chauvet, E. Synergistic effects of water temperature and dissolved nutrients on litter decomposition and associated fungi. *Glob. Chang. Biol.* **2011**, *17*, 551–564. [CrossRef]
24. Fernandes, I.; Pascoal, C.; Guimaraes, H.; Pinto, R.; Sousa, I.; Cássio, F. Higher temperature reduces the effects of litter quality on decomposition by aquatic fungi. *Freshw. Biol.* **2012**, *57*, 2306–2317. [CrossRef]
25. Fernandes, I.; Seena, S.; Pascoal, C.; Cássio, F. Elevated temperature may intensify the positive effects of nutrients on microbial decomposition in streams. *Freshw. Biol.* **2014**, *59*, 2390–2399. [CrossRef]
26. Bergfur, J.; Friberg, N. Trade-offs between fungal and bacterial respiration along gradients in temperature, nutrients and substrata: Experiments with stream derived microbial communities. *Fungal Ecol.* **2012**, *5*, 46–52. [CrossRef]
27. López-Rojo, N.; Martínez, A.; Pérez, J.; Basaguren, A.; Pozo, J.; Boyero, L. Leaf traits drive plant diversity effects on litter decomposition and FPOM production in streams. *PLoS ONE* **2018**, *13*, e0198243. [CrossRef]
28. Mbaka, J.G.; Schäfer, R.B. Effect of small impoundments on leaf litter decomposition in streams. *River Res. Appl.* **2015**, *5*, 907–913. [CrossRef]
29. Conte, M.; Ennaanay, D.; Mendoza, G.; Walter, T.M.; Wolny, S.; Freyberg, D.; Nelson, E.; Solorzano, L. Retention of nutrients and sediment by vegetation. In *Natural Capital*; Oxford University Press: Oxford, UK, 2011; pp. 89–110.
30. Migliorini, G.H.; Romero, G.Q. Warming and leaf litter functional diversity, not litter quality, drive decomposition in a freshwater ecosystem. *Sci. Rep.* **2020**, *10*, 20333. [CrossRef]

Article

Enhancing the Output of Climate Models: A Weather Generator for Climate Change Impact Studies

Pietro Croce, Paolo Formichi * and Filippo Landi

Department of Civil and Industrial Engineering-Structural Division, University of Pisa, Largo Lucio Lazzarino 1, 56122 Pisa, Italy; p.croce@ing.unipi.it (P.C.); filippo.landi@ing.unipi.it (F.L.)
* Correspondence: p.formichi@ing.unipi.it; Tel.: +39-3483391336

Abstract: Evaluation of effects of climate change on climate variable extremes is a key topic in civil and structural engineering, strongly affecting adaptation strategy for resilience. Appropriate procedures to assess the evolution over time of climatic actions are needed to deal with the inherent uncertainty of climate projections, also in view of providing more sound and robust predictions at the local scale. In this paper, an ad hoc weather generator is presented that is able to provide a quantification of climate model inherent uncertainties. Similar to other weather generators, the proposed algorithm allows the virtualization of the climatic data projection process, overcoming the usual limitations due to the restricted number of available climate model runs, requiring huge computational time. However, differently from other weather generation procedures, this new tool directly samples from the output of Regional Climate Models (RCMs), avoiding the introduction of additional hypotheses about the stochastic properties of the distributions of climate variables. Analyzing the ensemble of so-generated series, future changes of climatic actions can be assessed, and the associated uncertainties duly estimated, as a function of considered greenhouse gases emission scenarios. The efficiency of the proposed weather generator is discussed evaluating performance metrics and referring to a relevant case study: the evaluation of extremes of minimum and maximum temperature, precipitation, and ground snow load in a central Eastern region of Italy, which is part of the Mediterranean climatic zone. Starting from the model ensemble of six RCMs, factors of change uncertainty maps for the investigated region are derived concerning extreme daily temperatures, daily precipitation, and ground snow loads, underlying the potentialities of the proposed approach.

Keywords: climate change; extremes; factor of change; climatic actions; climatic variables; regional climate models; weather generator; climate effects

Citation: Croce, P.; Formichi, P.; Landi, F. Enhancing the Output of Climate Models: A Weather Generator for Climate Change Impact Studies. *Atmosphere* **2021**, *12*, 1074. https://doi.org/10.3390/atmos12081074

Academic Editors: Baojie He, Ayyoob Sharifi, Chi Feng and Jun Yang

Received: 31 July 2021
Accepted: 18 August 2021
Published: 21 August 2021

1. Introduction

Climate change is becoming more and more relevant in many sciences and engineering disciplines, including civil and structural engineering. Thus, its implications and potential future risks are increasingly debated [1–4]. The potential increase in frequency and intensity of natural hazards significantly influences not only the design and the service level of new structures and infrastructures but also the structural safety and the assessment of existing ones, with relevant consequences in terms of direct and indirect costs [5], and in planning of maintenance or strengthening interventions [4,6–8]. It must be remarked that especially for critical infrastructures, huge indirect costs can derive from the increase of out of service time [9,10]. In assessing the influences of climate changes on the performances of engineering works, attention should be devoted both to the peculiarities of the investigated structures and to the objectives of the study. In fact, while in many works, the most pertinent information is the evolution over time of the mean value of the relevant climatic variables [11], in other works, the focus is on the extremes [5]. A number of cases are typically governed by mean values: for example, estimates of electrical power productivity of wind turbines [12] or hydroelectric power plants [13], assessment of navigability of rivers and channels [14], evaluation of water table level, availability of fresh water resources [15], estimation of sea level rise [11], agricultural and zootechnics production [16], and deterioration effects on structures and finishes [17,18]. Other cases are mainly governed by extreme values: for example, design of civil engineering and geotechnical works [4,6–8], assessment of minimum vital outflow of rivers and torrents, structural design of wind turbines [19], evaluation of minimum height of embankments [20], out of service time of navigable channels [14], icing effects on cables and electrical power lines [21,22], flooding [23], and scour of bridge piles [4].

In modern codes, such as the Eurocodes [24–27], design values of effects of climatic variables, snow and rain precipitations, shade air temperature, wind velocity, etc., are evaluated starting from the so-called representative values of climatic variables themselves, which are characterized by a given probability of exceedance, p, in a reference time interval [24,28]. In the Eurocodes [24–27], four main representative values are usually defined [24]:

- The characteristic value, for which $p = 2\%$ in one year;
- The combination value, leading, together with the characteristic value of another variable action of different nature, to a combined effect characterized by $p = 2\%$ in one year;
- The frequent value, roughly exceeded from 100 to 300 times in one year;
- The quasi permanent value, exceeded for more than 50% of the design working life of the construction.

Assuming suitable extreme value distributions, the aforementioned representative values are commonly evaluated under the hypothesis of climate being stationary over time [4,29]. As this assumption is clearly violated when climate change effects occur, relevant assessment of representative values should duly take into account non-stationarity, also in view of the definition of suitable adaptation strategies for the design of resilient buildings and infrastructures [30,31].

The evolution over time of climatic variables and climatic actions is usually assessed on the basis of the outputs of General (or Global) Climate Models (GCMs) or Regional Climate Models (RCMs). In comparison with GCMs, the RCMs are associated to more refined grid resolution, typically ranging from 10 to 20 km, so allowing more detailed local estimates. In any case, owing to the complexity of the problem and the computational time needed for a single run (simulation), independently on the scale of the model, i.e., the typical dimensions of individual cells, the number of reliable climatic models and the total number of available runs are so limited that further elaborations are needed, aiming to increase the representativeness of the outcomes of climate models. A promising method to enhance the significance of the outputs of the climate models is based on the implementation of suitable weather generators. The basic idea of the method is the artificial generation of additional outputs of climatic models, via random extraction of daily values of relevant climatic variables from appropriate homogenous populations of climatic variables, whose statistical properties are suitably assessed from the available runs of various climatic models. In this way, it is also possible to evaluate model uncertainties due to the inherent climate variability, which is fundamental information for the interpretation of climate projection, especially at smaller scales [32]. Weather generators, which have been proposed and largely developed in the water and hydraulic engineering field [33], are nowadays largely adopted for statistical downscaling in climate change investigations [34–37].

Evidently, the assembly of the homogenous populations to be randomly sampled is the core of the procedure, potentially impacting the outcomes. More specifically, assumptions about the probabilistic description of such populations determine post-sample resulting populations complying with the starting hypotheses, so that uncertainties associated with probabilistic models cannot be directly considered. To overcome this issue, a novel weather generation technique is proposed, making direct use of the outputs of climate models. The proposed procedure is particularly suitable for studying the evolution of climatic variables over time as a function of climate change, starting from the output of Regional Climate

Models (RCMs). The procedure allows evaluating the dependence of climate projections on most relevant sources of uncertainty, i.e., emission scenarios, climate models, and internal variability [38].

Nonparametric resampling techniques were already proposed for generating daily temperature and precipitation from historical data [39,40]. In these methods, resampling was carried out considering the nearest neighbors for the simulated day in terms of a weighted Euclidean distance [39] or conditioning on indices of atmospheric circulation [40].

In the present study, the generation is elaborated on the basis of climate model outputs according to an original and simplified procedure, with the aim of reproducing the internal variability of climate model runs, in such a way to estimate the uncertainty in the expected changes of climate statistics.

The efficiency and performances of the proposed weather generator are demonstrated discussing a relevant case study: the variation over time, until the year 2100, of characteristic values of significant climatic actions, daily maximum and minimum temperature, precipitations, and ground snow load, in the central Eastern part of Italy, which is located in the Mediterranean climatic area. In the study, the weather generated series are analyzed in terms of factor of changes (FC) [41], described in the following section, which are a very effective way to describe the relative variations of the considered variable.

The aim of the case study is to illustrate the suitability of the proposed methodology for the treatment of climate model outputs and is not intended to provide detailed guidance for adaptation strategies in the field of climatic actions on structures. In fact, a discussion is currently ongoing on the use of IPCC baseline scenarios, which are central to climate science and adaptation policy [42]. Scenarios are affected by deep uncertainties that are inherent in assumptions about significant factors: future technology developments, lifestyle changes, policy formulations, economic and demographic trends, and so on [43]. Notwithstanding this deep uncertainty, no equal probabilities are generally assigned by expert to the total range of future emissions. A discussion about this relevant issue can be found in [43]. In the present study, to assess the potentialities of the method, a sensitivity analysis is performed considering two possible scenarios belonging to the Representative Concentration Pathways (RCPs) set [44]: RCP4.5 and RCP8.5. Of course, the definition of the most "realistic" RCP is out of the scope of the present work.

2. Methodology

2.1. Weather Generators

As already mentioned, stochastic weather generators are suitable random extraction algorithms for the automatic generation of time-series of climatic variables. The main feature of these algorithms is their ability to generate day-by-day data series of climatic variables, starting from suitable populations of input data, based on observations, if any, and projections provided by climate models. Of course, weather generators cannot be confused with weather prediction algorithms: in fact, while weather projections are based on numerical methods for the solution of the physical equations describing the earth climate and its dependency on time, weather generators are implements for the origination of additional and virtual time-dependent results or forecasts, based on synthetic representations of available information, regardless of whether the information itself is obtained from observation or simulations [34].

Weather generators were firstly proposed and applied for risk assessment purposes in hydrological and agricultural fields [33,45]. Their main intended use was the simulation of series of climatic data sufficiently extended over time, allowing to fill the lack of reliable meteorological data, and to extend the available information to unmonitored sites, by resorting to artificial measurements. More recently, the field of application of weather generators has been considerably enlarged. In fact, starting from the outputs of individual runs of the climate models and taking into account the relevant statistical parameters of each climatic variable, they are more and more adopted to obtain supplementary virtual runs of the models. Although a sensible difference is still present between the predictions of regional climate models and local observations [46], weather generators can be very useful tools to enhance the climate projections and to achieve more sound information about the influence of climate changes. More precisely, once the evolution of the governing climatic parameters is derived from the climate model outputs, site-specific climate change scenarios can be assessed applying suitable weather generation algorithms. Truly, distinctive features of weather generators are much wider: for example, they can be efficiently applied to evaluate how the uncertainties on the projections of effects of climate change influence climate variable statistics and calculate the prediction intervals associated with the available climate model outputs. To obtain this information, two different approaches can be envisaged, depending on the way the changes over time are described: the variations over time of featuring statistical parameters of relevant climatic variables can be studied considering trends in the weather series, or, alternatively, by means of the already recalled factor of change approach, considering the variations of the statistical properties of the climate variables. The latter approach is based on the assumption, corroborated by numerical and experimental evidence, that influences of climate change are more soundly evaluated considering the variation of statistical parameters, rather than the whole weather series, which are very sensitive to the "natural variability" of climate.

The factors of change, $FC_{rep}(t)$, express the relationship between a representative value of the considered variable at the time t, $F_{rep}(t)$, and the corresponding representative value, $F_{rep}(t=0)$, at the time $t=0$. Factors of changes are generally expressed in terms of quotients, in the non-dimensional form:

$$FC_{rep}(t) = \frac{F_{rep}(t)}{F_{rep}(t=0)}. \tag{1}$$

However, when the observed climate variables are measured on interval scales, such as for temperature, they are expressed in terms of differences, in the dimensional form:

$$FC_{rep}(t) = F_{rep}(t) - F_{rep}(t=0). \tag{2}$$

The factor of change approach consists of the following:

- Evaluation of factors of change (FCs): climate model outputs and associated climate variable statistics concerning a future time interval are compared with those obtained for the control period $t=0$;
- Implementation of a suitable weather generator (WG): random samples are generated modifying the relevant statistical properties of the parameters, which are used by the WG algorithm, according to the previously detected FCs. As discussed before, the scale of local observations is different from that of climate model outputs; therefore, the WG cannot be run directly using the statistics of climate model outputs. Adopting the FC approach, the discrepancy between RCM outputs and observations is by-passed [37];
- Assessment of climate change effects on representative values: the assessment is done directly using the generated future weather series, or deriving their influence on statistical properties of their distributions.

It must be underlined that to be effective, weather generators require the preliminary parametrization of complex climate processes, which depend on many climatic variables, whose variations cannot be directly estimated from the climate model outputs. In fact, climate projections may not be available for all the components of the WG. Moreover, the factors of change so derived largely ignore the inherent uncertainty of the climate model itself. Aiming to enhance the output of climate models, improving the predictions of the climate change effects, an innovative and easy to apply weather generator has been developed. The proposed generator allows not only to simulate additional runs of climatic models directly sampling from the available outputs but also to evaluate the factors of change of the considered climatic variable and the variation over time of the main stochastic properties of extreme statistics.

2.2. A Weather Generator for the Virtualization of the Outputs of Regional Climate Model

The innovative implementation of the weather generator algorithm proposed here aims to virtualize the outputs of Regional Climate Models. In "traditional" weather generation, random data are extracted from appropriate reference probability distribution functions (*pdf*s) or cumulative distribution functions (*CDF*s), whose relevant statistical parameters are derived fitting the statistics of climate variables derived from the climate model outputs. As anticipated, this kind of generation is sensitive to the assumptions, so that "errors" or "misinterpretations" of available information cannot be corrected in the subsequent phases. For this reason, the key feature of the proposed method is that data are generated directly sampling from the climate model outputs. The scheme of implementation of the algorithm and the fundamental steps of the procedure are summarized in the block diagram reported in Figure 1.

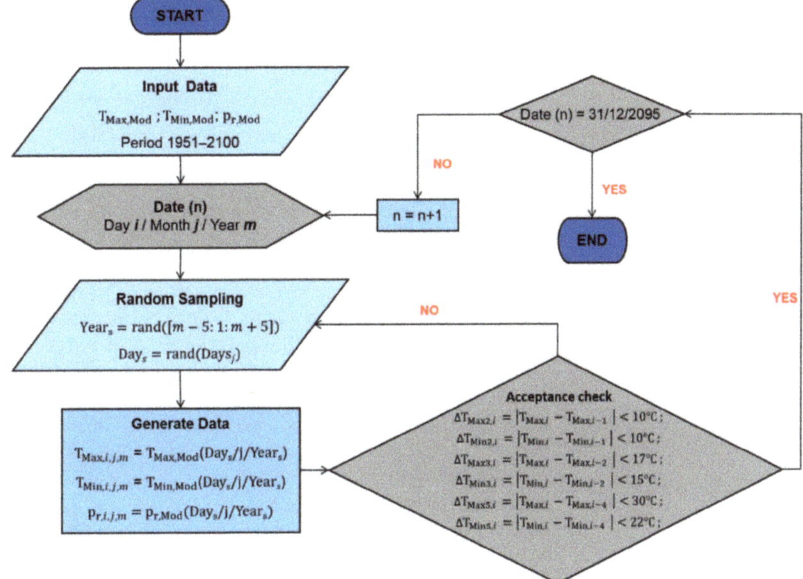

Figure 1. Implementation scheme of the weather generator algorithm.

The basic information needed to start the generation algorithm is the output of the considered climate model. The typical climate model output is a series of climatic data, e.g., daily maximum and minimum shade air temperatures ($T_{\mathrm{Max,Mod}}$ and $T_{\mathrm{Min,Mod}}$) and daily precipitation ($p_{\mathrm{r,Mod}}$), as those provided by high-resolution RCMs, selected among those included in EURO-CORDEX [47,48].

The procedure described in Figure 1 is applied for each considered model, to avoid inconsistencies in the generation, which could arise by mixing models characterized by different biases.

The generation of the daily information about the *i*-th day of the *j*-th month of the *m*-th year ($T_{\mathrm{Max},i,j,m}$, $T_{\mathrm{Min},i,j,m}$ and $p_{\mathrm{r},i,j,m}$) is performed randomly extracting the data from the output of the climate model concerning whichever day of the *j*-th months belonging to the considered year, and to the five consecutive years preceding or following the considered year, so that the time interval considered is 11 years long, from year $m-5$ to year $m+5$. This sampling period of 11 years has been chosen in such a way to properly modulate the contrasting needs of having a period, at the same time, long enough to assure adequately numerous data in the population, and short enough to exclude significant influence of climate change on climatic variables. Differently from [39,40], no specific probability is assigned to the set of days of the identified sampling period: in fact, a uniformly distributed sampling is performed.

To avoid the generation of unrealistic data, six additional constraints have been imposed for the acceptance of randomly generated daily data, as summarized in Equations (3)–(8); obviously, data not fulfilling the constraints are discarded and newly generated. Constraints refer to maximum and minimum temperatures variations in two consecutive days (Equations (3) and (4)), in three consecutive days (Equations (5) and (6)), and in five consecutive days (Equations (7) and (8)). The values, which are obviously depending on the country, or the geographical area considered in the study, need to be assessed analyzing actual data. For the relevant case study considered later in the paper, the following inequalities to be simultaneously fulfilled by the generated data in the i-th day have been established:

$$\Delta T_{\mathrm{Max2},i,j,m} = \left| T_{\mathrm{Max},i,j,m} - T_{\mathrm{Max},i-1,j,m} \right| < 10\ ^{\circ}\mathrm{C}, \quad (3)$$

$$\Delta T_{\mathrm{Min2},i,j,m} = \left| T_{\mathrm{Min},i,j,m} - T_{\mathrm{Min},i-1,j,m} \right| < 10\ ^{\circ}\mathrm{C}, \quad (4)$$

$$\Delta T_{\mathrm{Max3},i,j,m} = \left| T_{\mathrm{Max},i,j,m} - T_{\mathrm{Max},i-2,j,m} \right| < 17\ ^{\circ}\mathrm{C}, \quad (5)$$

$$\Delta T_{\mathrm{Min3},i,j,m} = \left| T_{\mathrm{Min},i,j,m} - T_{\mathrm{Min},i-2,j,m} \right| < 15\ ^{\circ}\mathrm{C}, \quad (6)$$

$$\Delta T_{\mathrm{Max5},i,j,m} = \left| T_{\mathrm{Max},i,j,m} - T_{\mathrm{Max},i-4,j,m} \right| < 30\ ^{\circ}\mathrm{C}, \quad (7)$$

$$\Delta T_{\mathrm{Min5},i,j,m} = \left| T_{\mathrm{Min},i,j,m} - T_{\mathrm{Min},i-4,j,m} \right| < 22\ ^{\circ}\mathrm{C}. \quad (8)$$

The generation is carried out at each cell disregarding spatial correlation with the neighboring ones as the analysis of the extremes for the definition of climatic actions in structural design is not affected by their spatial behavior. Anyhow, in view of studies focusing on variables that are sensitive to spatial correlation, such as for example the total precipitation on large river basins, the procedure can be easily modified by implementing iterative steps to satisfy additional constraints between neighboring cells. With this aim, several suitable procedures can be envisaged, for example based on empirical limitations or on more refined analytical approaches. Empirical limitations could be defined considering recorded daily variations of climatic variables between adjacent cells. In this way, generated values are accepted only if the limitations are satisfied; if not, the sampling is iterated. A more refined approach could be based on the checking of the correlation length; in implementing the weather generation algorithm, the correlation length of the generated data series is checked against the correlation length of the original climate model output. Estimates of correlation functions and correlation lengths can be obtained from the experimental semi-variograms [49–51] of the original RCM output and of the generated series. When the estimated correlation length of the generated series is outside the confidence interval [51] of the original one, the generation is repeated.

By using the above-mentioned algorithm, a huge number of consistent series of daily maximum and minimum shade air temperatures ($T_{\mathrm{Max},i,j,m}$ and $T_{\mathrm{Min},i,j,m}$) but also precipitations ($p_{\mathrm{r},i,j,m}$) have been generated for the period 1956–2095, therefore not only covering the projected period but also the historical one. For the case study illustrated in the following, 1000 series are generated for each investigated climate model and scenario.

As far as the consistency of the generated precipitation series is concerned, specific checks at the cell level have been performed. The mean and coefficient of variation of daily precipitation, the fraction of wet days, and the mean wet-day amount calculated from the generated series are compared with those directly obtained analyzing the climate model output. As expected, the results show a very good agreement. As an example, in Table 1, the calculated parameters for the two series are compared for one cell in the study region (see Section 3.1) and for the period 1981–1990. Furthermore, it must be underlined that for the aims of the study, the focus is on the analysis of the extremes.

Performance metrics for the weather generation algorithms are evaluated by making use of the Taylor diagram [52], which relates the root mean square error (RMSE), the pattern correlation, and the standard deviation; the Taylor diagram is generally adopted to assess the ability of climate model simulations to reproduce observed climate statistics [53]. In this study, Taylor diagram and performance statistics are evaluated comparing the generated series and with the original climate model outputs. As an example, the results obtained for monthly mean statistics of daily maximum temperature are reported in Figure 2 with reference to the same cell, model and time period considered for the comparison in Table 1. As expected, a very good correlation is obtained between the simulations and the reference data as well as very low RMSE.

Table 1. Mean, coefficient of variation, CoV, fraction of wet days (with 0.3 mm or more), f_{wet}, and mean wet-day amount, m_{wet}, for the RCM ECEARTH-HIRHAM5 and the generated series at cell 101 for the period 1981–1990.

Parameters	RCM	Weather Generated Series
Mean (mm/day)	2.26	2.24
CoV	3.30	3.26
f_{wet} (%)	31.29	31.27
m_{wet} (mm/day)	7.18	7.12

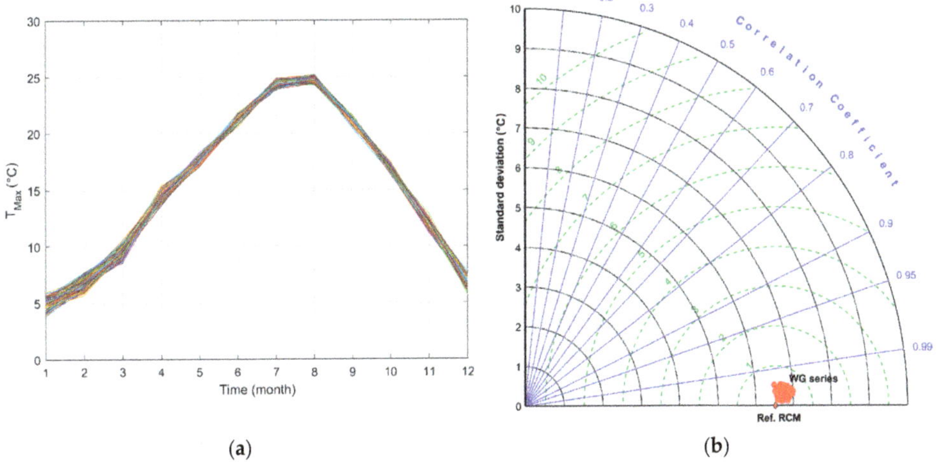

Figure 2. Monthly mean statistics of T_{Max} (**a**), Taylor diagram (**b**) considering the RCM ECEARTH-HIRHAM5 at cell 101 and the period 1981–1990. In (**b**), each red dot represents a generated series made with the investigated model, whereas the red diamond represents the reference one.

The analysis of the whole set of generated series, each one simulating one run of the climate model, allows evaluating the factors of change for convenient time intervals, as well as the confidence and the prediction intervals for the estimated changes over time.

2.3. Factors of Change and Extreme Values Theory

Analysis of climate extremes and evaluation of the values associated to a given probability of exceedance is generally performed resorting to the classical Extreme Value Theory, providing that the stationarity hypothesis is fulfilled. The basic concept of stationarity from which engineers work assumes that climate is variable but with variations whose properties are constant with time and that occur around an unchanging mean state [54]; i.e., given a time interval of fixed length, the statistical properties of the climatic variable are independent of the position of the time interval on the time axis. This assumption of stationarity is still common practice for the design criteria of new structures and infrastructures, and it is accepted for the time intervals generally considered in the elaboration of extremes of climatic variables, 40–50 years. However, considering the speed of current climate change, this assumption can become questionable; therefore, concepts and models able to take into account also variations of annual extremes over time are increasingly studied [55]. On the basis of these observations, an ad hoc procedure has been set up to evaluate the evolution of the extremes of climatic variables over time, in such a way to assess not only the influence of climate change on the extreme themselves, but also the consequences in terms of structural design.

The basic idea of this ad hoc procedure is that the influence of climate change on representative values of climatic variables can be captured studying the evolution of characteristic values over time. To perform this kind of assessment, it has been assumed that as long as the considered time period is shorter than 40 years, the stationarity hypothesis holds, and the internal influence of climate change can be disregarded; consequently, the tendency can be evaluated considering 40-year-long time periods (time windows) and assessing the variation over time of both the representative values of climatic variables and the statistical parameters of extreme value distributions. In the present study, the time shift between two consecutive time windows has been fixed to ten years.

As in structural design, characteristic values of climatic actions usually correspond to an annual probability of exceedance $p = 2\%$, this value has been assumed for the elaboration of extremes in each time window, also in view of using the outcomes of the study for the updating of climatic actions in specific sites, or, more generally, for updating of maps of climatic actions, such as those provided in Eurocodes; see e.g., [56] for a review of ground snow loads maps in Europe.

According to the methodology just sketched out, the artificially generated series of climatic data, extended from 1956 until 2095, have been subdivided into eleven time windows. For each time window, an annual extreme value analysis has been performed with the block maxima approach, assuming a Gumbel distribution and said extreme value type I distribution, whose CDF is:

$$F(x < X, i) = \exp\left\{-\exp\left[-\left(\frac{x-\mu(i)}{\sigma(i)}\right)\right]\right\} \quad i = 1, \dots 11, \quad \mu(i) \in \mathbb{R} \text{ and } \sigma(i) > 0, \tag{9}$$

where $\mu(i)$ and $\sigma(i)$ are the parameters of the distribution, depending on the considered time window, i, to be estimated by means of a suitable method, for example, the least square method (LSM). The Gumbel distribution is assumed to be consistent with the prevailing model adopted in Europe for climatic actions in structural design [56,57], but the choice of the most appropriate distribution function is not straightforward and may depend on the investigated climate variable, the location, and the record length [29].

Thus, the characteristic values of the climatic variable, $c_k(i)$, in the i-th time window are

$$c_k(i) = \mu(i) + \sigma(i)\,\{-\ln[-\ln(1-p)]\}\,, \tag{10}$$

where p is the annual probability of exceedance, for example $p = 0.02$, as in Eurocodes.

An underestimation of extremes is generally detected from the analysis of climate model outputs [32,46,58–60]; therefore, to estimate future trends of climatic actions, a suitable calibration strategy should be adopted. Nevertheless, in the framework of the factor of change approach adopted in the present study, this kind of calibration is not needed, owing to the relative nature of the factor of change, which largely compensates the biases in climate projections if they will be maintained unchanged over time. In fact, the factor of change approach, which is widely used in climate impact research [58,61], represents a sound and easy calibration methodology.

The rationale of factors of change approach is the assumption that the actual changes over time of the observed climate variable y_i are practically coincident with the ones predicted by the climate model. On this basis, x_i^p is the reference characteristic value of the considered climatic variable derived from the observations, y_i^p is its reference characteristic value obtained from the projections, x_i^f is the future characteristic value including the actual effects of climate change, and y_i^f is its future characteristic value obtained from the projections; thus, the following relationship can be established:

$$y_i^f \approx y_i^p + \left(x_i^f - x_i^p\right), \tag{11}$$

when factors of change are expressed in terms of differences, and

$$y_i^f \approx y_i^p \frac{x_i^f}{x_i^p}, \tag{12}$$

when factors of change are expressed in terms of quotient.

Actually, assuming that factors of change are nearly insensitive to absolute errors in the estimations of representative values, the proposed weather generation algorithm can provide sound estimates of factors of change of the investigated climate variable. The characteristic values obtained from the extreme value analysis of each generated series are compared considering the i-th time window, at the time $40(i-1)$ years, and the reference characteristic value obtained for the first time window ($i = 0$). For extreme temperatures, the factor of change is defined in terms of differences

$$\Delta T_{\mathrm{Max,k}}(i) = T_{\mathrm{Max,k}}[40(i-1)] - T_{\mathrm{Max,k}}(0) \tag{13}$$

$$\Delta T_{\mathrm{Min,k}}(i) = T_{\mathrm{Min,k}}[40(i-1)] - T_{\mathrm{Min,k}}(0) \tag{14}$$

while a product factor of change in terms of quotient can be assessed for extreme precipitations

$$FC_{p_{\mathrm{r,k}}}(i) = \frac{p_{\mathrm{r,k}}[40(i-1)]}{p_{\mathrm{r,k}}(0)}, \tag{15}$$

and ground snow loads

$$FC_{s_{\mathrm{k}}}(i) = \frac{s_{\mathrm{k}}[40(i-1)]}{s_{\mathrm{k}}(0)}. \tag{16}$$

The factors of change are evaluated for the whole set of generated series; thus, a significant ensemble is collected, allowing a probabilistic description of the expected changes for the extremes of the investigated climate variables.

2.4. Factors of Change Maps

The analysis of the relevant outcomes obtained from the elaboration of generated series of climatic data allows deriving a significant collection of factors of change, satisfactorily describing the variation over time of the main statistical parameter characterizing the relevant climatic variables. In this way, prediction intervals corresponding to different percentiles of the factors of change ensemble can be determined, in such a way that changes in extreme values and pertinent uncertainty ranges can be highlighted. In the authors' opinion, for the sake of practical applications, values corresponding to the 25th percentile and to the 75th percentile are particularly significant; therefore, in the following, they have been assumed as systematic reference, for the assessment of prediction intervals.

To show the potential and the effectiveness of the method, it has been applied to study the evolution over time of factors of changes in a relevant case study, regarding a central Eastern area of Italy, which belongs, as already said, to the so-called Mediterranean climatic zone, for which relevant factors of change maps have been derived, as illustrated in the following sections.

To make the reading of these maps easier, the results are represented as bivariate color maps [62], for each individual cell of a given area. This representation allows simultaneously drawing two limit percentiles in the same map, so giving an impressive illustration not only of the evolution over time of extremes values but also of the pertinent uncertainty interval. Moreover, these maps can be a useful platform to set up suitably modified climatic maps to be used in structural codes and standards, duly taking into account higher-level adaptation strategies. A significant example of such factors of change maps is illustrated in Figure 3, concerning the daily maximum temperatures in the cells of the Italian region considered in the case study. In the bivariate color map in Figure 3 and in the following ones as well, the color associated to each cell indicates, on the horizontal and vertical axis of the legend respectively, the lower and upper limits of the 25–75% prediction interval. For example, in Figure 3, the color corresponding to 1.75 °C on the horizontal axis and to 3.75 °C on the vertical axis is denoting a cell for which the 25–75% prediction limits are 1.75 °C and 3.75 °C.

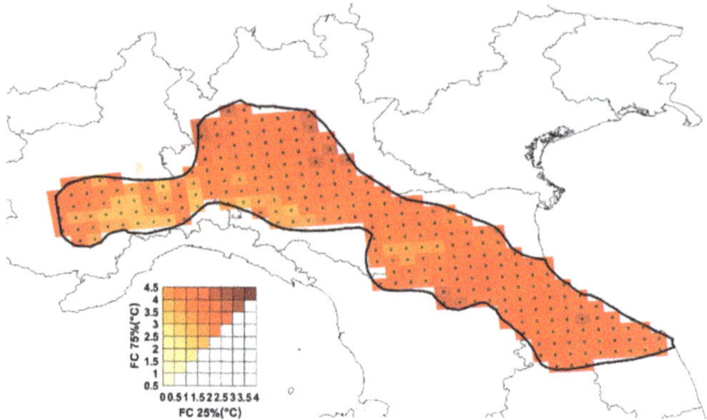

Figure 3. Example of bivariate factor of change map for $T_{Max,k}$.

3. Results

3.1. Study Area and Datasets

The procedure illustrated in the previous paragraph is applied here to investigate climate change impact on extreme temperatures and precipitation for the above-mentioned Italian region. Regarding snow loads, the studied area includes sites belonging to Zones 3 and 4 of the Mediterranean climatic region, as defined in EN1991-1-3:2003 [25]. These zones are illustrated in Figure 4a, together with the 272 cells for which high-resolution climate projections are available, being included in the EUR11 12.5 × 12.5 km grid of the Regional Climate Model of the EURO-CORDEX Figure 4b) [47,48].

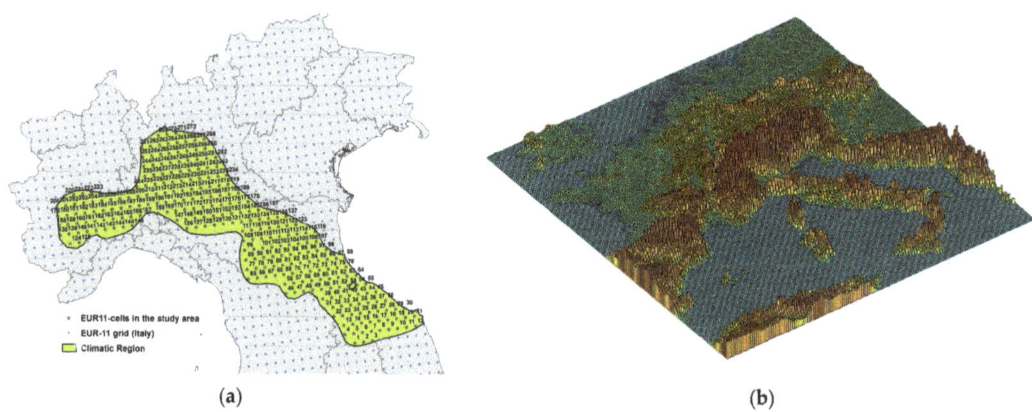

Figure 4. Investigated area in the Italian Mediterranean region (**a**), illustration of the Italian topography at EUR11-grid resolution (**b**).

3.2. Dataset of Daily Maximum and Minimum Temperature

The dataset of climate projections considered in the study covers the period 1951–2100. The above-mentioned period is split in two parts: a control period, from 1951 until 2005, and a future period, 2006–2100. The control period is used for the so-called *Historical Experiment*, aiming to assess the capability of the climate model to reproduce past weather: in this run, the observed changes in the greenhouse gas content of the atmosphere are forced in the model, checking the agreement between the predictions of the model and actual observations. The future period is used for projections in the *RCPs Experiment*, where the runs of the model are forced according to various scenarios, which are characterized by the assumed atmospheric composition changes described by various representative concentration pathways (RCPs) [54]. The scenarios mostly considered in climate risk studies are RCP4.5 and RCP8.5, for which the majority of climate projections are available. They correspond to medium and maximum greenhouse gas emission pathways, and they have been adopted in the present study as well, even if the discussion on climate scenarios is rapidly evolving and at present, RCP4.5 appears to offer more realistic baselines [63], while the use of RCP8.5 is debated by the scientific community [42,64], often appearing excessively pessimistic. Moreover, the discussion is even more complex, since it is influenced not only by the input greenhouse gasses emissions but also by the output of the models.

Regarding the historical experiment, climatic data provided by several meteorological institutes, namely the Laboratoire des Sciences du Climat et de l'Environnement, Institut Pierre Simon Laplace, IPSL-INERIS, the Max Planck Institute, MPI-CSC, the Royal Netherlands Meteorological Institute (KNMI), and the Danish Meteorological Institute, were processed. The same research institutes also provided the climate projections considered in the assessment of variations of climatic variables induced by climate change. A synthetic description and the main characteristics of each model are summarized in Table 2.

Table 2. Synthetic description of the climate models considered in the multi-model ensemble.

Institute_ID	RCM Name	Driving_GCM Name	Driving_Ensemble Member	Period
DMI	HIRHAM5	EC-EARTH	r3i1p1	1951–2100
CLMcom	CCLM4-8-17	CNRM-CM5-LR	r1i1p1	1951–2100
CLMcom	CCLM4-8-17	EC-EARTH	r12i1p1	1951–2100
KNMI	RACMO22E	EC-EARTH	r1i1p1	1951–2100
MPI-CSC	REMO2009	MPI-ESM-LR	r1i1p1	1951–2100
IPSL-INERIS	WRF331F	IPSL-CM5A-MR	r1i1p1	1951–2100

The results presented in the following sections are derived applying the proposed weather generation algorithm starting from the multi-model ensemble, which is obtained combining the individual climate models and adopting a constant weight for each of them, so hypothesizing that each model is equally suitable to reproduce the effects of climate change over time. The rationale of considering a model ensemble is that each model of the ensemble can be generally considered as the result of a reasonable, independent sampling of future climate. Truly, a further refinement could be introduced, suitably weighing the climate models, i.e., modifying the individual weight of each model, based on their ability to reproduce past observations [65], but this topic involves very complicated and delicate aspects, which cannot be easily tackled in a general way.

3.3. Effects of Climate Change on Extreme Temperatures

By means of the new weather generator algorithm previously described, factors of change and suitable prediction intervals have been assessed, in comparison with the reference time window, 1956–1995 for characteristic values of daily maximum and minimum temperature, $T_{Max,k}$ and $T_{Min,k}$, in each individual cell of the studied region. Of course, as the considered climatic variable is the temperature, factors of change have been derived in terms of differences [66], according to Equations (13) and (14).

A first example of the obtained results is shown in Figure 5, where factors of change in terms of differences are reported for characteristic values of maximum temperature in one cell of the investigated region (cell 101). Trends are illustrated for the two considered scenarios together with the associated uncertainty, i.e., the prediction interval 25–75%.

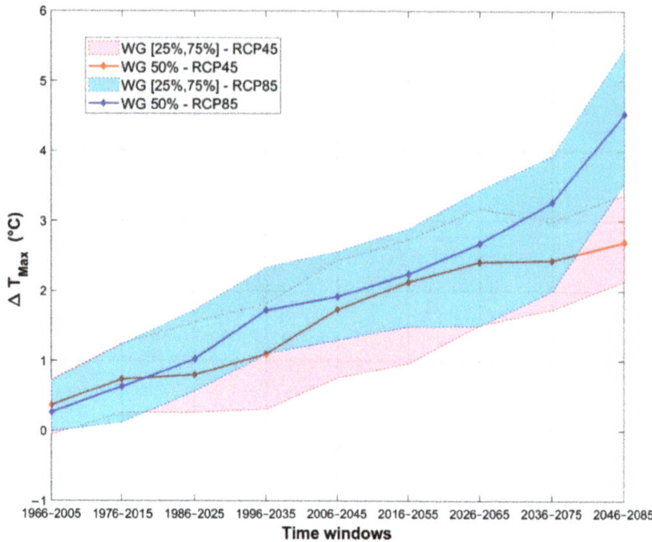

Figure 5. Factors of change for $T_{\mathrm{Max,k}}$ at cell 101—prediction interval 25–75%.

The main outcomes of the study are summarized in the factors of change maps, as illustrated in Figures 6 and 7, for the RCP4.5 and RCP8.5 scenarios, respectively. In the bivariate maps, the characteristic values of daily maximum air shade temperature ($T_{\mathrm{Max,k}}$) in four significant time windows (1976–2015, 1996–2032, 2016–2055, and 2035–2075) are illustrated, referring to the 25th and the 75th percentile.

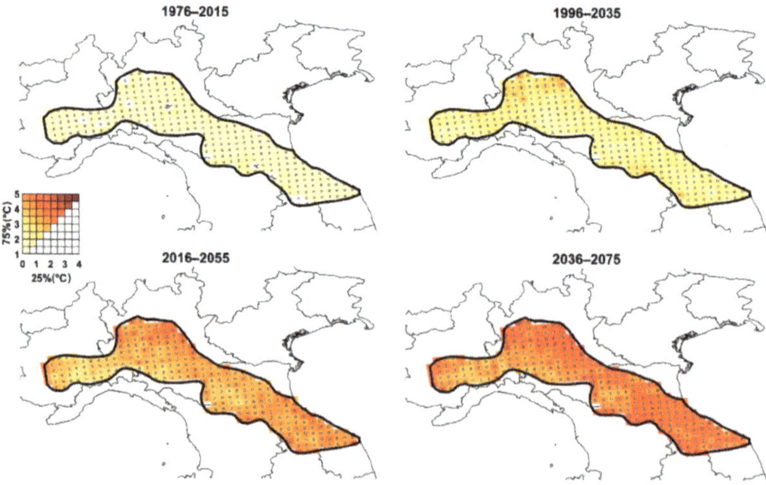

Figure 6. Factors of change for $T_{\mathrm{Max,k}}$ in the time windows 1976–2015, 1996–2035, 2016–2055, and 2036–2075 in comparison with the reference time interval 1956–1995—prediction interval (25–75%) map (Scenario RCP4.5).

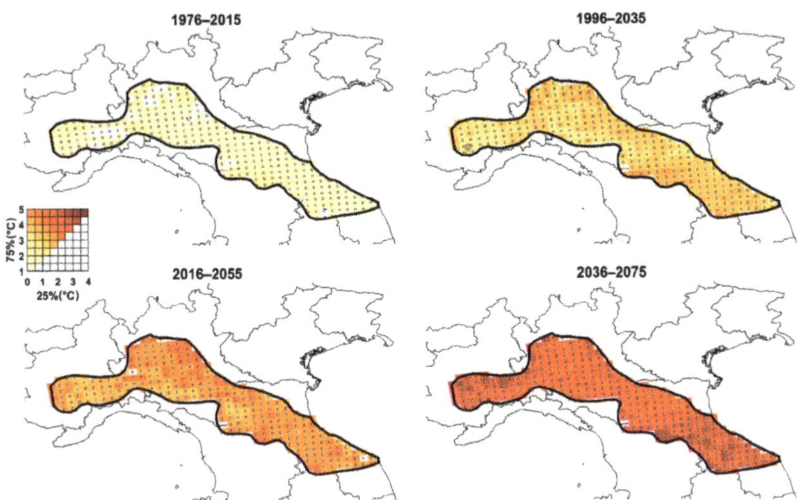

Figure 7. Factors of change for $T_{Max,k}$ in the time windows 1976–2015, 1996–2035, 2016–2055, and 2036–2075 in comparison with the reference time interval 1956–1995—prediction interval (25–75%) map (Scenario RCP8.5).

Table 3 summarizes the average factors of change obtained in the investigated region according to the RCP4.5 and RCP8.5 emission scenarios. In the table, results corresponding to prediction percentiles 25%, 50%, and 75% are given for each time window considered in the study.

Table 3. Mean of factors of change (°C) percentiles for $T_{Max,k}$ in the study region.

Time Window	RCP4.5			RCP8.5		
	25%	50%	75%	25%	50%	75%
1966–2005	0.04	0.41	0.82	0.03	0.36	0.72
1976–2015	0.31	0.87	1.42	0.35	0.88	1.42
1986–2025	0.45	1.16	1.87	0.81	1.43	2.06
1996–2035	0.65	1.49	2.25	1.26	1.93	2.67
2006–2045	0.89	2.01	3.10	1.44	2.19	2.85
2016–2055	1.33	2.33	3.31	1.77	2.51	3.19
2026–2065	1.58	2.63	3.63	2.00	2.76	3.54
2036–2075	1.87	2.75	3.55	2.47	3.37	4.17
2046–2085	2.18	2.83	3.67	3.93	5.10	6.08

Characteristic values of daily minimum air shade temperature ($T_{Min,k}$) are illustrated in a similar way in Figures 8 and 9, while analogously to what was previously done for maximum temperature, the factors of change percentiles (25%, 50%, 75%) averaged over the region are summarized in Table 4 for each time window.

The results confirm that there is a high probability that extreme temperatures significantly rise over time. This tendency accords with the expectations about mean temperatures [67], but it emerges more markedly, confirming the conclusions obtained analyzing actual measurements [29].

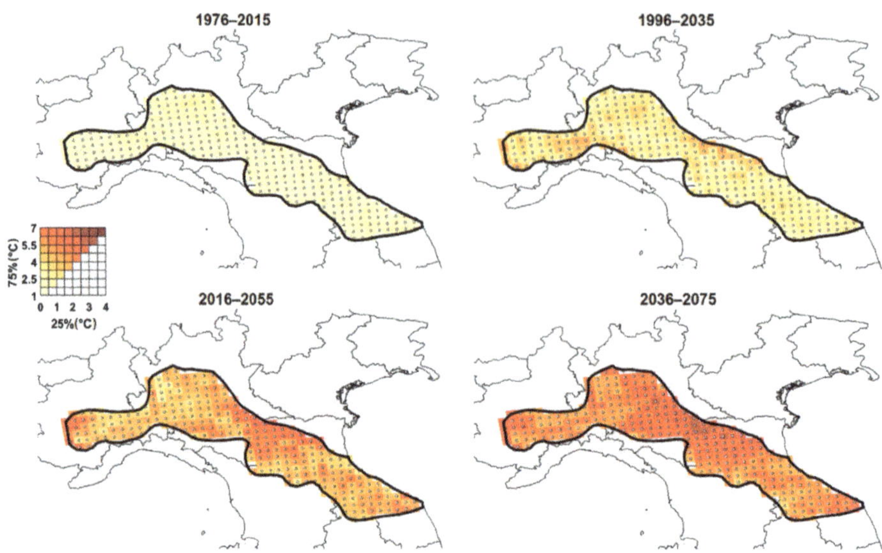

Figure 8. Factors of change for $T_{Min,k}$ in the time windows 1976–2015, 1996–2035, 2016–2055, and 2036–2075 in comparison with the reference time interval 1956–1995—prediction interval (25–75%) map (Scenario RCP4.5).

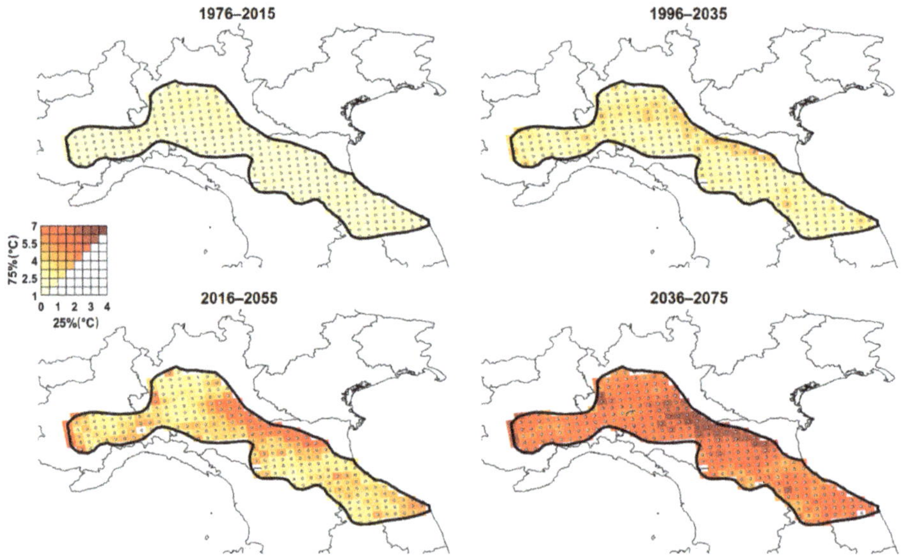

Figure 9. Factors of change for $T_{Min,k}$ in the time windows 1976–2015, 1996–2035, 2016–2055, and 2036–2075 in comparison with the reference time interval 1956–1995—prediction interval (25–75%) map (Scenario RCP8.5).

For example, considering the time window 2036–2075, the lower and upper limits of the prediction interval of the increase of the characteristic value of the maximum temperature are 1.87 °C and 3.55 °C, respectively, with an average value of 2.75 °C, considering the RCP4.5 scenario, and 2.47 °C and 4.17 °C, respectively, with an average value of 3.37 °C, considering the RCP8.5 scenario. In the same time window 2036–2075, a more pronounced rise is expected for $T_{Min,k}$: the lower and upper limits of the prediction interval of the increase of the characteristic value of the minimum temperature are 2.47 and 4.17 °C, respectively, with an average value of 3.37 °C, considering the RCP4.5 scenario, and 2.29 and 5.23 °C, respectively, with an average value of 3.52 °C, considering the RCP8.5 scenario.

Table 4. Mean of factors of change (°C) percentiles for $T_{Min,k}$ in the study region.

Time Window	RCP4.5			RCP8.5		
	25%	50%	75%	25%	50%	75%
1966–2005	−0.30	0.14	0.73	−0.28	0.15	0.73
1976–2015	−0.23	0.56	1.60	−0.16	0.63	1.66
1986–2025	−0.12	0.98	2.48	−0.29	0.84	2.44
1996–2035	0.31	1.57	3.16	−0.08	1.23	2.92
2006–2045	0.65	2.01	3.89	0.24	1.63	3.39
2016–2055	1.01	2.59	4.52	0.83	2.13	3.67
2026–2065	1.48	3.03	4.87	1.55	2.75	4.19
2036–2075	1.59	3.18	5.22	2.29	3.52	5.23
2046–2085	2.40	3.81	6.02	2.83	4.42	8.55

Clearly, on the basis of these observations, suitable updates of current temperature maps provided in the Italian National Annex to EN1991-1-5:2003 [27] seem to be necessary.

3.4. *Effects of Climate Change on Extreme Precipitation*

Regarding the effects of climate change on rainfalls, there is observational evidence that the frequency and intensity of heavy rainfalls is increasing in several regions all around the world. This evidence is further validated by the outcomes of many recent research works [68,69]. This observation is in close agreement with the classical thermodynamic law, also known as Clausius–Clapeyron law, stating that warmer the air, the higher its capacity to hold water vapor [70,71]. Looking only at the increase of atmospheric moisture content, a scaling rate of warming around 6–7% K^{-1} can be predicted, according to the recalled Clausius–Clapeyron law.

Considering an annual probability of exceedance $p = 2\%$, factors of change of characteristic values of daily precipitation ($p_{r,k}$) and associated prediction intervals have been also derived by means of the proposed weather generator.

Factors of change maps for characteristic values of precipitation ($p_{r,k}$) and associated prediction intervals for time windows 1976–2015, 1996–2035, 2016–2055, and 2035–2075 are presented in the bivariate maps reported in Figures 10 and 11, referring to the RCP4.5 and RCP8.5 scenarios, respectively. Evidently, in this case, as in the next one concerning snow loads, factors of changes are derived in form of quotients, according to Equation (15).

Factors of change percentiles (25%, 50%, 75%), averaged over the region, are summarized in Table 5 for each time window.

Table 5. Mean of factors of change percentiles for $p_{r,k}$ in the investigated region.

Time Window	RCP4.5			RCP8.5		
	25%	50%	75%	25%	50%	75%
1966–2005	0.96	1.01	1.06	0.96	1.01	1.06
1976–2015	0.94	1.02	1.12	0.94	1.02	1.12
1986–2025	0.91	1.03	1.19	0.91	1.03	1.18
1996–2035	0.89	1.05	1.24	0.89	1.05	1.23
2006–2045	0.90	1.06	1.25	0.91	1.08	1.25
2016–2055	0.91	1.07	1.25	0.94	1.11	1.32
2026–2065	0.92	1.07	1.25	0.97	1.16	1.39
2036–2075	0.93	1.09	1.28	1.01	1.20	1.47
2046–2085	0.97	1.13	1.36	1.04	1.25	1.56

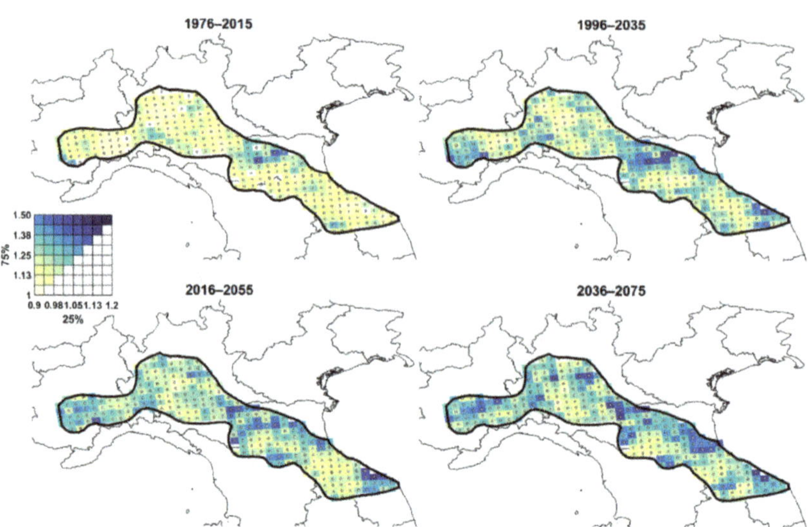

Figure 10. Factors of change for $p_{r,k}$ in the time windows 1976–2015, 1996–2035, 2016–2055, and 2036–2075 in comparison with the reference time interval 1956–1995—prediction interval (25–75%) map (Scenario RCP4.5).

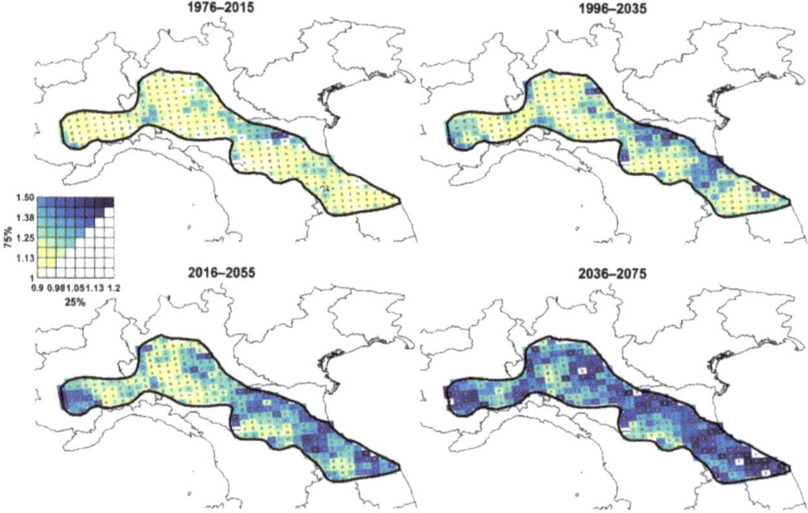

Figure 11. Factors of change for $p_{r,k}$ in the time windows 1976–2015, 1996–2035, 2016–2055, and 2036–2075 in comparison with the reference time interval 1956–1995—prediction interval (25–75%) map (Scenario RCP8.5).

3.5. Ground Snow Loads

Starting from the generated samples of daily temperatures and precipitation (T_{max}, T_{min} and p_r), they have been assessed the characteristic values of the snow load on ground. Since the estimation of ground snow load requires reconstructing the complicated processes governing snowfall, snow accumulation, total or partial melting of the snow cover, and so on, the application of a weather generation algorithm has been supplemented with the procedure developed by the authors to simulate snow load on the ground on the basis of precipitation and temperature data [6,7], so obtaining characteristic values of snow load in each 40-year-long time window.

For the investigated region, factors of change maps, concerning characteristic ground snow load, s_k, in time windows 1976–2015, 1996–2035, 2016–2055, and 2036–2075, in comparison with the reference time window 1956–1995, are presented in Figures 12 and 13, for the RCP4.5 and RCP8.5 scenarios, respectively.

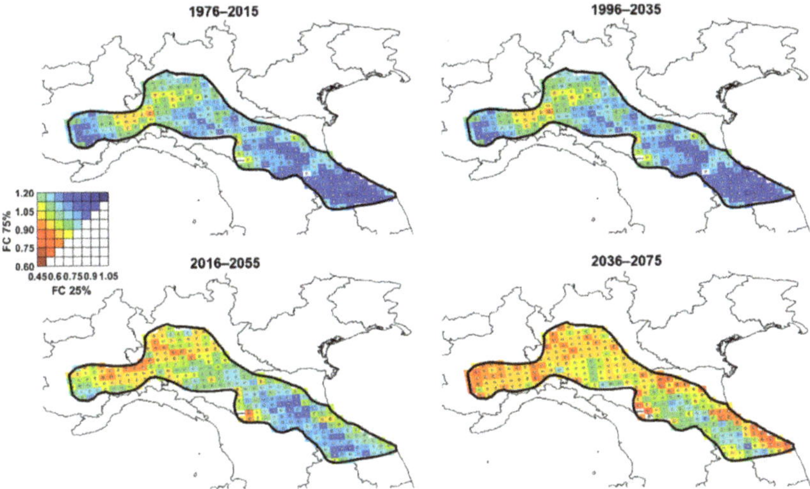

Figure 12. Factors of change for s_k in the time windows 1976–2015, 1996–2035, 2016–2055, and 2036–2075 in comparison with the reference time interval 1956–1995—prediction interval (25–75%) map (Scenario RCP4.5).

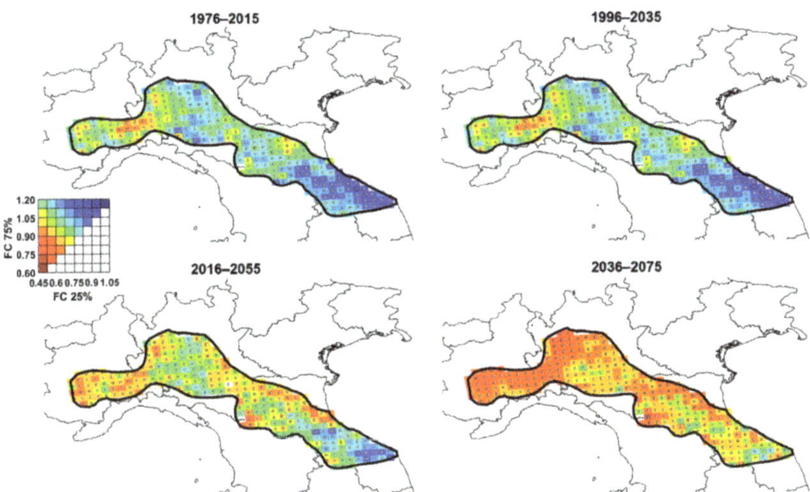

Figure 13. Factors of change for s_k in the time windows 1976–2015, 1996–2035, 2016–2055, and 2036–2075 in comparison with the reference time interval 1956–1995—prediction interval (25–75%) map (Scenario RCP8.5).

The outcomes confirm that the investigated region is characterized by a general decreasing trend, even if locally different behavior can be detected, depending on the grid cell, on the investigated time window, and on the considered scenario. In fact, as already highlighted by O'Gorman in [72], the reduction of snowfall fraction and the increase of heavy precipitation caused by global warming, may locally lead to a contrasting response in terms of snowfall variation. On the one hand, the temperature rise facilitates the melting of snow and increases the fraction of precipitation falling as rain; on the other hand, it is associated with a rise of precipitation rate during extreme events, potentially leading to heavy snowfalls and higher ground snow load. Thus, the overall outcome, in terms of increase or decrease of the ground snow load, is dependent on the ratio of these two competing factors [73].

Nevertheless, according the obtained results, in the near future (1996–2035), a constant or increasing trend ($FC_{50\%} > 0.95$) can be expected in around 15% of the region in the RCP4.5 scenario, and in around 7% of the region in the RCP8.5 scenario. However, a decrease is expected in the long-term future (2036–2075) for the whole region.

As previously done for precipitation, the results are averaged over all grid cells to better visualize changes in ground snow load, reducing the influence of unforced variability at the grid box level. The resulting factors of change in the considered time windows, averaged on the region, are reported for the given prediction percentiles (25%, 50%, and 75%) in Table 6. The results confirm that the ground snow loads in the investigated region generally decrease, by approximately 10% to 15% in the time interval 1991–2030 and 25% to 30% in the future (2041–2080). Moreover, the obtained results are practically independent on the considered RCP scenario.

Table 6. Mean of factors of change percentiles for s_k in the study region.

Time Window	RCP4.5			RCP8.5		
	25%	50%	75%	25%	50%	75%
1966–2005	0.91	0.98	1.03	0.91	0.98	1.03
1976–2015	0.85	0.95	1.04	0.84	0.94	1.04
1986–2025	0.80	0.92	1.04	0.78	0.90	1.02
1996–2035	0.76	0.88	1.01	0.72	0.84	0.99
2006–2045	0.73	0.85	0.98	0.69	0.81	0.95
2016–2055	0.69	0.82	0.95	0.67	0.78	0.92
2026–2065	0.67	0.79	0.92	0.65	0.76	0.88
2036–2075	0.64	0.76	0.89	0.61	0.72	0.85
2046–2085	0.62	0.73	0.85	0.56	0.68	0.80

4. Discussion

As already mentioned in the introduction, the aim of the present study is to provide a methodology for the treatment of climate model outputs by means of a weather generation technique combined with the factor of change approach. The results obtained for the case study presented in Section 3 show the suitability of the methodology but are not intended to represent a detailed guidance for adaptation planning.

Dealing with regional climate projections, the skill of the ensemble to reproduce multidecadal changes in the study region should be first assessed. This can be carried out by comparing the variation of climate statistics provided by the climate model ensemble and by the available high-resolution observational dataset for the historical period. As an example, factors of change for daily temperatures and precipitation can be evaluated for a part of the investigated region from the analysis of the Eraclito-ERG5 dataset [74]. The Eraclito-ERG5 is a high-quality gridded observational dataset for the Emilia Romagna region in Italy, covering the period 1961–2020 and characterized by a horizontal resolution of about 5 km × 5 km.

In Figures 14–16, factors of change for characteristic values of daily maximum and minimum temperature, and daily precipitation are reported for the time window 1981–2020 with respect to the period 1961–2000. Trends are generally consistent with those previously presented in Section 3; local differences can be justified remarking on the one hand that the database of observations and climate projections have different resolutions, on the other that the time windows are not perfectly coincident. Of course, the output of climate models with this extremely refined resolution could significantly improve the evaluation of climate change impacts. The wish is that they will soon be available.

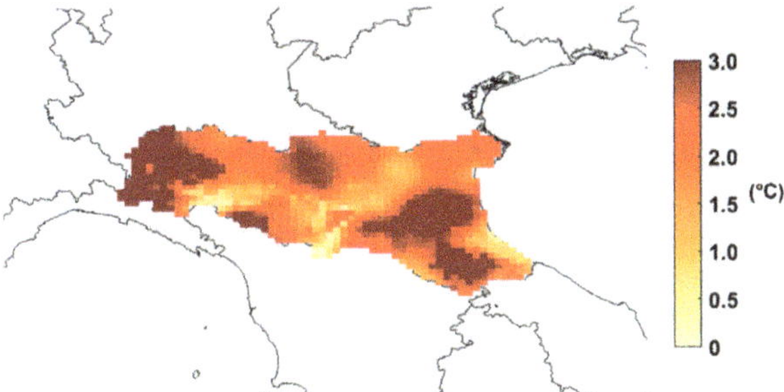

Figure 14. Factors of change for $T_{max,k}$ for the time window 1981–2020 in comparison with the reference time interval 1961–2000.

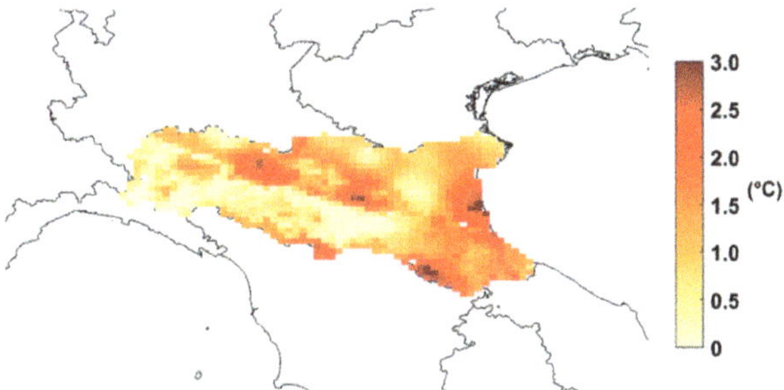

Figure 15. Factors of change for $T_{min,k}$ for the time window 1981–2020 in comparison with the reference time interval 1961–2000.

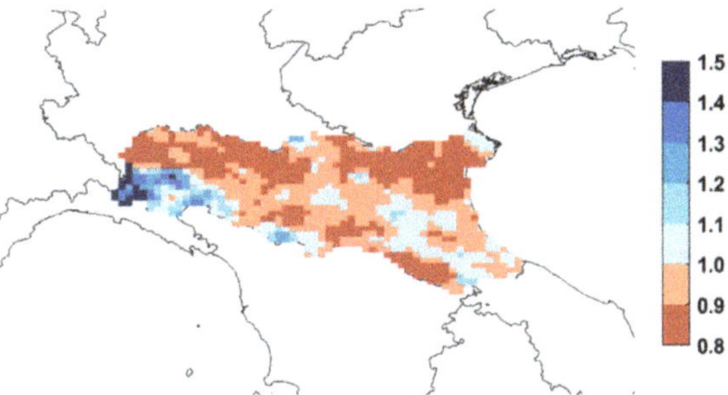

Figure 16. Factors of change for $p_{r,k}$ for the time window 1981–2020 in comparison with the reference time interval 1961–2000.

5. Conclusions

In the paper, a new procedure for the analysis of climate model outputs is presented in view of a probabilistic assessment of future trends of climatic actions.

The proposed methodology is based on an innovative weather generator, which investigates the internal variability of climate models and allows improving the statistical representativeness of the climate model ensemble, preserving the consistency of the original climate model output. In fact, performance metrics have shown a very good agreement between the statistics of the generated series and the original ones for daily temperatures and precipitation.

Thus, the generated series can be used to investigate changes in the statistics of climatic variables, e.g., temperatures and precipitation, providing a quantification of the uncertainty associated with the predicted changes.

The results, presented for different climatic actions, in terms of confidence maps for the study region in Italy, confirm that the technique is very promising and can be successfully applied for the adaptation of climatic actions maps given in codes and standards for structural design. For example, starting from the obtained FC maps, current thermal and ground snow load maps based on past observations and provided in the Italian National Annex to Eurocode EN1991-1-5 [27] and EN1991-1-3 [25] respectively can be updated considering climate change impacts [75]. As the soundness of the output of the method is a function of the quality and of the resolution of climate models, it will be improved as soon as the next generation of climate projections will be available.

Author Contributions: Conceptualization, P.C., P.F. and F.L.; methodology, P.C., P.F. and F.L.; software, P.C., P.F. and F.L.; validation, P.C., P.F. and F.L.; data curation, P.C., P.F. and F.L.; writing—original draft preparation, P.C., P.F. and F.L.; writing—review and editing, P.C., P.F. and F.L.; visualization, P.C., P.F. and F.L.; resources, P.F. All authors have read and agreed to the published version of the manuscript.

Funding: This research received no external funding.

Institutional Review Board Statement: Not applicable.

Informed Consent Statement: Not applicable.

Data Availability Statement: The data presented in this study are available on request from the corresponding author. The data are not publicly available as they cannot be used for commercial purposes.

Acknowledgments: The authors acknowledge the World Climate Research Working Groups on Regional Climate, and on Coupled Modelling. The authors thank the climate modeling groups listed in Table 2 for providing the model outcomes via the infrastructure of the Earth System Grid Federation (https://esgf-data.dkrz.de/search/cordex-dkrz/, accessed on 18 August 2021).

Conflicts of Interest: The authors declare no conflict of interest.

References

1. Bastidas-Arteaga, E.; Stewart, M. *Climate Adaptation Engineering*; Butterworth-Heinemann: Oxford, UK, 2019.
2. Madsen, H.O. Managing structural safety and reliability in adaptation to climate change. In *Safety, Reliability, Risk and Life-Cycle Performance of Structures and Infrastructure*; CRC Press: Boca Raton, FL, USA, 2013; pp. 81–88.
3. Stewart, M.G.; Deng, X. Climate impact risks and climate adaptation engineering for built infrastructure. *ASCE-ASME J. Risk Uncertain. Eng. Syst. Part A Civ. Eng.* **2015**, *1*, 04014001. [CrossRef]
4. Croce, P.; Formichi, P.; Landi, F. Climate change: Impacts on climatic actions and structural reliability. *Appl. Sci.* **2019**, *9*, 5416. [CrossRef]
5. Forzieri, G.; Bianchi, A.; Batista e Silva, F.B.; Herrera, M.A.M.; Leblois, A.; Lavalle, C.; Aerts, J.C.J.H.; Feyen, L. Escalating impacts of climate extremes on critical infrastructures in Europe. *Glob. Environ. Chang.* **2018**, *48*, 97–107. [CrossRef] [PubMed]
6. Croce, P.; Formichi, P.; Landi, F. Structural safety and design under climate change. In *20th Congress of IABSE, New York City 2019: The Evolving Metropolis*; International Association for Bridge and Structural Engineering (IABSE): Zurich, Switzerland, 2019; pp. 1130–1135.
7. Croce, P.; Formichi, P.; Landi, F.; Marsili, F. Climate change: Impact on snow loads on structures. *Cold Reg. Sci. Technol.* **2018**, *150*, 35–50. [CrossRef]
8. Croce, P.; Formichi, P.; Landi, F.; Mercogliano, P.; Bucchignani, E.; Dosio, A.; Dimova, S. The snow load in Europe and the climate change. *Clim. Risk Manag.* **2018**, *20*, 138–154. [CrossRef]
9. Forzieri, G.; Bianchi, A.; Herrera, M.A.M.; Batista e Silva, F.; Lavalle, C.; Feyen, L. *Resilience of Large Investments and Critical Infrastructures in Europe to Climate Change*; EUR27906; Publications Office of the European Union: Luxembourg, 2016. [CrossRef]
10. Organization for Economic Co-Operation and Development (OECD). *Climate-Resilient Infrastructure*; OECD ENVIRONMENT POLICY PAPER NO. 14; OECD Publishing: Paris, France, 2018.

11. Intergovernmental Panel on Climate Change (IPCC). *Climate Change 2021: The Physical Science Basis. Contribution of Working Group I to the Fifth Assessment Report of the Intergovernmental Panel on Climate Change—Summary for Policymakers*; Cambridge University Press: Cambridge, UK, 2021.
12. Pryor, S.C.; Barthelmie, R.J.; Bukovsky, M.S.; Leung, L.R.; Sakaguchi, K. Climate change impacts on wind power generation. *Nat. Rev. Earth Environ.* **2020**, *1*, 627–643. [CrossRef]
13. Wagner, T.; Themeßl, M.; Schüppel, A.; Gobiet, A.; Stigler, H.; Birk, S. Impacts of climate change on stream flow and hydro power generation in the Alpine region. *Environ. Earth Sci.* **2017**, *76*, 4. [CrossRef]
14. Christodoulou, A.; Christidis, P.; Bisselink, B. Forecasting the impacts of climate change on inland waterways. *Transp. Res. Part D Transp. Environ.* **2020**, *82*, 102159. [CrossRef]
15. Konapala, G.; Mishra, A.K.; Wada, Y.; Mann, M.E. Climate change will affect global water availability through compounding changes in seasonal precipitation and evaporation. *Nat. Commun.* **2020**, *11*, 3044. [CrossRef]
16. Mbow, C.C.; Rosenzweig, L.G.; Barioni, T.G.; Benton, M.; Herrero, M.; Krishnapillai, E.; Liwenga, P.; Pradhan, M.G.; Rivera-Ferre, T.; Sapkota, F.N.; et al. Food security. In *Climate Change and Land: An IPCC Special Report on Climate Change, Desertification, Land Degradation, Sustainable Land Management, Food Security, and Greenhouse Gas Fluxes in Terrestrial Ecosystems*; Cambridge University Press: Cambridge, UK, 2019.
17. Stewart, M.G.; Wang, X.; Nguyen, M. Climate change impact and risks of concrete infrastructure deterioration. *Eng. Struct.* **2011**, *33*, 1326–1337. [CrossRef]
18. Bastidas-Arteaga, E.; Schoefs, F.; Stewart, M.G.; Wang, X. Influence of global warming on durability of corroding RC structures: A probabilistic approach. *Eng. Struct.* **2013**, *51*, 259–266. [CrossRef]
19. Bisoi, S.; Haldar, S. Impact of climate change on design of offshore wind turbine considering dynamic soil–structure interaction. *J. Offshore Mech. Arct. Eng.* **2017**, *139*, 061903. [CrossRef]
20. Hoekstra, A.Y.; De Kok, J.L. Adapting to climate change: A comparison of two strategies for dike heightening design. *Nat. Hazards* **2008**, *47*, 217–228. [CrossRef]
21. Lutz, J.; Dobler, A.; Nygaard, B.E.; Mc Innes, H.; Haugen, J.E. Future projections of icing on power lines over Norway. In Proceedings of the International Workshop on Atmospheric Icing of Structures IWAIS 2019, Reykjavík, Iceland, 23–28 June 2019.
22. Faggian, P.; Bonanno, R.; Pirovano, G. Research activities to cope with wet snow impacts on overhead power lines in future climate over Italy. In Proceedings of the International Workshop on Atmospheric Icing of Structures IWAIS 2019, Reykjavík, Iceland, 23–28 June 2019.
23. Hirabayashi, Y.; Mahendran, R.; Koirala, S.; Konoshima, L.; Yamazaki, D.; Watanabe, S.; Kim, H.; Kanae, S. Global flood risk under climate change. *Nat. Clim. Chang.* **2013**, *3*, 816–821. [CrossRef]
24. European Committee for Standardization (CEN). *EN 1990. Eurocode—Basis of Structural Design*; CEN: Brussels, Belgium, 2002.
25. European Committee for Standardization (CEN). *EN 1991-1-3. Eurocode 1: Actions on Structures—Part 1–3: General Actions—Snow Loads*; CEN: Brussels, Belgium, 2003.
26. European Committee for Standardization (CEN). *EN 1991-1-4. Eurocode 1: Actions on Structures—Part 1–4: General Actions—Wind Actions*; CEN: Brussels, Belgium, 2005.
27. European Committee for Standardization (CEN). *EN 1991-1-5. Eurocode 1: Actions on Structures—Part 1–5: General Actions—Thermal Actions*; CEN: Brussels, Belgium, 2003.
28. International Organization for Standardization (ISO). *ISO 2394 General Principles on Reliability for Structures*; ISO: Geneva, Switzerland, 2015.
29. Croce, P.; Formichi, P.; Landi, F. Evaluation of current trends of climatic actions in Europe based on observations and regional reanalysis. *Remote Sens.* **2021**, *13*, 2025. [CrossRef]
30. European Commission. *EU Strategy on Adaptation to Climate Change, COM (2013) 216*; European Commission: Brussel, Belgium, 2013.
31. European Commission. *Adapting Infrastructure to Climate Change, SWD (2013) 137*; European Commission: Brussel, Belgium, 2013.
32. Brekke, L.D.; Barsugli, J.J. Uncertainties in projections of future changes in extremes. In *Extremes in a Changing Climate*; AghaKouchak, A., Easterling, D., Hsu, K., Schubert, S., Sorooshian, S., Eds.; Springer: Dordrecht, The Netherlands, 2013; pp. 309–346. [CrossRef]
33. Wilks, D.; Wilby, R. The weather generation game: A review of stochastic weather models. *Prog. Phys. Geogr.* **1999**, *23*, 329–357. [CrossRef]
34. Semenov, M.A.; Barrow, E.M. Use of a stochastic weather generator in the development of climate change scenarios. *Clim. Chang.* **1997**, *35*, 397–414. [CrossRef]
35. Fowler, H.J.; Blenkinsop, S.; Tebaldi, C. Linking climate change modelling to impacts studies: Recent advances in downscaling techniques for hydrological modelling. *Int. J. Climatol.* **2007**, *27*, 1547–1578. [CrossRef]
36. Kilsby, C.G.; Jones, P.D.; Burton, A.; Ford, A.C.; Fowler, H.J.; Harpham, C.; James, P.; Smith, A.; Wilby, R.L. A daily weather generator for use in climate change studies. *Environ. Model. Softw.* **2007**, *22*, 1705–1719. [CrossRef]
37. Fatichi, S.; Ivanov, V.Y.; Caporali, E. Simulation of future climate scenarios with a weather generator. *Adv. Water Resour.* **2011**, *34*, 448–467. [CrossRef]
38. Hawkins, E.; Sutton, R.T. The potential to narrow uncertainty in regional climate predictions. *Bull. Am. Meteorol. Soc.* **2009**, *90*, 1095–1107. [CrossRef]

39. Buishand, T.A.; Brandsma, T. Multi-site simulation of daily precipitation and temperature in the Rhine basin by nearest-neighbor resampling. *Wat. Resour. Res.* **2001**, *37*, 2761–2776. [CrossRef]
40. Buishand, T.A.; Brandsma, T. Multi-site simulation of daily precipitation and temperature conditional on the atmospheric circulation. *Clim. Res.* **2003**, *25*, 121–133. [CrossRef]
41. Anandhi, A.; Frei, A.; Pierson, D.C.; Schneiderman, E.M.; Zion, M.S.; Lounsbury, D.; Matonse, A.H. Examination of change factor methodologies for climate change impact assessment. *Water Resour. Res.* **2011**, *47*, W03501. [CrossRef]
42. Burgess, M.G.; Ritchie, J.; Shapland, J.; Pielke Jr, R. IPCC baseline scenarios have over-projected CO_2 emissions and economic growth. *Environ. Res. Lett.* **2021**, *16*, 014016. [CrossRef]
43. Ho, E.; Budescu, D.V.; Bosetti, V.; van Vuuren, D.P.; Keller, K. Not all carbon dioxide emission scenarios are equally likely: A subjective expert assessment. *Clim. Chang.* **2019**, *155*, 545–561. [CrossRef]
44. Van Vuuren, D.P.; Edmonds, J.; Kainuma, M.; Riahi, K.; Thomson, A.; Hibbard, K.; Hurtt, G.C.; Kram, T.; Krey, V.; Lamarque, J.F.; et al. The representative concentration pathways: An overview. *Clim. Chang.* **2011**, *109*, 5–31. [CrossRef]
45. Hutchinson, M.F. Methods of generation of weather sequences. In *Agricultural Evironments*; Bunting, A.H., Ed.; CAB International: Wallingford, UK, 1986; pp. 149–157.
46. Croce, P.; Formichi, P.; Landi, F.; Marsili, F. A novel probabilistic methodology for the local assessment of future trends of climatic actions. *Beton-und Stahlbetonbau* **2018**, *113*, 110–115. [CrossRef]
47. Jacob, D.; Petersen, J.; Eggert, B.; Alias, A.; Christensen, O.B.; Bouwer, L.M.; Braun, A.; Colette, A.; Déqué, M.; Georgievski, G.; et al. EURO-CORDEX: New high-resolution climate change projections for European impact research. *Reg. Environ. Chang.* **2014**, *14*, 563–578. [CrossRef]
48. Kotlarski, S.; Keuler, K.; Christensen, O.B.; Colette, A.; Déqué, M.; Gobiet, A.; Goergen, K.; Jacob, D.; Lüthi, D.; van Meijgaard, E.; et al. Regional Climate Modelling on European Scale: A joint standard evaluation of the EURO-CORDEX ensemble. *Geosci. Model Dev.* **2014**, *7*, 1297–1333. [CrossRef]
49. Cressie, N.A.C. *Statistics for Spatial Data, Revised Edition*; John Wiley & Sons Inc.: Hoboken, NJ, USA, 1994.
50. Cooley, D.; Nychka, D.; Naveau, P. Bayesian spatial modeling of extreme precipitation return levels. *J. Am. Stat. Assoc.* **2007**, *102*, 824–840. [CrossRef]
51. BjØrnstad, O.N.; Falck, W. Nonparametric spatial covariance functions: Estimation and testing. *Environ. Ecol. Stat.* **2001**, *8*, 53–70. [CrossRef]
52. Taylor, K.E. Summarizing multiple aspects of model performance in a single diagram. *J. Geophys. Res.* **2001**, *106*, 7183–7192. [CrossRef]
53. Gleckler, P.J.; Taylor, K.E.; Doutriaux, C. Performance metrics for climate models. *J. Geophys. Res.* **2008**, *113*, D06104. [CrossRef]
54. Klein Tank, A.M.; Zwiers, F.W.; Zhang, X. *Guidelines on Analysis of Extremes in a Changing Climate in Support of Informed Decisions for Adaptation*; Tech. Rep. WCDMP-No. 72; World Meteorological Organization (WMO): Geneva, Switzerland, 2009.
55. Cooley, D. Return periods and return levels under climate change. In *Extremes in a Changing Climate*; AghaKouchak, A., Easterling, D., Hsu, K., Schubert, S., Sorooshian, S., Eds.; Springer: Dordrecht, The Netherlands, 2013; pp. 97–114. [CrossRef]
56. Croce, P.; Formichi, P.; Landi, F.; Marsili, F. Harmonized European ground snow load map: Analysis and comparison of national provisions. *Cold Reg. Sci. Technol.* **2019**, *168*, 102875. [CrossRef]
57. Formichi, P.; Danciu, L.; Akkar, S.; Kale, O.; Malakatas, N.; Croce, P.; Nikolov, D.; Gocheva, A.; Luechinger, P.; Fardis, M.; et al. Eurocodes: Background and applications. Elaboration of maps for climatic and seismic actions for structural design with the Eurocodes. *JRC Sci. Policy Rep.* **2016**. [CrossRef]
58. Maraun, D. Bias correcting climate change simulations—A critical review. *Curr. Clim. Chang. Rep.* **2016**, *2*, 211–220. [CrossRef]
59. Maraun, D.; Widmann, M. *Statistical Downscaling and Bias Correction for Climate Research*; Cambridge University Press: Cambridge, UK, 2018.
60. Berg, P.; Christensen, O.B.; Klehmet, K.; Lenderink, G.; Olsson, J.; Teichmann, C.; Yang, W. Precipitation extremes in a EURO-CORDEX 0.11° ensemble at hourly resolution. *Nat. Hazards Earth Syst. Sci. Discuss.* **2018**, *362*. [CrossRef]
61. Gleick, P.H. Methods for evaluating the regional hydrologic impacts of global climatic changes. *J. Hydrol.* **1986**, *88*, 97–116. [CrossRef]
62. Teuling, A.J.; Stöckli, R.; Seneviratne, S.I. Bivariate colour maps for visualizing climate data. *Int. J. Climatol.* **2011**, *31*, 1408–1412. [CrossRef]
63. Hausfather, Z.; Peters, G. Emissions—The 'business as usual' story is misleading. *Nature* **2020**, *577*, 618–620. [CrossRef]
64. Schwalm, C.R.; Glendon, S.; Duffy, P.B. RCP8.5 tracks cumulative CO_2 emissions. *Proc. Natl. Acad. Sci. USA* **2020**, *117*, 19656–19657. [CrossRef]
65. Tebaldi, C.; Knutti, R. The use of the multi-model ensemble in probabilistic climate projections. *Philos. Trans. R. Soc. A Math. Phys. Eng. Sci.* **2007**, *365*, 2053–2075. [CrossRef]
66. Croce, P.; Formichi, P.; Landi, F.; Marsili, F. Evaluating the effect of climate change on thermal actions on structures. In *Life-Cycle Analysis and Assessment in Civil Engineering: Towards an Integrated Vision*; Caspeele, R., Taerwe, L., Frangopol, D.M., Eds.; Taylor & Francis Group: Oxfordshire, UK, 2019; pp. 1751–1758, ISBN 978-1-138-62633-1.
67. Intergovernmental Panel on Climate Change (IPCC). *Climate Change 2013: The Physical Science Basis. Contribution of Working Group I to the Fifth Assessment Report of the Intergovernmental Panel on Climate Change*; Cambridge University Press: Cambridge, UK; New York, NY, USA, 2013.

68. Westra, S.; Alexander, L.V.; Zwiers, F.W. Global increasing trends in annual maximum daily precipitation. *J. Clim.* **2013**, *26*, 3904–3918. [CrossRef]
69. Donat, M.G.; Lowry, A.L.; Alexander, L.V.; O'Gorman, P.A.; Maher, N. More extreme precipitation in the world's dry and wet regions. *Nat. Clim. Chang.* **2016**, *6*, 508–513. [CrossRef]
70. Fischer, E.M.; Knutti, R. Observed heavy precipitation increase confirms theory and early models. *Nat. Clim. Chang.* **2016**, *6*, 986–991. [CrossRef]
71. O'Gorman, P.A. Precipitation extremes under climate change. *Curr. Clim. Chang. Rep.* **2015**, *1*, 49–59. [CrossRef]
72. O'Gorman, P.A. Contrasting responses of mean and extreme snowfall to climate change. *Nature* **2014**, *515*, 416–418. [CrossRef]
73. Räisänen, J. Warmer climate: Less or more snow? *Clim. Dyn.* **2008**, *30*, 307–319. [CrossRef]
74. Antolini, G.; Auteri, L.; Pavan, V.; Tomei, F.; Tomozeiu, R.; Marletto, V. A daily high-resolution gridded climatic data set for Emilia-Romagna, Italy, during 1961–2010. *Int. J. Climatol.* **2016**, *36*, 1970–1986. [CrossRef]
75. Croce, P.; Formichi, P.; Landi, F. Implication of climate change on climatic actions on structures: The update of climatic load maps. In *IABSE Symposium Wrocław 2020, Synergy of Culture and Civil Engineering—History and Challenges—Report*; IABSE: Zurich, Switzerland, 2020; pp. 877–884.

Article

Panama's Current Climate Replicability in a Non-Hydrostatic Regional Climate Model Nested in an Atmospheric General Circulation Model

Reinhardt Pinzón [1,2,3,*], Noriko N. Ishizaki [4], Hidetaka Sasaki [5] and Tosiyuki Nakaegawa [5,*]

1 Centro de Investigaciones Hidráulicas e Hidrotécnicas (CIHH), Group HPC-Cluster-Iberogun, Universidad Tecnológica de Panamá (UTP), Panamá City P.O. Box 0819-07289, Panama
2 Sistema Nacional de Investigación (SNI), SENACYT, Panamá City P.O. Box 0816-02852, Panama
3 Centro de Estudios Multidisciplinarios de Ingeniería Ciencias y Tecnología (CEMCIT-AIP), El Dorado, Panamá City P.O. Box 0819-07289, Panama
4 National Institute for Environmental Studies, Tsukuba 305-8506, Japan; ishizaki.noriko@nies.go.jp
5 Meteorological Research Institute, Tsukuba 305-0052, Japan; hsasaki@mri-jma.go.jp
* Correspondence: reinhardt.pinzon@utp.ac.pa (R.P.); tnakaega@mri-jma.go.jp (T.N.); Tel.: +507-560-3762 (R.P.); +81-29-853-8538 (T.N.)

Abstract: To simulate the current climate, a 20-year integration of a non-hydrostatic regional climate model (NHRCM) with grid spacing of 5 and 2 km (NHRCM05 and NHRCM02, respectively) was nested within the AGCM. The three models did a similarly good job of simulating surface air temperature, and the spatial horizontal resolution did not affect these statistics. NHRCM02 did a good job of reproducing seasonal variations in surface air temperature. NHRCM05 overestimated annual mean precipitation in the western part of Panama and eastern part of the Pacific Ocean. NHRCM05 is responsible for this overestimation because it is not seen in MRI-AGCM. NHRCM02 simulated annual mean precipitation better than NHRCM05, probably due to a convection-permitting model without a convection scheme, such as the Kain and Fritsch scheme. Therefore, the finer horizontal resolution of NHRCM02 did a better job of replicating the current climatological mean geographical distributions and seasonal changes of surface air temperature and precipitation.

Keywords: MRI-AGCM; Panama; precipitation; NHRCM; surface air temperature; present climate; nesting method; simulations; mean square error

Citation: Pinzón, R.; Ishizaki, N.N.; Sasaki, H.; Nakaegawa, T. Panama's Current Climate Replicability in a Non-Hydrostatic Regional Climate Model Nested in an Atmospheric General Circulation Model. *Atmosphere* **2021**, *12*, 1543. https://doi.org/10.3390/atmos12121543

Academic Editors: Baojie He, Ayyoob Sharifi, Chi Feng and Jun Yang

Received: 30 September 2021
Accepted: 18 November 2021
Published: 23 November 2021

1. Introduction

Scientists and policymakers alike are requesting high-resolution estimates of global warming and its consequences in order to map out the changes and consequences in greater detail. Dynamical downscaling of global climate models is a valuable strategy for generating comprehensive estimates of regional climate changes owing to global warming [1,2]. As an example, previous works projected the future climate in and around Panama with a global climate model (GCM) with grid spacing of 20 km [3–7].

Even with fine grid spacing, however, this model did not adequately replicate extreme events and did not resolve the main mountain ranges or river basins, all of which must be depicted if climate projections are to be relevant for water resource and flood planning, agriculture, and other applications. To overcome these issues, a non-hydrostatic regional climate model (NHRCM) was used in order to simulate current climate in Panama.

Only a few studies on high-resolution models (e.g., [8–10]) have been published in Central America. Climate change projections for Central America and Mexico were initially performed with a regional climate model [8]. Dynamical downscaling with horizonal resolution of 27, 9, and 3 km over Panama using the Weather Research and Forecasting (WRF) model was performed to forecast one-day precipitation with different physical parameterizations [10]. This is the only study on forecasting for Panama, but it is not a study on a climate time-scale. In this study, we nested our NHRCM within the results of AGCM's current climate experiment, integrated it for all four seasons over a 20-year period from 1980 to 1999, and analyzed its performance to make the model useful for climate studies surrounding Panama.

2. Materials and Methods

2.1. Experiment Design

This study used a multiple nesting approach, as shown in Figure 1a. The lateral boundary data on the outskirts were derived from current climate simulations, employing an AGCM with a grid spacing of 20 km [11]. The AGCM is based on the Japan Meteorological Agency's (JMA) previous operational model for numerical weather prediction [12]. The integration period was 1979 to 2003. This simulation was designed to perform dynamical downscaling with accurate current climate replication to avoid garbage-in-garbage-out, since coupled GCMs replicate current climate with distinct and/or systematic biases, especially at regional scale. The AGCM is externally forced with observation-based time series of sea surface temperature, sea ice concentration, greenhouse gas concentration, sulfate aerosol concentration, ozone gas concentration, and volcanic aerosol concentration. The AGCM does a good job of replicating the current climate in Central America [4] and in most of the world [13].

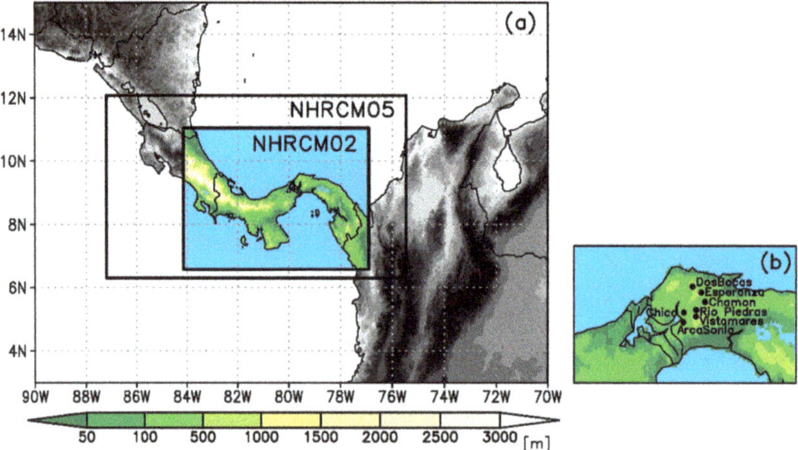

Figure 1. (**a**) Target area for current climate simulations. Inner domain (colored) represents area for NHRCM02, outer domain (grayscale) represents area for NHRCM05. (**b**) Eight ground-based precipitations stations observed by Panama Canal Authority.

NHRCMs with grid spacing of 5 and 2 km (NHRCM05 and NHRCM02, respectively) were nested within the AGCM. The AGCM lacks cloud water, cloud ice, and other cloud characteristic variables, so both NHRCM05 and NHRCM02 were used to produce them for the lateral border of the two smaller domains with 5 and 2 km grid spacing. Calculations were simplified by skewing the inner domain, which still spans almost the entire Panama Isthmus. The setup of NHRCM employed in this study can be found in Table 1. NHRCM was externally forced with observation-based time series of sea surface temperature and greenhouse gas concentration. In this work, the time slice integration method was applied. Each integration took place between 1 December and 4 December of the following year. The first month of results was discarded in favor of spin-up. Since 1980, this cycle has repeatedly occurred 20 times. Comparisons of climate between NHRCM05 and NHRCM02 are the focus in this study; comparisons between NHRCMs and MRI-AGCM are provided in the Figure 1a.

Table 1. Model specifications chosen in this study. Minimal details are given in main text, and full details are described in references.

Model	MRI-AGCM	NHRCM	
Grid space	20 km	5 km	2 km
Boundary condition	-	MRI-AGCM 20 km	
Spectral nudging	-	Applied	Not applied
Convection scheme	Yoshimura et al. (2018) [14]	Kain and Fritsch's scheme (1990) [15]	Not applied
Boundary layer	Mellor-Yamada (MY; 1974) Level2 [16]	MYNN2.5; Nakanishi and Niino 2004 [17]	
Radiation process	JMA (2007) [18]	Yabu et al. (2005) [19] and Kitagawa (2000) [20]	
Land surface model	SiB ver.0919 [21]	iSiB [22]	
Sea surface temperatures and sea ice	COBE SST (1° × 1°) [23]		
External atmospheric forcing	Greenhouse gases, sulfur and volcanic aerosol, and ozone gases	Greenhouse gases	

2.2. Surface Air Temperature and CRU TS v4.05 Data

The The gridded Climatic Research Unit (CRU) Time-series (TS) data version 4.05 dataset was used, which is the fourth version of the gridded product established by the Climate Research Unit of the University of East Anglia [24]. It includes a number of variables (precipitation, surface air temperature, mean temperature, etc.) compiled in a global (excluding Antarctica) 0.5 × 0.5 grid (56 km) from 1901 to 2017, obtained by the interpolation of monthly data collected from the World Meteorological Organization's archives.

2.3. Precipitation Data

For the observation dataset, we used GSMaP Gauge v.5 global satellite mapping of precipitation data [25,26] (https://sharaku.eorc.jaxa.jp/GSMaP/guide.html#09; accessed on 2 February 2020). Ground truth data are used to adjust the satellite data bias in this dataset. It has a time resolution of one hour and a spatial resolution of 0.1°. For the target season, this dataset is only accessible for the period from 2000 to 2010. There are differences in the reference period between the simulations and GSMaP. Monthly precipitation in CRU TS with a 0.5 × 0.5 grid was used to see if there was an issue with the reference period difference.

Eight ground-based precipitation stations observed by the Panama Canal Authority were also used (Figure 1b). The eight stations are located in and around the upper Chagres River Basin, a part of the Panama Canal Basin [27], which is a very important area for water resources of canal operations; the coordinates of the observation stations range from 79.40° E to 79.51° E and from 9.19° N to 9.45° N, corresponding to about a single grid box of MRI-AGCM. The elevation ranges from 110 to 2100 m.

3. Results and Discussion

3.1. Surface Air Temperature

The models were validated by comparing the CRU to the simulations averaged over CRU grids on land. Figure 2 depicts the differences in annual mean climatological surface air temperature between the simulation and observations. Except for several grids, the bias of the surface air temperature simulated with NHRCM02 lies between −1 and 1 °C (Figure 2d). A large negative bias of −3 °C was found in the mountainous terrain of the western border. NHRCM05 did a good job of simulating surface air temperature, although the biases in NHRCM05 are large compared to those in NHRCM (Figure 2c).

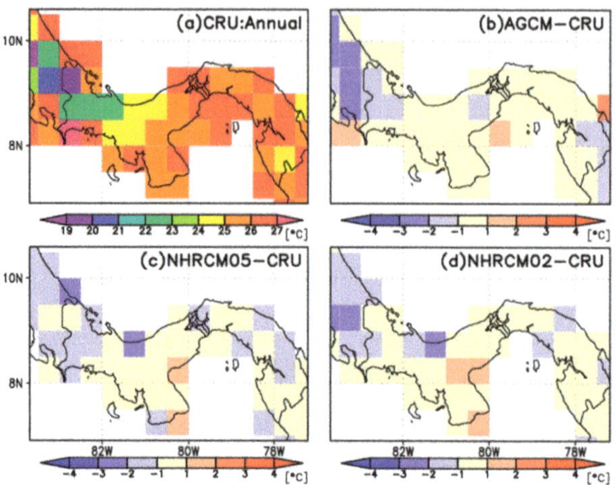

Figure 2. Geographic evaluation of 20-year mean annual surface air temperature derived from (**a**) CRU (CRU_TS4.05), and biases in (**b**) MRI-AGCM, (**c**) NHRCM05, and (**d**) NHRCM02. Values in biases are difference between simulations and CRU observations. See Table 2 for quantitative evaluation.

As the grid size becomes finer, the NHRCMs have negative biases, opposite to those of MRI-AGCM. NHRCM02 has small biases that are geographically distributed in Panama, and the area average over the country is minimal (Table 2). For the 20-year mean annual surface air temperature, the spatial correlation between the NHRCMs and CRU was 0.90 and 0.92 for 5 and 2 km grid spacing, respectively. The root mean square errors (RMSEs) for NHRCMs are almost the same. Therefore, NHRCM02 has better performance in simulating surface air temperature, and the spatial horizontal resolution contributes slightly to the replicability of geographic distribution of surface air temperature.

Table 2. Quantitative evaluation of 20-year mean annual surface air temperature averaged over Panama in three simulations against observation. See Figure 2 for geographic evaluation.

	Bias (°C)	RMSE (°C)	Correlation
AGCM	0.55	1.05	0.93
NHRCM05	−0.66	1.09	0.90
NHRCM02	−0.51	1.08	0.92

NHRCM02 did a good job of reproducing seasonal variations in surface air temperature (see Figure 3). The NHRCMs had cool biases in the rainy season, from May to November, in comparison to CRU, whereas in the dry season, from January to March, the simulated surface air temperatures of NHRCMs were almost the same as those of CRU. All models accurately simulated maximum mean monthly surface air temperatures at the end of the dry season, but with a 1-month behind/advance maximum, while the minimum surface air temperature in NHRCMs occurred in October, with two months advance, in comparison to CRU.

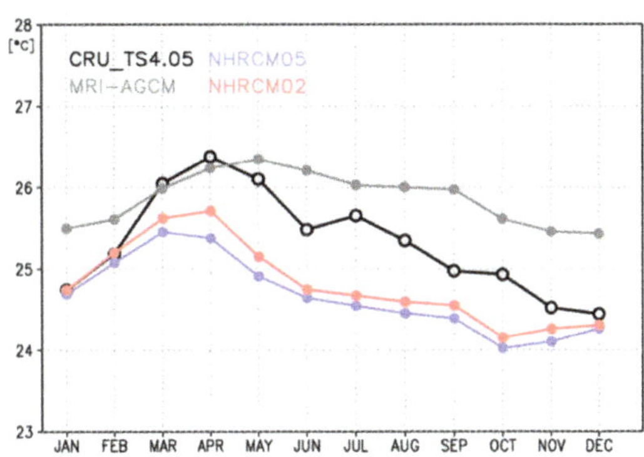

Figure 3. Seasonal changes in 20-year mean monthly surface air temperatures averaged over grids with altitude below 1000 m. Black, gray, blue, and red lines represent observations, MRI-AGCM, NHRCM05, and NHRCM02, respectively.

For seasonal changes in 20-year mean surface air temperature, NHRCM02 had a smaller bias and RMSE and higher temporal correlation than NHRCM05 (Table 3). The correlation between model simulation and CRU was 0.74 and 0.83 for NHRCM05 and NHRCM02, respectively, and NHRCM02 did a better job at simulating seasonal changes than NHRCM05.

Table 3. Same as in Table 2 but for seasonal changes in 20-year mean monthly surface air temperature in three simulations against observation. See Figure 3 for seasonal changes.

	Bias (°C)	RMSE (°C)	Correlation
AGCM	0.55	0.66	0.87
NHRCM05	−0.66	0.76	0.74
NHRCM02	−0.51	0.61	0.83

MRI-AGCM showed large negative biases in many grids, especially in the western border and the Caribbean Sea (Figure 2b). MRI-AGCM's average annual mean surface air temperature biases over Panama were negative by 0.55 °C (Table 2). As the grid size becomes finer, NHRCMs have negative biases, opposite to MRI-AGCM. The three models show very small varieties ranging from 1.05 to 1.09 °C. The spatial correlation for MRI-AGCM was slightly high compared to NHRCM02. The range of monthly surface air temperature of NHRCMs was better than that of MRI-AGCM (Figure 3). MRI-AGCM had the best temporal correlation of seasonal changes in the three models (Table 3), although the difference in the correlation coefficients is not statistically significant. Therefore, the nested NHRCMs reproduced the spatial details of surface air temperature as well as MRI-AGCM did. Different horizontal resolutions of datasets of observations and models means different representative altitudes. The altitude–surface air temperature dependence may affect the biases.

3.2. Precipitation

The models were validated by comparing the GSMaP to the simulations averaged over the GSMaP grid. The differences in annual mean climatological precipitation between simulations and observations are provided in Figure 4. NHRCM05 overestimated annual mean precipitation in the western part of Panama and eastern part of the Pacific Ocean side. Very large biases, exceeding a ratio of 3, were seen in the western part of the Caribbean Sea. NHRCM05 was responsible for this overestimation, as it was not seen in MRI-AGCM. Precipitation in NHRCM05 is understood as high sensitivity to mountainous terrains [28]. NHRCM02 simulated annual mean precipitation better than NHRCM05, probably due to a convection-permitting model without a convection scheme, such as Kain and Fritsch's scheme. The underestimation is seen in the eastern part of the Azuero Peninsula, the driest region in Panama.

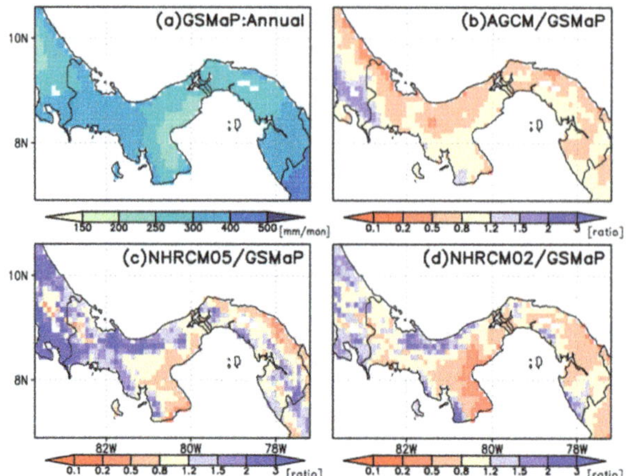

Figure 4. Same as in Figure 2 but for 20-year mean annual precipitation derived from (**a**) GSMaP v5, and biases in (**b**) MRI-AGCM, (**c**) NHRCM05, and (**d**) NHRCM02. Bias values are ratios of simulation to observation of GSMaP. See Table 4 for quantitative evaluation.

The finer version of NHRCM, NHRCM02, had smaller absolute bias and RMSE and higher spatial correlation than NHRCM05 (Table 4). These results are consistent with the fact that orographic precipitation is generally simulated in a model with a finer horizontal resolution.

Table 4. Same as in Table 2 but for quantitative evaluation of 20-year mean annual precipitation. See Figure 4 for geographic evaluation.

	Bias (mm/Month)	RMSE (mm/Month)	Correlation
AGCM	−44.9	95.8	0.40
NHRCM05	127.5	253.1	0.33
NHRCM02	4.0	131.5	0.38

Figure 5 shows the seasonal changes in precipitation in Panama. The two models simulated similar seasonal changes, except for the period October to February. There are three monthly precipitation stages: about 470 mm/month from May to August, about 410 mm/month from September to November, and dry conditions or less precipitation from December to April. The seasonal changes in Panama were fairly well represented by NHRCM02 (Figure 5), but the second stage was not. NHRCM05 overestimated precipitation, especially in the second and the first half of the second stages. NHRCM02 also overestimated precipitation in the same period because it used the lateral boundary simulated in NHRCM05.

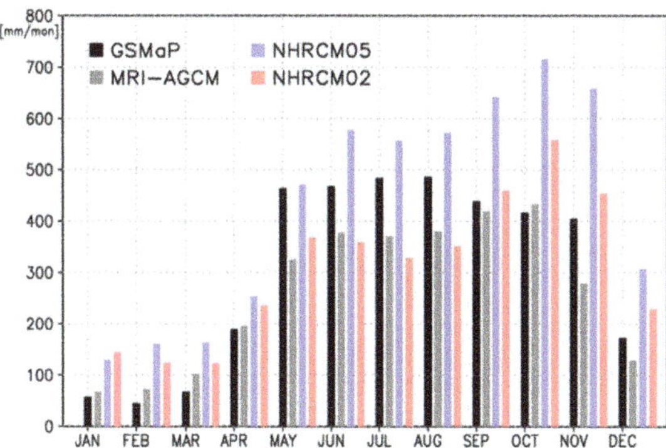

Figure 5. Seasonal changes of 20-year mean monthly precipitation in three simulations against observations.

A quantitative evaluation of seasonal changes is tabulated in Table 5. The biases are the same as in Table 4, since they do not vary when spatial or temporal means are computed on an annual time scale. Although NHRCMs are nested in the lateral boundary conditions of MRI-AGCM, they have opposite signs in the biases. NHRCM05 has larger RMSE than NHRCM02. The temporal correlation between NHRCMs and CRU was 0.92 for NHRCM05 and 0.85 for NHRCM02.

Table 5. Same as in Table 3 but for quantitative evaluation of seasonal changes of 20-year mean monthly precipitation in three simulations against observations. See Figure 5 for seasonal changes.

	Bias (mm/Month)	RMSE (mm/Month)	Correlation
AGCM	−44.86	77.14	0.95
NHRCM05	127.52	151.10	0.92
NHRCM02	4.04	95.03	0.85

Scatter plots of annual precipitation totals of the three models and GSMaP are depicted in Figure 6. The range of annual precipitation is confined between about 200 and 500 mm/month, while for NHRCM02 the range is widely distributed between 50 and 1000 mm/month. NHRCM05 only simulated about 1400 mm/month in the three grids. These values obviously seem to be overestimated, since the maximum simulated in NHRCM02 is about 1100 mm/month and that in GSMaP is about 480 mm/month. The seasonal variations of monthly climatological precipitation in NHRCM05 resemble those of the observations but with larger amounts. The time series of daily precipitation always show large amounts of precipitation, which contributes to the overestimations, but little event-specific heavy precipitation with outlier amounts is seen. The outlier of about 1400 mm/month in NHRCM05 is not an implausible value, and the so-called gridpoint storm on a sub-daily time scale [29] may have affected the overestimations and is a challenging study relevant to extreme precipitation events. Therefore, the higher horizontal resolution of 2 km and the convection-permitting model without a convection scheme is responsible for the low maximum simulated in NHRCM02.

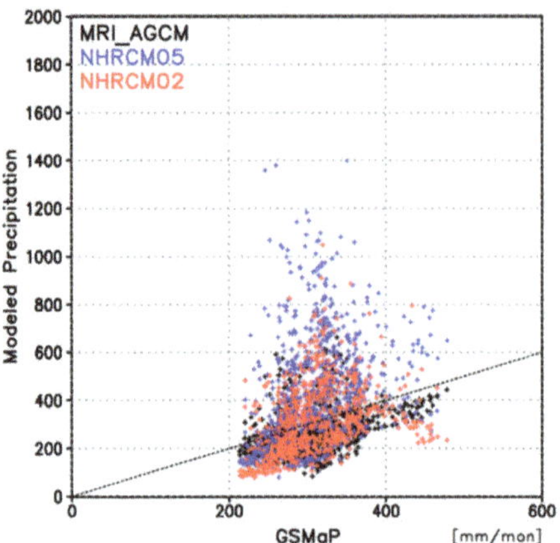

Figure 6. Scatter plot showing 20-year mean annual precipitation amount of GSMaP and NHRCM02. Black dots represent values (in mm/month) for entire domains including oversea areas, and blue dots represent values for land areas only.

The area in and around the upper Chagres River Basin, a part of the Panama Canal Basin (Figure 1b), was selected to evaluate the capability of NHRCM02 to simulate precipitation over complex terrain, since it has complex topography with elevation ranging from 110 to 2100 m. NHRCM02 well simulated the spatial pattern of precipitation, while NHRCM05 did not (Figure 7). The spatial correlation coefficient of precipitation in NHRCM02 and NHRCM05 with ground-based stations is 0.76 and −0.95, respectively. The negative correlation coefficient of NHRCM05 suggests that less than 5 km scale topography strongly affects local precipitation there. The size of the basin corresponds to that of a grid box of MRI-AGCM, and MRI-AGCM cannot capture the spatial pattern of precipitation in the basin. GSMaP has a correlation of 0.4, but the absolute precipitation is almost the same at the eight stations due to 0.1° horizontal resolution. This shows the benefit of the high horizontal resolution of NHRCM02.

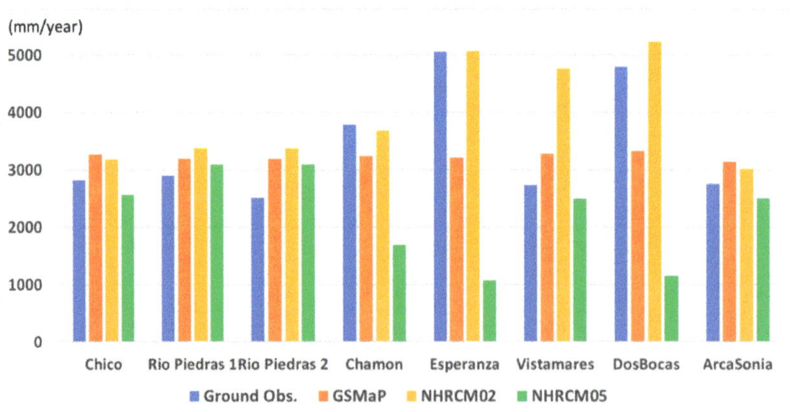

Figure 7. Comparison of annual precipitation at eight stations in upper Chagres River Basin among four datasets: ground observations, GSMaP, NHRCM02, and NHRCM05.

Overestimation at Vistamares in NHRCM02 is probably due to the complexity of the topography, which cannot be represented even at 2 km since the maximum and minimum annual precipitation in the eight grids surrounding Vistamares varies from about 2900 to 5000 mm. It may be enhanced due to climatological time-scale gridpoint storms fixed with the complexity of the topography.

MRI-AGCM has a smaller bias than NHRCM05, although NHRCM05 has finer horizontal resolution and uses the lateral boundary conditions obtained from MRI-AGCM (Figure 4). MRI-AGCM's annual precipitation biases averaged over Panama were negative at −44.9 mm/year and the smallest of the three models (Table 4). The spatial correlations of MRI-AGCM and NHRCM02 were similar and higher than that of NHRCM05. The seasonal changes in Panama in MRI-AGMC were similar to those in NHRCM02 (Figure 5). The temporal correlation between MRI-AGCM and CRU was 0.95 for AGCM, the highest of the three models (Table 5). The range of annual precipitation amount was smaller for MRI-AGCM than NHRCMs but larger than GSMaP (Figure 6). Only NHRCM02 replicated the small-scale precipitation distribution in the complex terrain (Figure 7). Therefore, NHRCM02 reproduced many characteristics of precipitation as well as MRI-AGCM did.

3.3. Uncertainty in Precipitation Dataset

There are uncertainties in various precipitation datasets [30]. The annual climatological precipitation in Panama varies from 2501 to 2924 mm/year [31]. GSMaP used in this study showed the largest annual climatological precipitation of 3185 mm/year. The range of annual precipitation is large, and the choice of dataset affects the reproducibility. The biases in NHRCMs become larger than those shown in Tables 4 and 5 when another dataset is used.

The geographic distribution of precipitation seems too smooth (Figure 4a) when we consider the topography in Panama (Figure 1), because orographic precipitation produces small-scale distribution. This is primarily the horizontal resolution of GSMaP. The spatial correlation of NHRCMs will be higher than that of MRI-AGCM when a precipitation dataset with a horizontal resolution higher than GSMaP is used.

The range of annual precipitation of GSMaP at each grid is small in comparison with the ranges of the three models (Figure 6). This is also due to the insufficient horizontal resolution of GSMaP. The spatial variability of precipitation is distinct in a mountainous area of the Panama Canal catchment [27]. Therefore, the horizontal resolution of a precipitation dataset may affect the reportability of precipitation, especially for finer horizontal resolution of models such as NHRCMs.

Another uncertainty in precipitation datasets stems from the specific time period, since precipitation has large variabilities in time. NHRCMs and GSMaP have no overlap period; the former period is 1981 to 2000, while the latter period is 2001 to 2010. The mean climatological precipitation was compared between the two periods using precipitation in CRU TS. The ratio of annual precipitation of the former to the latter is 1.03. The maximum monthly precipitation ratio is 1.19 in November, while the minimum is 0.95 in October. These differences allow us to compare climatological precipitation between the two reference periods.

4. Conclusions

We used an inner nested grid with spacing of 2 and 5 km to run NHRCMs for a 20-year integration of the current climate. As the grid size becomes finer, NHRCMs have smaller biases. NHRCM02 has small biases that are geographically distributed in Panama, and the area average over the country is minimal. NHRCM02 did a better job of simulating surface air temperature than NHRCM05, and the spatial horizontal resolution did not affect the biases and RMSE, but did affect the spatial correlation.

NHRCM02 simulated annual mean precipitation better than NHRCM05, probably due to a convection-permitting model without a convection scheme, such as the Kain and Fritsch scheme. The spatial correlations of NHRCM02 were slightly higher than those of NHRCM05. The seasonal changes of precipitation were well simulated in both models.

MRI-AGCM had good replicability of precipitation similar to NHRCM02 and outperformed NHRCM05. This suggests that 5 km horizontal resolution is not suitable for simulating climate in Panama and 2 km is required for correct simulation. The difference between NHRCM02 and NHRCM05 is whether it is a convection-permitting model or not, or, in other words, whether the convection scheme is switched on or not, which may play an important role in replicability. The range of annual precipitation in GSMaP at each grid is small compared with the three models. Therefore, the horizontal resolution of a precipitation dataset may affect the reportability of precipitation, especially for models with finer horizontal resolution such as NHRCMs. NHRCMs tend to simulate geographic precipitation too finely to compare with GSMaP. Therefore, the horizontal resolution of a precipitation dataset may affect the reportability of precipitation, especially for models with finer horizontal orientation such as NHRCMs.

There are many challenging topics for simulation of the current climate with a regional climate model in Panama as well as all over the world [32]. For example, the climatological aspect of the diurnal cycle in the tropics is a key phenomenon to be replicated in regional climate models [28]. The replicability of extreme events [5,11,29,33,34] in the current climate is an important feature in climate modeling, along with the replicability of climates. We expect impact assessment researchers and decisionmakers to make use of the NHRCM to analyze the details of projected global warming and related climate changes around Panama.

Author Contributions: Conceptualization, R.P. and T.N.; methodology, R.P., N.N.I., H.S. and T.N.; software, N.N.I. and H.S.; validation, R.P., N.N.I., H.S. and T.N.; resources, R.P. and T.N.; writing—original draft preparation R.P. and T.N.; writing—review and editing, R.P., N.N.I., H.S. and T.N.; visualization, N.N.I. All authors have read and agreed to the published version of the manuscript.

Funding: R.P. had scientific support from SENACYT through projects FID-2016-275 and EIE-2018-16, and through the Sistema Nacional de Investigación (SNI) of SENACYT. R.P. visited MRI with support from the Exchange Program for Research Institutions in Developing Countries of the Japanese Ministry of Land, Infrastructure, Transport and Tourism. T.N. received funds and support from the JSPS KAKENHI Grant Numbers-in-Aid for Specially Promoted Research (grant numbers 20K12154, 16H06291, and 21H05002) and support from Theme C of the TOUGOU program (JPMXD0717935498), funded by the Japanese Ministry of Education, Culture, Sports, Science and Technology.

Institutional Review Board Statement: Not applicable.

Informed Consent Statement: Not applicable.

Data Availability Statement: The data presented in this study are available on request from the corresponding author for. The data are not publicly available due to the size of data produced under ongoing research program.

Acknowledgments: The authors acknowledge administrative support provided by the Universidad Tecnológica de Panamá. We thank the editor and reviewers for their time and constructive comments, which have helped improve our manuscript.

Conflicts of Interest: The authors declare no conflict of interest.

References

1. Sasaki, H.; Kurihara, K.; Takayabu, I.; Uchiyama, T. Preliminary experiments of reproducing the present climate using the non-hydrostatic regional climate model. *SOLA* **2008**, *4*, 25–28. [CrossRef]
2. Sasaki, H.; Murata, A.; Hanafusa, M.; Oh'izumi, M.; Kurihara, K. Reproducibility of present climate in a non-hydrostatic regional climate model nested within an atmosphere general circulation model. *SOLA* **2011**, *7*, 173–176. [CrossRef]
3. Fábrega, J.; Nakaegawa, T.; Pinzón, R.; Nakayama, K.; Arakawa, O.; SOUSEI Theme-C Modeling Group. Hydroclimate projections for Panama in the 21st century. *Hydrol. Res. Lett.* **2013**, *7*, 23–29. [CrossRef]
4. Nakaegawa, T.; Kitoh, A.; Kusunoki, S.; Murakami, H.; Arakawa, O. Hydroclimate change over central America and the Caribbean in a global warming climate projected with 20-km and 60-km mesh MRI atmospheric general circulation models. *Pap. Meteorol. Geophys.* **2014**, *65*, 15–33. [CrossRef]

5. Nakaegawa, T.; Kitoh, A.; Murakami, H.; Kusunoki, S. Maximum 5-day rainfall total and the maximum number of consecutive dry days over central America in the future climate projected by an atmospheric general circulation model with three different horizontal resolutions. *Theor. Appl. Climatol.* **2014**, *116*, 155–168. [CrossRef]
6. Pinzón, R.E.; Hibino, K.; Takayabu, I.; Nakaegawa, T. Virtual experiencing future climate changes in Central America with MRI-AGCM: Climate analogues study. *Hydrol. Res. Lett.* **2017**, *11*, 106–113. [CrossRef]
7. Kusunoki, S.; Nakaegawa, T.; Pinzón, R.; Sanchez-Galan, J.E.; Fábrega, J.R. Future precipitation changes over Panama projected with the atmospheric global model MRI-AGCM3.2. *Clim. Dyn.* **2019**, *53*, 5019–5034. [CrossRef]
8. Karmalkar, A.V.; Bradley, R.S.; Diaz, H.F. Climate change in Central America and Mexico: Regional climate model validation and climate change projections. *Clim. Dyn.* **2011**, *37*, 605–629. [CrossRef]
9. Poleo, D.; Vindas, C.; Stoltz, W. Comparación y evaluación de diferentes esquemas de parametrización de cúmulos con WRF EMS aplicadas al caso del huracán otto. *Tóp. Meteorol. Oceanogr.* **2017**, *16*, 28–40. [CrossRef]
10. Sierra-Lorenzo, M.; Bezanilla-Morlot, A.; Centella-Artola, A.; León-Marcos, A.; Borrajero-Montejo, I.; Ferrer-Hernández, A.; Salazar-Gaitán, J.; Lau-Melo, A.; Picado-Traña, F.; Pérez-Fernández, J. Assessment of different WRF configurations performance for a rain event over Panama. *Atmos. Clim. Sci.* **2020**, *10*, 280–297. [CrossRef]
11. Kitoh, A.; Ose, T.; Kurihara, K.; Kusunoki, S.; Sugi, M.; KAKUSHIN Team-3 Modeling Group. Projection of changes in future weather extremes using super-high resolution global and regional atmospheric models in the KAKUSHIN Pro- gram: Results of preliminary experiments. *Hydrol. Res. Lett.* **2009**, *3*, 49–53. [CrossRef]
12. Mizuta, R.; Oouhi, K.; Yoshimura, H.; Noda, A.; Katayama, K.; Yukimoto, S.; Hosaka, M.; Kusunoki, S.; Kawai, H.; Nakagawa, M. 20 km-mesh global climate simulations using JMA-GSM model—Mean climate state. *J. Meteorol. Soc. Jpn.* **2006**, *84*, 165–185. [CrossRef]
13. Ito, R.; Nakaegawa, T.; Takayabu, I. Comparison of regional characteristics of land precipitation climatology projected by an MRI-AGCM multi-cumulus scheme and multi-SST ensemble with CMIP5 multi-model ensemble projections. *Prog. Earth Planet. Sci.* **2020**, *7*, 77. [CrossRef]
14. Yoshimura, H.; Mizuta, R.; Murakami, H. A spectral cumulus parameterization scheme interpolating between two convective updrafts with semi-Lagrangian calculation of transport by compensatory subsidence. *Mon. Weather Rev.* **2015**, *143*, 597–621. [CrossRef]
15. Kain, J.S.; Fritsch, J.M. A one-dimensional entraining/detraining plume model and its application in convective parameterization. *J. Atmos. Sci.* **1990**, *47*, 2784–2802. [CrossRef]
16. Mellor, G.L.; Yamada, T. A hierarchy of turbulence closure models for planetary boundary layers. *J. Atmos. Sci.* **1974**, *31*, 1791–1806. [CrossRef]
17. Nakanishi, M.; Niino, H. An improved Mellor–Yamada level-3 model with condensation physics: Its design and verification. *Bound. Layer Meteorol.* **2004**, *112*, 1–31. [CrossRef]
18. JMA. Outline of the Operational Numerical Weather Prediction at the Japan Meteorological Agency. March 2007. Available online: http://www.jma.go.jp/jma/jma-eng/jma-center/nwp/outline2007-nwp/index.htm (accessed on 10 November 2021).
19. Yabu, S.; Murai, S.; Kitagawa, H. Clear-sky radiation scheme. *Rep. Numer. Predict. Div.* **2005**, *51*, 53–64. (In Japanese)
20. Kitagawa, H. Radiation processes. *Rep. Numer. Predict. Div.* **2000**, *46*, 16–31. (In Japanese)
21. Hirai, M.; Sakashita, T.; Kitagawa, H.; Tsuyuki, T.; Hosaka, M.; Oh'izumi, M. Development and validation of a new land surface model for JMA's operational global model using the CEOP observation dataset. *J. Meteorol. Soc. Jpn.* **2007**, *85*, 1–24. [CrossRef]
22. Hirai, M.; Oh'izumi, M. Development of a new land-surface model for JMA-GSM. In *Extended Abstract of 20th Conference on Weather Analysis and Forecasting/16th Conference on Numerical Weather Prediction*; American Meteorological Society: Tokyo, Japan, 2004; Available online: https://ams.confex.com/ams/84Annual/webprogram/Paper68652.html (accessed on 10 November 2021).
23. Ishii, M.; Shouji, A.; Sugimoto, S.; Matsumoto, T. Objective analyses of sea-surface temperature and marine meteorological variables for the 20th century using ICOADS and the Kobe collection. *Int. J. Climatol.* **2005**, *25*, 865–879. [CrossRef]
24. Harris, I.; Osborn, T.J.; Jones, P.; Lister, D. Version 4 of the CRU TS monthly high-resolution gridded multivariate climate dataset. *Sci. Data* **2020**, *7*, 109. [CrossRef] [PubMed]
25. Kubota, T.; Shige, S.; Hashizume, H.; Aonashi, K.; Takahashi, N.; Seto, S.; Hirose, M.; Takayabu, Y.N.; Nakagawa, K.; Iwanami, K.; et al. Global precipitation map using satellite-borne microwave radiometers by the GSMaP project: Production and validation. *IEEE Trans. Geosci. Remote Sens.* **2007**, *45*, 2259–2275. [CrossRef]
26. Kubota, T.; Aonashi, K.; Ushio, K.; Shige, S.; Takayabu, Y.N.; Kachi, M.; Arai, Y.; Tashima, T.; Masaki, T.; Kawamoto, N.; et al. Global satellite mapping of precipitation (GSMaP) products in the GPM era. In *Satellite Precipitation Measurement*; Levizzani, V., Kidd, C., Kirschbaum, D.B., Kummerow, C.D., Nakamura, K., Turk, K.J., Eds.; Springer: Berlin, Germany, 2020. [CrossRef]
27. Nakaegawa, T.; Pinzon, R.; Fabrega, J.; Cuevas, J.A.; De Lima, H.A.; Cordoba, E.; Nakayama, K.; Lao, J.I.B.; Melo, A.L.; Gonzalez, D.A.; et al. Seasonal changes of the diurnal variation of precipitation in the upper Río Chagres basin, Panamá. *PLoS ONE* **2019**, *14*, e0224662. [CrossRef] [PubMed]
28. Takayabu, I.; Ishizaki, N.N.; Nakaegawa, T.; Sasaki, H.; Wongseree, W. Potential of representing the diurnal cycle of local-scale precipitation in northeastern Thailand using 5-km and 2-km grid regional climate models. *Hydrol. Res. Lett.* **2021**, *15*, 1–8. [CrossRef]
29. Chan, S.C.; Kendon, E.J.; Fowler, H.J.; Blenkinsop, S.; Roberts, N.M.; Ferro, C.A.T. The value of high-resolution met office regional climate models in the simulation of multihourly precipitation extremes. *J. Clim.* **2014**, *27*, 6155–6174. [CrossRef]

30. Ebert, E.E.; Janowiak, J.E.; Kidd, C. Comparison of near-real-time precipitation estimates from satellite observations and numerical models. *Bull. Am. Meteorol. Soc.* **2007**, *88*, 47–64. [CrossRef]
31. Nakaegawa, T.; Arakawa, O.; Kamiguchi, K. Investigation of climatological onset and withdrawal of the rainy season in panama based on a daily gridded precipitation dataset with a high horizontal resolution. *J. Clim.* **2015**, *28*, 2745–2763. [CrossRef]
32. Prein, A.F.; Rasmussen, R.; Stephens, G. Challenges and advances in convection-permitting climate modeling. *Bull. Am. Meteorol. Soc.* **2017**, *98*, 1027–1030. [CrossRef]
33. Croce, P.; Formichi, P.; Landi, F. Evaluation of current trends of climatic actions in Europe based on observations and regional reanalysis. *Remote Sens.* **2021**, *13*, 2025. [CrossRef]
34. Forzieri, G.; Bianchi, A.; E Silva, F.B.; Herrera, M.A.M.; Leblois, A.; LaValle, C.; Aerts, J.C.; Feyen, L. Escalating impacts of climate extremes on critical infrastructures in Europe. *Glob. Environ. Chang.* **2018**, *48*, 97–107. [CrossRef] [PubMed]

MDPI

Article

Green's Function for Static Klein–Gordon Equation Stated on a Rectangular Region and Its Application in Meteorology Data Assimilation

Hao Cheng [1,2], Xiyu Mu [3], Hua Jiang [2], Ming Wei [1] and Guoqing Liu [2,*]

[1] Collaborative Innovation Center on Forecast and Evaluation of Meteorological Disasters, Nanjing University of Information Science & Technology, No. 219 Ningliu Road, Nanjing 210044, China; chenghao@njtech.edu.cn (H.C.); njueducn@126.com (M.W.)
[2] College of Mathematics & Physics, Nanjing Tech University, No. 30 Puzhu Road(S), Nanjing 211816, China; quantumcat@163.com
[3] Jiangsu Institute of Meteorological Sciences, No. 20 Kunlun Road, Nanjing 210009, China; xiyumufish@gmail.com
* Correspondence: guoqing@njtech.edu.cn

Abstract: The meteorology data assimilation applications often encounter variational problems with unknown weights, where the corresponding Euler equation is an elliptic partial differential equation. This research focused on retrieving the weights in remote sensing data assimilation by means of the computer-friendly form of the Green's function obtained by eigenfunction expansion for the boundary value problem of the static Klein–Gordon equation on a rectangular region. With the help of the proposed retrieving method, the assimilation problem of estimating regional precipitation with weather radar and rain-gauge is solved in the Green's function method. Results show that high accuracy of the proposed method makes it a good candidate for data assimilation problems in operational use.

Keywords: Green's function; Klein–Gordon equation; rectangle area; uniformly convergence; regional precipitation; data assimilation

Citation: Cheng, H.; Mu, X.; Jiang, H.; Wei, M.; Liu, G. Green's Function for Static Klein–Gordon Equation Stated on a Rectangular Region and Its Application in Meteorology Data Assimilation. *Atmosphere* **2021**, *12*, 1602. https://doi.org/10.3390/atmos12121602

Academic Editors: Baojie He, Ayyoob Sharifi, Chi Feng and Jun Yang

Received: 27 October 2021
Accepted: 29 November 2021
Published: 30 November 2021

Publisher's Note: MDPI stays neutral with regard to jurisdictional claims in published maps and institutional affiliations.

1. Introduction

In modern numerical weather prediction, data assimilation is an effective method to improve data accuracy by fusing datasets with different precisions. Data assimilation applications often encounter variational problems, where the corresponding Euler equation is an elliptic partial differential equation (PDE), usually a static Klein–Gordon equation (SKGE). In applications, there are two obstacles: unknown boundary conditions, and unknown weights. Such problems of retrieving the unknown boundary condition and weights are generally known as inverse problems. Most previous research on these problems has concentrated on approximating parameter estimation [1–5] to some extent. Wei, et al. (1998, 2003) calculated the weights with the matrix theory and the finite difference method of partial differential equation [6,7].

In solving elliptic equations by Green's functions, Melnikov (2011) constructed the corresponding Green's functions for the SKGE of several unbounded domains into convergent series, which were more suitable for the numerical implementation [8]. Using Fourier transformation, Aseeri et al. (2015) solved the cubic Klein-Gordon equation and examined strong scaling of the code [9]. Vibrating systems examples were described by Klein–Gordon equation and the correspondence between the classical and quantum settings of this equation was discussed in [10]. Muravey (2015) provided explicit formulas for the Green's functions of an elliptic PDE in an infinite strip and a half-plane [11]. However, due to convergence of series, the explicit Green's function on a rectangular region remains unsolved. Cheng, et al. (2017) found the representation in double series of Green's function for SKGE on a rectangular region. Although it is uniformly convergent, the huge amount of computing complexity along with the double series hinders the practical application [12].

Needless to say, the retrieval of the unknown weights of an elliptic PDE requires the explicit solution based on the Green's function, which is in the form of an infinite series. In fact, retrieving the weights is more difficult than retrieving the unknown boundary condition, because the weights appear in the expression of Green's function. Considering the approximation calculation, the infinite series of the Green's function must be acquired in a computer-friendly form. This is the motivation of our research.

This current work obtains a new series representation of the Green's function for the SKGE using eigenfunction expansion. The Green's function is decomposed into a singular component and a regular component. The singular component, with a logarithmic singularity, is expressed as an explicit elementary function, which is computer-friendly for Green's function application. The regular component is expressed as an infinite series. The uniform convergence of this infinite series is investigated. With help of the computer-friendly form Green's function, the weights are calculated by comparing the observations in the interior region with the formula solution of Green's function in the corresponding positions. Real precipitation data assimilation results show that Green's function method is a good candidate for data assimilation problems in operational use.

2. The Data Assimilation and Its Variational Problem

In weather prediction areas, a particular physical parameter u_R is provided from remote observation. The corresponding physical quantity $\widetilde{u}$ retrieved from u_R according to the empirical formula is usually with large error. Some high-precision observation u_G in the same area can be used to improve the estimation accuracy, but the number of high-precision observations is limited. More reasonable constraints are also needed to improve the estimation accuracy. According to the smoothness and the dependence on the observation, an unknown function $u(x,y)$ is built to satisfy the following functional extremum problem

$$\iint\limits_{\Omega} \left[\left(\frac{\partial u}{\partial x}\right)^2 + \left(\frac{\partial u}{\partial y}\right)^2 + \mu_R(u-u_R)^2 + \mu_G(u-u_G)^2 \right] dxdy = \min \tag{1}$$

where Ω stands for a special field, $\mu_R \geq 0,\ \mu_G \geq 0$ are weights corresponding to different error items. In addition, the values of $u(x,y)$ on the boundary Γ of area Ω are given by

$$u(x,y)|_{\Gamma} = \phi(x,y) \tag{2}$$

For this functional extremum problem with the boundary constraint, the corresponding Euler's equation is presented as follows [13,14]:

$$\frac{\partial^2 u}{\partial x^2} + \frac{\partial^2 u}{\partial y^2} - \mu u = \hat{u}\,, \tag{3}$$

where $\hat{u} = \mu_R u_R + \mu_G u_G$, $\mu = \mu_R + \mu_G$. This is a typical Dirichlet problem for an elliptic equation stated on a special region. Theoretically, the elliptic equation can be solved by the Green's function. However, as mentioned before, the weights are usually unknown in real data assimilation problem.

The Green's function of a PDE depends on not only the form of the equation but also the boundary condition. It is important to find an explicit solution of the Green's function [12]. In order to get an analytical expression of the corresponding Green's function, we consider the special situation of a rectangular area. The solution of Equation (3) has area of

$$\Omega = \{(x,y)|0 \leq x \leq a, 0 \leq y \leq b\}.$$

For simplicity of discussion, in this paragraph, let $k^2 = \mu, f = \hat{u}$, we discuss the Klein–Gordon equation as follows.

$$\begin{cases} \nabla^2 u - k^2 u = f \\ u|_\Gamma = \phi \end{cases} \tag{4}$$

As Hilbert has shown, a boundary problem is solvable if a generalized Green's function is introduced. Therefore, we should find the corresponding Green's function—i.e., $G(x,y;\xi,\eta)$—which satisfies [13]

$$\begin{cases} \nabla^2 G - k^2 G = \delta(x-\xi)\delta(y-\eta) \\ G|_\Gamma = 0 \end{cases} \tag{5}$$

for any point (ξ,η) in domain Ω. Here, δ is a Dirac function.

On the other hand, if $u(x,y), G(x,y;\xi,\eta) \in C^2(\Omega)$, from the Green's function, we have

$$\iint\limits_{\Omega} \left[u\nabla^2 G - G\nabla^2 u\right] dxdy = \int_\Gamma \left(u\frac{\partial G}{\partial \vec{n}} - G\frac{\partial u}{\partial \vec{n}}\right) ds. \tag{6}$$

where $\vec{n}$ is normal vector of the boundary Γ. Let u satisfy Equation (4), substituting u into Equation (6) gives

$$\iint\limits_{\Omega} \left[u\nabla^2 G - G\left(k^2 u + f\right)\right] dxdy = \int_\Gamma \left(u\frac{\partial G}{\partial \vec{n}} - G\frac{\partial u}{\partial \vec{n}}\right) ds.$$

Then, for a generalized Green's function satisfying Equation (5), the above equation can be simplified as

$$\begin{aligned} \iint_\Omega \left[u(\nabla^2 G - k^2 G) - Gf\right] dxdy &= \iint_\Omega u\delta(x-\xi, y-\eta) dxdy - \iint_\Omega Gf dxdy \\ &= \int_\Gamma \left(u\frac{\partial G}{\partial \vec{n}} - G\frac{\partial u}{\partial \vec{n}}\right) ds. \end{aligned}$$

The solution of Equation (4) therefore becomes

$$u(\xi,\eta) = \iint\limits_{\Omega} G(x,y;\xi,\eta) f(x,y) dxdy + \int_\Gamma \phi(x,y) \frac{\partial G(x,y;\xi,\eta)}{\partial \vec{n}} ds \tag{7}$$

The next key problem is to get the solution of Equation (5), the generalized Green's function $G(x,y;\xi,\eta)$.

It is well known that non-homogeneous boundary conditions can be transformed into homogeneous boundary conditions. For simplicity, we assume that $\phi(x,y) = 0$ here. Then we get

$$u(\xi,\eta) = \iint\limits_{\Omega} G(x,y;\xi,\eta) f(x,y) dxdy = \int_0^b \int_0^a G(x,y;\xi,\eta) f(x,y) dxdy \tag{8}$$

where the Green's function $G(x,y;\xi,\eta)$, according to the Appendix A, is defined by

$$(x,y;\xi,\eta) = \frac{2}{b} \sum_{n=1}^{\infty} g_n(x,\xi) \sin vy \sin v\eta, v = \frac{n\pi}{b} \tag{9}$$

The coefficients $g_n(x,\xi)$ for $x < \xi$ are of the form

$$\begin{aligned} g_n(x,\xi) = \tfrac{1}{4h\sin hha} [&\exp(h(x-\xi-a)) - \exp(h(x+\xi-a)) \\ &+ \exp(-h(x-\xi-a)) - \exp(-h(x+\xi-a))] \end{aligned} \tag{10}$$

where $h = \sqrt{v^2 + k^2}$. Substituting (9), (10) into (8) yields

$$\begin{aligned} u(\xi,\eta) = & \frac{2}{b}\int_0^b\int_0^\xi \sum_{n=1}^{\infty} \frac{\sinh h(\xi-a)\sinh hx}{h\sinh ha} \sin vy \sin v\eta f(x,y)dxdy \\ & +\frac{2}{b}\int_0^b\int_\xi^a \sum_{n=1}^{\infty} \frac{\sinh h\xi \sinh h(x-a)}{h\sinh ha} \sin vy \sin v\eta f(x,y)dxdy \end{aligned} \tag{11}$$

where $f = \mu_R u_R + \mu_G u_G$.

3. Retrieving the Weights by Formal Solution

Obviously, formal solution expression (8) contains unknown parameters, μ_R, μ_G. The mechanism for retrieving is to minimize the error between the observations and the analytical solution in the observation points. It is well known that the analytical solution can be obtained if the Green's function of the PDE is known. In the previous section, a computer-friendly representation of the Green's function is developed, the approximation calculation of analytical solution can be implemented by directly truncating the infinite series.

As precipitation data u_G measured by rain gauge is more accurate, we try to select the weights that satisfy

$$J(\mu_R, \mu_G) = \sum_{(\xi,\eta)\in\Omega_G} (u(\xi,\eta) - u_G(\xi,\eta))^2 = min. \tag{12}$$

where Ω_G represents the rain gauge point set. According to the knowledge of calculus, we compute the partial derivatives as follows

$$\begin{aligned} \frac{\partial J}{\partial \mu_R} &= 2\sum_{(\xi,\eta)\in\Omega_G} (u(\xi,\eta) - u_G(\xi,\eta))\frac{\partial u(\xi,\eta)}{\partial \mu_R} \\ &= 2\sum_{(\xi,\eta)\in\Omega_G} (u - u_G)\Big[\int_0^a\int_0^b G(x,y;\xi,\eta)\frac{\partial f}{\partial \mu_R}dxdy \\ &\quad + \int_0^a\int_0^b \frac{\partial G(x,y;\xi,\eta)}{\partial \mu_R} f dxdy\Big], \\ \frac{\partial J}{\partial \mu_G} &= 2\sum_{(\xi,\eta)\in\Omega_G} (u(\xi,\eta) - u_G(\xi,\eta))\frac{\partial u(\xi,\eta)}{\partial \mu_G} \\ &= 2\sum_{(\xi,\eta)\in\Omega_G} (u - u_G)\Big[\int_0^a\int_0^b G(x,y;\xi,\eta)\frac{\partial f}{\partial \mu_G}dxdy \\ &\quad + \int_0^a\int_0^b \frac{\partial G(x,y;\xi,\eta)}{\partial \mu_G} f dxdy\Big] \end{aligned}$$

where $\frac{\partial f}{\partial \mu_R} = u_R$, $\frac{\partial f}{\partial \mu_G} = u_G$,

$$\begin{aligned} \frac{\partial G(x,y;\xi,\eta)}{\partial \mu_R} &= \frac{2}{b}\sum \frac{\partial g_n(x,\xi)}{\partial \mu_R} \sin \nu y \sin \nu\eta, \\ \frac{\partial G(x,y;\xi,\eta)}{\partial \mu_G} &= \frac{2}{b}\sum \frac{\partial g_n(x,\xi)}{\partial \mu_G} \sin \nu y \sin \nu\eta \end{aligned}$$

with

$$\begin{aligned} \frac{\partial g_n(x,\xi)}{\partial \mu_R} &= \frac{\partial g_n(x,\xi)}{\partial h}\frac{\partial h}{\partial \mu_R} = \frac{\partial g_n(x,\xi)}{\partial h}\frac{1}{2h}, \\ \frac{\partial g_n(x,\xi)}{\partial \mu_G} &= \frac{\partial g_n(x,\xi)}{\partial h}\frac{\partial h}{\partial \mu_G} = \frac{\partial g_n(x,\xi)}{\partial h}\frac{1}{2h}, \\ \frac{\partial g_n(x,\xi)}{\partial h} &= \frac{\partial}{\partial h}\left(\frac{\sinh h(\xi-a)\sinh hx}{h\sinh ha}\right) \\ &= \frac{[(\xi-a)\cosh h(\xi-a)+x\cosh hx]h\sinh ha}{h^2\sinh^2 ha} \\ &\quad - \frac{(\sinh ha+h\cosh ha)\sinh h(\xi-a)\sinh hx}{h^2\sinh^2 ha}. \end{aligned}$$

$$\begin{aligned}
\frac{\partial g_n(x,\xi)}{\partial \mu_R} &= \frac{\partial g_n(x,\xi)}{\partial h}\frac{\partial h}{\partial \mu_R} = \frac{\partial g_n(x,\xi)}{\partial h}\frac{1}{2h}, \\
\frac{\partial g_n(x,\xi)}{\partial \mu_G} &= \frac{\partial g_n(x,\xi)}{\partial h}\frac{\partial h}{\partial \mu_G} = \frac{\partial g_n(x,\xi)}{\partial h}\frac{1}{2h}, \\
\frac{\partial g_n(x,\xi)}{\partial h} &= \frac{\partial}{\partial h}\left(\frac{\sinh h(\xi-a)\sinh hx}{h\sinh ha}\right) \\
&= \frac{[(\xi-a)\cosh h(\xi-a)+x\cosh hx]h\sinh ha}{h^2\sinh^2 ha} \\
&\quad - \frac{(\sinh ha+h\cosh ha)\sinh h(\xi-a)\sinh hx}{h^2\sinh^2 ha}.
\end{aligned}$$

Finally, the steepest descent algorithm is designed for solving $J\left(\vec{z}\right) \triangleq J(\mu_R, \mu_G) = \min$.
Algorithm:

Step 1. Choose initial point $\vec{z}^{(0)} = \left(\mu_R^{(0)}, \mu_G^{(0)}\right)$, $l = 0$, precise requirement $\varepsilon > 0$;

Step 2. Calculate $\nabla J\left(\vec{z}^{(l)}\right) = \left(\frac{\partial J}{\partial \mu_R}, \frac{\partial J}{\partial \mu_G}\right)\vec{z}^{(l)}$, $\vec{z}^{(l)} = \left(\mu_R^{(l)}, \mu_G^{(l)}\right)$. If $\| \nabla J\left(\vec{z}^{(l)}\right) \| < \varepsilon$, stop. Otherwise, let $\vec{d}^{(l)} = -\nabla J\left(\vec{z}^{(l)}\right)$;

Step 3. Let $\vec{z}^{(l+1)} = \vec{z}^{(l)} + \lambda_l \vec{d}^{(l)}$, find λ_l from $J\left(\vec{z}^{(l+1)}\right) = \min$;

Step 4. Let $l = l + 1$, turn to Step 2.

According to the Appendix B, convergence analysis and truncation error analysis, the series representation of Green's function is uniformly converged, and the calculation of infinite series can be truncated.

4. Numerical Experiments

Simulations are conducted here to find the rate of convergence, and so investigate the performance of the proposed Green's function in Equation (7). Figure 1 plots $G(x, y; \xi, \eta)$ on the rectangular region ($0 \le x \le 80, 0 \le y \le 60$) selected from the radar detection range for (ξ, η) equals to (40, 30). Note that G equals 0 on the boundaries of the region and tends to $-\infty$ when (x, y) tends to (ξ, η). The convergence performance of the Green's function on many points in the rectangular area is tested to perform the simulation. Results show that all the points converge steadily to their true numerical values after a limited number of iterations. The convergence speed of the Green's function is much faster when (x, y) is far from (ξ, η).

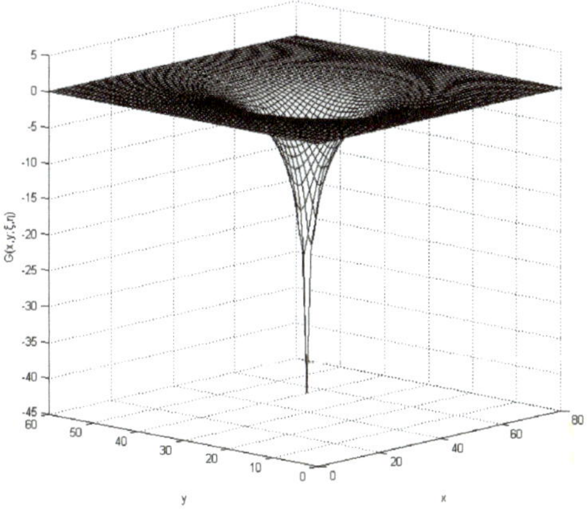

Figure 1. Plot of the Green's function $G(x, y; \xi, \eta)$ on a rectangular region.

In the following example, the Green's function method in Equation (8) is used to retrieve the rainfall value at some points in a rectangular area. The results are compared with the values from the Z-I relationship. The observation data are made by Changzhou radar at 18:38 UTC on 11 June 2016 (Figure 2). There are 624 rain gauges in the observation area. In the data assimilation problem, the 624 rain-gauges data are used to correct radar–retrieved rainfall. The optimal estimation weights are $\mu_R = 12, \mu_G = 96$.

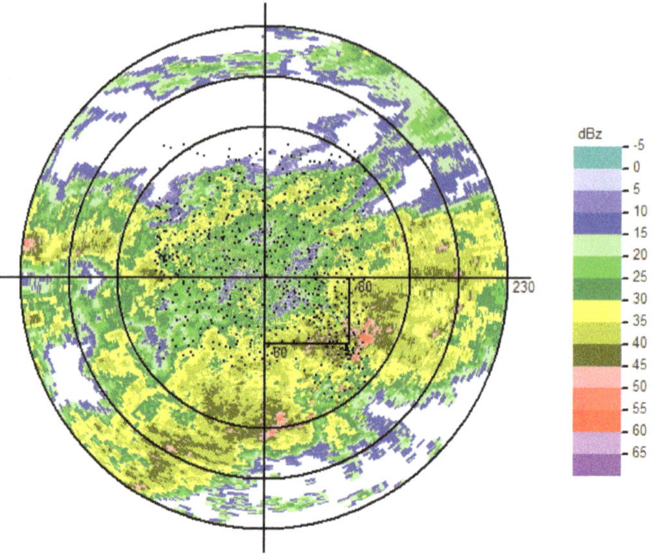

Figure 2. Radar observation and rain-gauge point position.

A rectangular region of 80 × 60 km^2 (Figure 3) and 103 rain gauges inside it are chosen to perform the Green's function method. Because it is a rectangular region, using partial derivative to replace directional derivative, we get the curve integral on the four boundaries by difference method. Assuming boundary condition equals the radar observation, rainfall of every rain-gauge point in the rectangular region is obtained by the Green's function method.

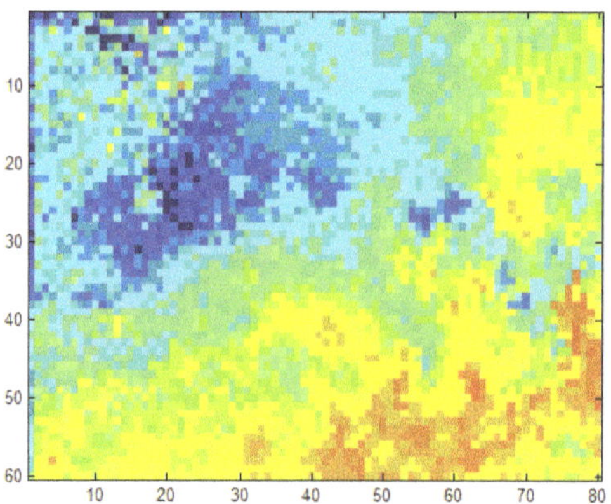

Figure 3. Rectangular Region (80 × 60 km^2).

In practical application, if the number of rain gauges is insufficient, virtual observations on positions without rain gauges can be obtained by linear interpolation. In our experiment, there are few rain gauges near the bottom of the rectangular area. At (50, 50), for example, the observed value 0.953 is obtained by interpolation of five nearby observation points (42, 55), (47, 41), (57, 46), (50, 45) and (53, 48), which can be used for subsequent parameter estimation or comparative analysis.

In the calculation, 75 rain gauges are selected to estimate the weights, and the remaining quarter of the rain gauges are used for comparative analysis. The results show that the retrieved values possess better correlation properties of the rain-gauge observations than radar observations (Figure 4). Correlation coefficient has grown from 0.6149 to 0.7844. Most retrieved values of the 103 points are closer to the rain-gauge observations than the values from Z-I relationship of radar and rainfall. Under the condition of optimal weights selection, the rainfall field obtained by variational inversion has higher accuracy. In this case, Z-I relationship is $Z = 296 * I^{1.24}$ in which Z is the radar reflectivity, I is the rainfall u_R. Table 1 shows some of these values.

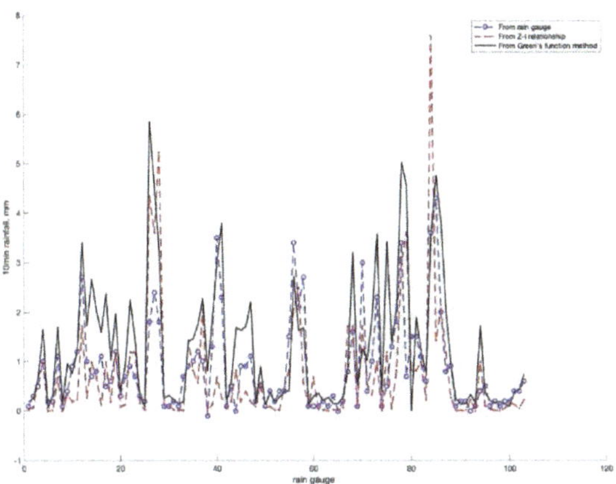

Figure 4. 10 min rainfall from different methods.

Table 1. Interior point values. Pos is interior point coordinates (unit: km). Obs is rain-gauge observation in 10 min (unit: mm). Rad is radar observation on the point (unit: dBz). Rtr is retrieved data by Z-I relationship of radar and rainfall (unit: mm). Joi is the joint value by the Green's function method (unit: mm).

Pos	**(17, 47)**	**(59, 21)**	**(62, 58)**	**(78, 36)**	**(67, 44)**	**(50, 28)**	**(28, 22)**	**(52, 25)**
Obs	1.2	0.7	1.8	2.7	2.3	0.9	0.2	0.6
Rad	30	27	47	35	38	31	11	30
Rtr	1.2322	0.9353	3.2013	1.6780	3.5835	0.9404	0.1644	0.7484
Joi	0.2224	0.1274	5.2259	0.5629	0.9826	0.2678	0.0065	0.2224

5. Conclusions

This study solves the data assimilation problems of two kinds of data, u_R and u_G, through a variational model. In order to retrieve the weighting parameters μ_R, μ_G from some of the observations in the interior region, the computer-friendly form of the Green's function is obtained by eigenfunction expansion for the boundary value problem of the static Klein–Gordon equation on a rectangular region. Convergence analysis in Appendix B proves that the series representation of Green's function is computer-friendly, and can be used to approximate computations by a direct truncation.

The numerical experiment's result shows that the series representation of Green's function converges steadily and rapidly. The Green's function method is used to retrieve the weights and assimilate precipitation data from weather radar and rain-gauges in the experiments. Results show that the Green's function method is a good candidate for data assimilation problems in operational use.

Author Contributions: Conceptualization, H.C., M.W. and G.L.; methodology, H.C. and G.L.; software, H.C. and H.J.; validation, H.C. and X.M.; writing—original draft preparation, H.C. and G.L.; writing—review and editing, H.C. and G.L. All authors have read and agreed to the published version of the manuscript.

Funding: This work was partially supported by National Nature Science Foundation of China (No. 61501224).

Acknowledgments: This work was partially supported by National Nature Science Foundation of China (No. 61501224). We are grateful to anonymous reviewers and an associate editor whose comments led to many improvements of the article.

Conflicts of Interest: The authors declare no conflict of interest.

Appendix A. The Green's Function for the SKGE Stated on a Rectangular Region

For the elliptic two-dimensional SKGE $(\nabla^2 - k^2)u(x,y) = f(x,y)$, $(x,y) \in \Omega$, the corresponding homogeneous boundary condition is $\begin{cases} u(x,0) = u(x,b) = 0 \\ u(0,y) = u(a,y) = 0 \end{cases}$. Focusing on the Dirichlet problem set up on a rectangle $\Omega = \{(x,y)\}|0 \le x \le a, 0 \le y \le b$, we study the following series representation of its Green's function

$$(x,y;\xi,\eta) = \frac{2}{b}\sum_{n=1}^{\infty} g_n(x,\xi)\sin vy \sin v\eta, v = \frac{n\pi}{b}. \quad \text{(A1)}$$

The coefficients $g(x,\xi)$ for $x < \xi$ are of the form

$$\begin{aligned} g_n(x,\xi) = \tfrac{1}{4h\sin hha} \quad & [\exp(h(x-\xi-a)) - \exp(h(x+\xi-a)) \\ & +\exp(-h(x-\xi-a)) - \exp(-h(x+\xi-a))] \end{aligned} \quad \text{(A2)}$$

where $h = \sqrt{v^2 + k^2}$. In fact, it is not difficult to find that

$$g_n(x,\xi)|_{x=0} = \frac{\exp(-h(\xi+a)) - \exp(h(\xi-a)) + \exp(h(\xi+a)) - \exp(-h(\xi-a))}{4h\sin hha} \neq 0$$

and $g_n(x,\xi)|_{x=0} \to \infty$ as $h \to \infty$, which yields the non-computer-friendly series representation. However, we find that the correct form is as follows.

$$\begin{aligned} g_n(x,\xi) &= \tfrac{\sin hh(\xi-a)\sin hhx}{h\sin hha} \\ &= \tfrac{1}{4h\sin hha} \quad [\exp(h(x+\xi-a)) - \exp(h(x-\xi+a)) \\ & \qquad +\exp(-h(x+\xi-a)) - \exp(-h(x-\xi+a))] \end{aligned} \quad \text{(A3)}$$

The following focuses on deducing Equation (A3). Substituting Equation (A1) into the SKGE gives the coefficient function $g_n(x,\xi)$ satisfying the differential equation

$$g''_n(x,\xi) - \left(v^2 + k^2\right)g_n(x,\xi) = \delta(x-\xi) \quad \text{(A4)}$$

with homogeneous boundary condition $g_n(0,\xi) = 0, g_n(a,\xi) = 0$. For x equal to anything but ξ, $g''_n(x,\xi) - (v^2 + k^2)g_n(x,\xi) = 0$. Therefore,

$$g_n(x,\xi) = \begin{cases} c_1\exp(hx) + c_2\exp(-hx), & x < \xi \\ d_1\exp(h(x-a)) + d_2\exp(-h(x-a)), & x \ge \xi \end{cases}$$

To determine the above constants c_1, c_2, d_1, d_2, integrate differential Equation (A4) from $\xi - \varepsilon$ to $\xi + \varepsilon$, where ε is infinitesimal. The result is $\frac{dg_n}{dx}\Big|_{\xi-\varepsilon}^{\xi+\varepsilon} = 1$. Integrating again gives $g_n|_{\xi-\varepsilon}^{\xi+\varepsilon} = 0$. That is, $g_n(x,\xi)$ as a function of x is continuous at $x = \xi$, but its first derivative jumps by +1 at that point, leading to

$$c_1 \exp(h\xi) + c_2 \exp(-h\xi) = d_1 \exp(h(\xi - a)) + d_2 \exp(-h(\xi - a))$$

and

$$d_1 \exp(h(\xi - a)) - d_2 \exp(-h(\xi - a)) - (c_1 \exp(h\xi) - c_2 \exp(-h\xi)) = 1/h.$$

Considering the homogeneous boundary conditions gives $\begin{cases} c_1 + c_2 = 0 \\ d_1 + d_2 = 0 \end{cases}$, from which

$$\begin{aligned} c_1 &= \frac{1}{2h} \frac{\sin hh(\xi-a)}{\sin hh\xi \cos hh(\xi-a) - \sin hh(\xi-a)\cos hh\xi} \\ &= \frac{1}{2h} \frac{\sin hh(\xi-a)}{\sin hha} \end{aligned}$$

and

$$\begin{aligned} d_1 &= \frac{1}{2h\cos hh(\xi-a)} \frac{\sin hh(\xi-a)\cos hh\xi}{\sin hh\xi \cos hh(\xi-a) - \sin hh(\xi-a)\cos hh\xi} \\ &= \frac{\sin hh(\xi-a)\cos hh\xi}{2h\cos hh(\xi-a)\sin hha} = \frac{1}{2h}\frac{\sin hh\xi}{\sin hha} \end{aligned}$$

Hence,

$$g_n(x,\xi) = \begin{cases} \frac{\sin hh(\xi-a)\sin hhx}{h\sin hha}, x < \xi \\ \frac{\sin hh\xi \sin hh(x-a)}{h\sin hha}, x \geq \xi \end{cases}. \tag{A5}$$

This yields the desired Green's function $G(x,y;\xi,\eta)$ on the rectangle $\Omega = \{(x,y)\}|$ $0 \leq x \leq a, 0 \leq y \leq b$ in the form

$$(x,y;\xi,\eta) = \begin{cases} \frac{2}{b}\sum\limits_{n=1}^{\infty} \frac{\sin hh(\xi-a)\sin hhx}{h\sin hha} \sin vy \sin v\eta,\ x < \xi \\ \frac{2}{b}\sum\limits_{n=1}^{\infty} \frac{\sin hh\xi \sin hh(x-a)}{h\sin hha} \sin vy \sin v\eta,\ x \geq \xi \end{cases}. \tag{A6}$$

In order to prove the convergence, rewrite the branch of the function $g_n(x,\xi)$, which is valid for $x < \xi$, in the form

$$\begin{aligned} g_n(x,\xi) &= \frac{1}{h}\frac{\sin hh(\xi-a)\sin hhx}{\sin hha} - \frac{\exp(h(x+\xi-2a))}{2h} \\ &\quad + \left[\frac{\exp(h(x+\xi-2a))}{2h} - \frac{\exp(v(x+\xi-2a))}{2v}\right] + \frac{\exp(v(x+\xi-2a))}{2v} \\ &= H_n(x,\xi) + M_n(x,\xi) + \frac{\exp(v(x+\xi-2a))}{2v} \end{aligned} \tag{A7}$$

where

$$H_n(x,\xi) = \frac{1}{h}\frac{\sin hh(\xi-a)\sin hhx}{\sin hha} - \frac{\exp(h(x+\xi-2a))}{2h},$$
$$M_n(x,\xi) = \frac{\exp(h(x+\xi-2a))}{2h} - \frac{\exp(v(x+\xi-2a))}{2v}.$$

This lets Equation (A1) be transformed into the following expression

$$\begin{aligned} (x,y;\xi,\eta) &= \frac{2}{b}\sum_{n=1}^{\infty} H_n(x,\xi)\sin vy \sin v\eta + \frac{2}{b}\sum_{n=1}^{\infty} M_n(x,\xi)\sin vy\sin v\eta \\ &\quad + \frac{1}{b}\sum_{n=1}^{\infty}\frac{\exp(v(x+\xi-2a))}{v}\sin vy \sin v\eta \end{aligned} \tag{A8}$$

with

$$\begin{aligned}
&\sum_{n=1}^{\infty} \frac{\exp(v(x+\xi-2a))}{v} \sin vy \sin v\eta \\
&= \frac{1}{2} \sum_{n=0}^{\infty} \frac{\exp(v(x+\xi-2a))}{v} \cos v(y-\eta) \ \frac{1}{2} \sum_{n=0}^{\infty} \frac{\exp(v(x+\xi-2a))}{v} \cos v(y+\eta) \\
&= \frac{1}{4\pi} \left[\ln \frac{1-2\exp(c(x+\xi-2a))\cos(y+\eta)+\exp(2c(x+\xi-2a))}{1-2\exp(c(x+\xi-2a))\cos(y-\eta)+\exp(2c(x+\xi-2a))} \right]
\end{aligned} \tag{A9}$$

where $c = \pi/b$. Substituting Equation (A9) into Equation (A8) then gives the new series representation of the Green's function for the Dirichlet problem set up for the SKGE on the rectangle Ω:

$$\begin{aligned}
&G(x,y;\xi,\eta) = \frac{2}{b} \sum_{n=1}^{\infty} H_n(x,\xi) \sin vy \sin v\eta + \frac{2}{b} \sum_{n=1}^{\infty} M_n(x,\xi) \sin vy \sin v\eta \\
&+ \frac{1}{4\pi} \left[\ln \frac{1-2\exp(c(x+\xi-2a))\cos(y+\eta)+\exp(2c(x+\xi-2a))}{1-2\exp(c(x+\xi-2a))\cos(y-\eta)+\exp(2c(x+\xi-2a))} \right], x < \xi
\end{aligned} \tag{A10}$$

The representation of $G(x,y;\xi,\eta)$ valid for $x \geq \xi$ can be derived from Equation (A10), by interchanging the variables x and ξ in the first of the two series only, because the second series is invariant to the swap.

Appendix B. Convergence Analysis and Truncation Error Analysis

For the effective computational implementation of the representation in Equation (A10), we study the two series separately in the following paragraphs. First, consider the series $\sum_{n=1}^{\infty} H_n(x,\xi) \sin vy \sin v\eta$. As it is difficult to prove directly the uniform convergence of the first of the two series in Equation (A10), decompose the series into two parts, one of which is a complete summation. Indeed, the complete summation can remove the disadvantage affecting the uniformly convergent criterion. This is the basic idea of the uniformly convergent proof of the infinite series. Elementary algebra shows

$$\begin{aligned}
H_n(x,\xi) &= \frac{1}{h} \frac{\sin hh(\xi-a) \sin hhx}{\sin hha} - \frac{\exp(h(x+\xi-2a))}{2h} \\
&= \frac{2\sin hh(\xi-a)\sin hhx - \exp(h(x+\xi-2a))\sin hha}{2h\sin hha} \\
&= \frac{1}{4h\sin hha}[(\exp(h(\xi-a)) - \exp(-h(\xi-a)))(\exp(hx) - \exp(-hx)) \\
&\qquad - \exp(-h(x+\xi-2a))(\exp(ha) - \exp(-ha))] \\
&= \frac{1}{4h\sin hha}[-\exp(h(x-\xi+a)) - \exp(-h(x-\xi+a)) \\
&\qquad + \exp(-h(x+\xi-a)) + \exp(h(x+\xi-3a))] \\
&= \frac{1}{2h(1-\exp(-2ha))}[-\exp(h(x-\xi)) - \exp(-h(x-\xi+2a)) \\
&\qquad + \exp(-h(x+\xi)) + \exp(h(x+\xi-4a))] \\
&= \frac{\exp(-h(x+\xi)) - \exp(h(x-\xi))}{2h(1-\exp(-2ha))} \\
&\qquad + \frac{\exp(h(x+\xi-4a)) - \exp(-h(x-\xi+2a))}{2h(1-\exp(-2ha))} \\
&\qquad = \frac{\exp(-h(x+\xi)) - \exp(h(x-\xi))}{2h} + H'_n(x,\xi),
\end{aligned}$$

where

$$\begin{aligned}
H'_n(x,\xi) = \frac{1}{2h(1-\exp(-2ha))} \quad &[\exp(h(x+\xi-4a)) - \exp(-h(x-\xi+2a)) \\
&+ \exp(-h(x+\xi+2a)) - \exp(h(x-\xi-2a))] \, .
\end{aligned} \tag{A11}$$

and

$$\sum_{n=1}^{\infty} \frac{\exp(-h(x+\xi)) - \exp(h(x-\xi))}{2h} \sin vy \sin v\eta$$

$$= -\frac{1}{4} \sum_{n=1}^{\infty} \frac{\exp(-h(x+\xi)) - \exp(h(x-\xi))}{h} [\cos v(y+\eta) - \cos v(y-\eta)]$$

$$= \frac{1}{8c} \left[\ln \frac{1 - 2\exp(c(x-\xi))\cos(y+\eta) + \exp(2c(x-\xi))}{1 - 2\exp(c(x-\xi))\cos(y-\eta) + \exp(2c(x-\xi))} - \ln \frac{1 - 2\exp(-c(x+\xi))\cos(y+\eta) + \exp(-2c(x+\xi))}{1 - 2\exp(-c(x+\xi))\cos(y-\eta) + \exp(-2c(x+\xi))} \right].$$

Then the Green's function in the form of Equation (A10) can be rewritten as

$$\begin{aligned}(x, y; \xi, \eta) = \frac{1}{4\pi} & \left[\ln \frac{1 - 2\exp(c(x+\xi-2a))\cos(y+\eta) + \exp(2c(x+\xi-2a))}{1 - 2\exp(c(x+\xi-2a))\cos(y-\eta) + \exp(2c(x+\xi-2a))} \right. \\ & + \ln \frac{1 - 2\exp(c(x-\xi))\cos(y+\eta) + \exp(2c(x-\xi))}{1 - 2\exp(c(x-\xi))\cos(y-\eta) + \exp(2c(x-\xi))} \\ & \left. - \ln \frac{1 - 2\exp(-c(x+\xi))\cos(y+\eta) + \exp(-2c(x+\xi))}{1 - 2\exp(-c(x+\xi))\cos(y-\eta) + \exp(-2c(x+\xi))} \right] \\ + \frac{2}{b} \sum_{n=1}^{\infty} H'_n(x, \xi) \sin vy \sin v\eta & + \frac{2}{b} \sum_{n=1}^{\infty} M_n(x, \xi) \sin vy \sin v\eta \end{aligned} \tag{A12}$$

To explore the convergence of $\sum_{n=1}^{\infty} H'_n(x, \xi) \sin vy \sin v\eta$ and analyze its truncation errors requires estimation of the asymptotic property of the common term $H'_n(x, \xi)$.

As $0 \le x < \xi \le a$, we have

$$\begin{aligned} |H'_n(x, \xi)| &= \left| \frac{1}{2h(1 - \exp(-2ha))} [\exp(h(x+\xi-4a)) - \exp(-h(x-\xi+2a)) \right. \\ &\qquad \left. + \exp(-h(x+\xi+2a)) - \exp(h(x-\xi-2a))] \right| \\ &= \left| \frac{\exp(-ha)}{2h(1 - \exp(-2ha))} [\exp(h(x+\xi-3a)) - \exp(-h(x-\xi+a)) \right. \\ &\qquad \left. + \exp(-h(x+\xi+a)) - \exp(h(x-\xi-a))] \right| \\ &\le \frac{\exp(-ha)}{2h(1 - \exp(-2ha))} [\exp(h(x+\xi-3a)) - \exp(-h(x-\xi+a)) \\ &\qquad + \exp(-h(x+\xi+a)) - \exp(h(x-\xi-a))] \\ &\le \frac{2\exp(-ha)}{h(1 - \exp(-2ha))} = O\left(\frac{1}{h\exp(ha)} \right), \; h \to \infty. \end{aligned}$$

Hence, $|H'_n(x, \xi) \sin vy \sin v\eta| \le \frac{2\exp(-ha)}{h(1 - \exp(-2ha))}$. The uniform convergence of the series $\sum_{n=1}^{\infty} \frac{2\exp(-ha)}{h(1 - \exp(-2ha))}$, from the series convergence criterion, clearly leads to the uniform convergence of the series $\sum_{n=1}^{\infty} H'_n(x, \xi) \sin vy \sin v\eta$.

In order to compute approximately the Green's function, analyze the modulus of its remainder.

$$\begin{aligned} \left| \sum_{n=N+1}^{\infty} H'_n \sin vy \sin v\eta \right| &\le \sum_{n=N+1}^{\infty} \frac{2}{(1 - \exp(-2ha)) h \exp(ha)} \\ &\le \frac{2}{(N+1)(1 - \exp(-2(N+1)a))} \sum_{n=N+1}^{\infty} \frac{1}{\exp(na)} \\ &= \frac{2\exp(-(N+1)a)}{(N+1)(1 - \exp(-2(N+1)a))(1 - \exp(-a))} \end{aligned}$$

Hence, the series $\sum_{n=1}^{\infty} H'_n(x, \xi) \sin vy \sin v\eta$ converges uniformly, and can be accurately computed by a direct truncation. The above estimate helps to find an approximate value of the truncating parameter N required to attain a desired accuracy.

Consider secondly the convergence of $\sum_{n=1}^{\infty} M_n(x,\xi)\sin vy \sin v\eta$, and the analysis of the modulus of its remainder,

$$
\begin{aligned}
|R_N(x,y;\xi,\eta)| &= \left|\sum_{n=N+1}^{\infty}\left[\frac{\exp(h(x+\xi-2a))}{2h}-\frac{\exp(v(x+\xi-2a))}{2v}\right]\sin vy\sin v\eta\right| \\
&\le \left|\sum_{n=N+1}^{\infty}\left[\frac{\exp(h(x+\xi-2a))}{2h}-\frac{\exp(v(x+\xi-2a))}{2v}\right]\right|
\end{aligned}
$$

The inequality regarding the truncation error is obtained as follows

$$
|R_N(x,y;\xi,\eta)| < \frac{b}{\pi}\left\{[1-\exp(k(x-\xi))]\sum_{n=N+1}^{\infty}\frac{1}{n}\exp\left(\frac{n\pi}{b}(x-\xi)\right)+k\frac{b}{\pi}\sum_{n=N+1}^{\infty}\frac{1}{n^2}\right\}. \tag{A13}
$$

Evidently,

$$
\sum_{n=N+1}^{\infty}\frac{1}{n^2} < \int_{N+1}^{\infty}\frac{1}{x^2}dx = \frac{1}{N+1} \tag{A14}
$$

and

$$
\ln(1+x) < x,\ x < 1. \tag{A15}
$$

Inequalities (A13)–(A15) allow estimation of the truncation error term:

$$
\begin{aligned}
|R_N(x,y;\xi,\eta)| &< \tfrac{b}{\pi}\left\{[1-\exp(k(x-\xi))]\sum_{n=N+1}^{\infty}\tfrac{1}{n}\exp(\tfrac{n\pi}{b}(x-\xi))+k\tfrac{b}{\pi}\sum_{n=N+1}^{\infty}\tfrac{1}{n^2}\right\} \\
&< \tfrac{b}{\pi}\left\{[1-\exp(k(x-\xi))]\ln\left[1+\exp(\tfrac{\pi}{b}(x-\xi))\right]\exp\left(\tfrac{N\pi}{b}(x-\xi)\right)+\tfrac{kb}{(N+1)\pi}\right\} \\
&< \tfrac{b}{\pi}\left\{[1-\exp(k(x-\xi))]\exp\left(\tfrac{(N+1)\pi}{b}(x-\xi)\right)+\tfrac{kb}{(N+1)\pi}\right\}.
\end{aligned}
$$

Obviously, the above inequality is dependent on the variables. To prove the uniform convergence and obtain a uniform estimation requires further estimation of the right side of the above inequality. Therefore, consider the extremum property of the function

$$
\psi(\theta) = [1-\exp(k\theta)]\exp\left(\frac{(N+1)\pi\theta}{b}\right),\ (\theta < 0).
$$

It is not difficult to find that the stationary point is $\theta = \frac{1}{k}\ln\left(\frac{(N+1)\pi/b}{(N+1)\pi/b+k}\right)$ and the corresponding maximum is

$$
\psi_{\max} = \frac{kb/(N+1)\pi}{1+kb/(N+1)\pi}\exp\left[-\frac{(N+1)\pi}{b}\frac{1}{k}\ln\left(1-\frac{kb}{(N+1)\pi+kb}\right)\right].
$$

When N is sufficiently large, we have

$$
\begin{aligned}
\psi_{\max} &< \tfrac{kb/(N+1)\pi}{1+kb/(N+1)\pi}\exp\left[-\tfrac{(N+1)\pi}{b}\tfrac{1}{k}\left(\tfrac{kb}{(N+1)\pi+kb}-\left(\tfrac{kb}{(N+1)\pi+kb}\right)^2/2\right)\right] \\
&< \tfrac{kb/(N+1)\pi}{1+kb/(N+1)\pi}\exp\left[-1+\tfrac{kb}{2(N+1)\pi+2kb}\right] < \tfrac{kb}{(N+1)\pi}.
\end{aligned}
$$

Substituting the above inequality into inequality (A13) gives

$$
|R_N(x,y;\xi,\eta)| < \frac{b}{\pi}\left[\frac{kb}{(N+1)\pi}+\frac{kb}{(N+1)\pi}\right] = \frac{2kb^2}{(N+1)\pi^2}. \tag{A16}
$$

Hence, the truncation error term converges to zero, and the function $\sum_{n=1}^{\infty} M_n(x,\xi)\sin vy\sin v\eta$ converges uniformly. The resulting uniform estimator of function $\sum_{n=1}^{\infty} M_n(x,\xi)\sin vy\sin v\eta$ is advantageous to the approximate calculation.

Then we find that the two series in Equation (A10) converge uniformly and can be accurately computed by a direct truncation. The above estimation can provide an appropriate value of the truncating parameter N required to attain a desired accuracy.

$$\begin{aligned} G(x,y;\xi,\eta) &= \tfrac{2}{b}\sum_{n=1}^{\infty} g_n(x,\xi)\sin\nu y\sin\nu\eta \\ &= \tfrac{1}{4\pi}\Big[\ln\tfrac{1-2\exp(c(x+\xi-2a))\cos(y+\eta)+\exp(2c(x+\xi-2a))}{1-2\exp(c(x+\xi-2a))\cos(y-\eta)+\exp(2c(x+\xi-2a))} \\ &\quad+\ln\tfrac{1-2\exp(c(x-\xi))\cos(y+\eta)+\exp(2c(x-\xi))}{1-2\exp(c(x-\xi))\cos(y-\eta)+\exp(2c(x-\xi))} \\ &\quad-\ln\tfrac{1-2\exp(-c(x+\xi))\cos(y+\eta)+\exp(-2c(x+\xi))}{1-2\exp(-c(x+\xi))\cos(y-\eta)+\exp(-2c(x+\xi))}\Big] \\ &+\tfrac{2}{b}\sum_{n=1}^{N} H_n'(x,\xi)\sin\nu y\sin\nu\eta+\tfrac{2}{b}\sum_{n=1}^{N} M_n(x,\xi)\sin\nu y\sin\nu\eta+O\left(\tfrac{1}{N}\right) \end{aligned} \tag{A17}$$

Similarly, interchanging the variables x with ξ in the first series gives the representation of $G(x,y;\xi,\eta)$ valid for $(x \geq \xi)$.

References

1. Talagrand, O.; Courtier, P. Variational assimilation of meteorological observations with the adjoint vorticity equation. I: Theory. *Q. J. R. Meteorol. Soc.* **1987**, *113*, 1311–1328. [CrossRef]
2. Al-Jamal, M.F. Numerical Solution of Elliptic Inverse Problems via the Equation Error Method. Ph.D. Thesis, Michigan Technological University, Houghton, MI, USA, 2012.
3. Mehraliyev, Y.T.; Kanca, F. An inverse boundary value problem for a second order elliptic equation in a rectangle. *Math. Modeling Anal.* **2014**, *19*, 241–256. [CrossRef]
4. Mehraliyev, Y.T. On solvability of an inverse boundary value problem for a fourth order elliptic equation. *J. Math. Syst. Sci.* **2013**, *3*, 560–566.
5. Orlovsky, D.G. Inverse problem for elliptic equation in Banach space with Bitsadze—Samarsky boundary value conditions. *J. Inverse ILL-Posed Probl.* **2013**, *21*, 141–157. [CrossRef]
6. Ming, W.; Renqing, D.; Wenzhong, G.; Takeda, T. Retrieval Single-Doppler Radar Wind with Variational Assimilation Method—Part I: Objective Selection of Functional Weighting Factors. *Adv. Atmos. Sci.* **1998**, *15*, 553–567. [CrossRef]
7. Ming, W.; Guo-qing, L.; Cheng-gang, W.; Wen-zhong, G.; Qin, X. Optimal Selection for the Weighted Coefficients of the Constrained Variational Problems. *Appl. Math. Mech.* **2003**, *24*, 926–944. [CrossRef]
8. Melnikov, Y.A. Construction of Green's functions for the two-dimensional static Klein-Gordon equation. *J. Part. Differ. Equ.* **2011**, *24*, 114–139.
9. Aseeri, S.; Batrašev, O.; Icardi, M.; Leu, B.; Liu, A.; Li, N.; Muite, B.K.; Müller, E.; Palen, B.; Quell, M. Solving the Klein-Gordon equation using Fourier spectral methods: A benchmark test for computer performance. In Proceedings of the Symposium on High Performance Computing (HPC'15), Montreal, QC, Canada, 17 June 2015; pp. 182–191.
10. Gravel, P.; Cauthier, C. Classical applications of the Klein-Gordon equation. *Am. J. Phys. Teach.* **2011**, *79*, 447–453. [CrossRef]
11. Muravey, D. The boundary value problem for a static 2D Klein-Gordon equation in the infinite strip and in the half-plane. *Mathematics* **2015**, *40*, 205–227.
12. Cheng, H.; Mu, X.; Jiang, H.; Liu, G.; Wei, M. Green's Function for the boundary value problem of thestatic Klein-Gordon equation stated on a rectangular region and its convergence analysis. *Bound. Value Probl.* **2017**, *2017*, 72. [CrossRef]
13. Courant, R.; Hilbert, D. *Methods of Mathematical Physics*; Interscience Publishers Inc.: New York, NY, USA, 1953.
14. Sasaki, Y.K. Some basic formulisms in numerical variational analysis. *Mon. Weather Rev.* **1970**, *98*, 875–883. [CrossRef]